Lecture Notes in Physics

Volume 975

The Lecture Notes in Physics

The series Lecture Notes in Physics (LNP), founded in 1969, reports new developments in physics research and teaching - quickly and informally, but with a high quality and the explicit aim to summarize and communicate current knowledge in an accessible way. Books published in this series are conceived as bridging material between advanced graduate textbooks and the forefront of research and to serve three purposes:

- to be a compact and modern up-to-date source of reference on a well-defined topic;
- to serve as an accessible introduction to the field to postgraduate students and nonspecialist researchers from related areas;
- to be a source of advanced teaching material for specialized seminars, courses and schools.

Both monographs and multi-author volumes will be considered for publication. Edited volumes should however consist of a very limited number of contributions only. Proceedings will not be considered for LNP.

Volumes published in LNP are disseminated both in print and in electronic formats, the electronic archive being available at springerlink.com. The series content is indexed, abstracted and referenced by many abstracting and information services, bibliographic networks, subscription agencies, library networks, and consortia.

Proposals should be sent to a member of the Editorial Board, or directly to the responsible editor at Springer:

Dr Lisa Scalone
Springer Nature
Physics
Tiergartenstrasse 17
69121 Heidelberg, Germany
lisa.scalone@springernature.com

More information about this series at http://www.springer.com/series/5304

Subhendra Mohanty

Astroparticle Physics and Cosmology

Perspectives in the Multimessenger Era

Subhendra Mohanty
Theory Division
Physical Research Laboratory
Ahmedabad, Gujarat, India

ISSN 0075-8450   ISSN 1616-6361   (electronic)
Lecture Notes in Physics
ISBN 978-3-030-56200-7   ISBN 978-3-030-56201-4   (eBook)
https://doi.org/10.1007/978-3-030-56201-4

This Springer imprint is published by the registered company Springer Nature Switzerland AG.
The registered company address is: Gewerbestrasse 11, 6330 Cham, Switzerland

Dedicated to
My Parents.

Preface

In the last decade, there have been several seminal discoveries starting with the Higgs boson at the LHC (2012); neutrinos with PeV energies at IceCube (2013); detection of gravitational waves from black hole and neutron star mergers by LIGO (2016) and the first picture of the black hole at the centre of the M87 galaxy by the EHT (2019). This comes after the success of the solar, atmospheric and reactor neutrino observation experiments in the last three decades and the cosmic microwave anisotropy measurement experiments (COBE (1989–1993), WMAP (2001–2011) and PLANCK(2009–2011)). Terrestrial dark matter detection experiments continue to put limits on the mass and cross-section of dark matter (if at all they are elementary particles). Astronomical observations are ongoing in a very large range of the electromagnetic spectrum (from radio to gamma rays) in addition to the observations of high energy neutrinos by IceCube. This ongoing multi-pronged observation of the universe will help us to answer the fundamental questions about the underlying theories that govern the universe.

In this book, we explore the theoretical consequences of these multi-messenger signals from the universe. A graduate student or researcher who is curious about learning a particular research topic may delve into the chapter of their choice to get an introduction to the subject. The treatment is more pedagogical and focussed compared with a review article. The main results discussed are worked out in the text.

In Chap. 1, we give a survey of the recent experimental observations, which started the multi-messenger era of astronomical observations like observations of gravitational waves, gamma rays, neutrinos and cosmic rays from the same source (e.g. blazers). We also list other important unsolved issues like final unobserved signal of the inflation paradigm, namely the observation of inflation-generated gravitational waves in the early universe via the measurement of the B-mode polarization of the CMB signal.

Dark matter is amongst the hottest areas of enquiry in particle physics and cosmology. In Chap. 2, we discuss the phenomenology of dark matter at cosmological and galactic scales. Direct detection experiments like Xenon-1T have ruled out a large swathe of the mass vs DM–nucleon cross-section parameter space challenging the conventional paradigm of the $\sim 100\,\mathrm{GeV}$ weakly interacting dark matter, where the observed relic density is naturally accounted for. New mechanisms for explaining the relic density like the freeze-in production of dark matter or the $3 \rightarrow 2$ annihilation process are required for evading the stringent constraints from direct detection experiments. Dark matter may be very light with a mass of 10^{-22} eV, or they may not be elementary particles at all and may be primordial black holes.

Large-scale galactic surveys such as the Sloan Digital Sky Survey, Dark Energy Survey and Baryon Oscillations Spectroscopy Survey are increasingly providing data about the distribution of matter in the universe and provide a scope for us to test theories of dark matter, dark energy, neutrino mass, etc. To test theory with observations from the survey, we need to understand perturbations of the metric and matter (dark matter, baryons, photons, neutrinos) in the framework of general relativity. We do this in Chap. 3.

In Chap. 4, we study the effect of perturbations on the cosmic microwave background anisotropy spectrum. The observations of COBE, WMAP, PLANCK, DESI, SPT, etc. turned cosmology into precision science. Future observations will tell us about the existence (or not) of B-mode polarization, which will test the theories of inflation. CMB can also be used to study the interactions of the dark sector like neutrino–dark matter interactions or neutrino–dark energy interactions. We lay down the basics of the theory of CMB anisotropy aimed at non-experts in the subject who would like to dive in and make predictions based on theories of their choice for ongoing and future observations.

In Chap. 5, we discuss models of inflation, which are consistent with the stringent limits on the tensor-to-scalar ratio and spectral index placed by PLANCK. There are a few well-motivated inflation models that make the cut, namely natural inflation, curvature-coupled Higgs inflation, R^2 Starobinsky inflation and the no-scale supergravity models. We discuss the pros and cons of each of these models and discuss their other phenomenological consequence and future prospects in CMB and LSS observations.

The discovery of the Higgs boson completed the last element in the particle spectrum envisaged in the standard model. The discovery of this first elementary scalar particle brings in several questions regarding the Higgs potential. In the standard model, the Higgs potential becomes negative at energy scales of $\sim 10^{11}$ GeV. Is the universe stable against tunnelling to a negative (Higgs potential) energy phase? What additional particles can make the Higgs potential stable or metastable (the timescale tunnelling is smaller than the lifetime of the universe)?

In Chap. 6, we address these questions. We develop the idea of effective potential of scalar fields. We discuss the Coleman–Weinberg (one loop at all order in coupling) corrections to the tree-level potential. We also derive the effective potential at finite temperature, which will decide the nature of phase transition during symmetry breaking as the universe cools. Phase transitions in the early universe can generate gravitational waves that may be observed in LIGO or future gravitational wave detectors.

In Chap. 7, we discuss gravitational waves. The gravitational wave energy radiated rate by binary stars is calculated using the (quicker) effective field theory technique and the result compared to the classical derivation. Energy loss rate of other possible light scalars like axions from black holes or neutron stars is also derived. The gravitational waveform expected at the detectors from binary mergers is derived. The stochastic gravitational waves from phase transitions, which may be observed in gravitational wave detectors, are worked out.

In Chap. 8, we study black holes. Black hole mergers have been observed through their gravitational signals, and the ring down of the black holes has been seen in the LIGO observations (2016). From the gravitational wave observations, the mass and spin of the black holes can be estimated. Moreover, recently there is a direct observation of the black hole shadow by the Event Horizon Telescope team of the super massive black hole at the centre of M87 galaxy. These observations are a good motivation for studying the details of the rotating black holes as well as more general types of black holes like the dilation-axion black holes predicted from string theory. The deviation from the Kerr metric may be observable in gravitational wave signals or more directly from the shape and size of the photon shadows of the super-massive galactic centre black holes. Kerr black holes may also generate high energy particles by the mechanism of super-radiance that we will discuss later in that chapter.

We hope that the reader will find the subjects covered interesting and be motivated to dive into the research work in this field.

I thank the students Surya Nayak, Prafulla Panda, Sarira Sahu, Anshu Gupta, Akhilesh Nautiyal, Suratna Das, Soumya Rao, Moumita Das, Tanushree Kaushik, Gaurav Tomar, Girish Chakravarti, Bhavesh Chauhan, Ashish Narang, Priyank Parashari, Prakrut Chaubal and Tanmay Poddar for their wonderful collaborations.

I thank the postdoctoral fellows I have worked with namely, Peter Stockinger, Kaushik Bhattacharya, Joydeep Chakrabortty, Gaurav Goswami, Ila Garg, Ujjal Kumar Dey, Naveen Singh, Najimuddin Khan, Sampurn Anand, Arindam Mazumdar, Soumya Sadhukhan, Soumya Jana, Sukannya Bhattacharya, Ayon Patra, Tripurari Srivastava, Abhass Kumar and Abhijit Saha for keeping me up-to-date about the new avenues of research over the years. Many of my collaborators have contributed plots and illustrations for this book. I thank them all for their immense help.

I thank my long-term collaborators Eduard Masso, Anjan Joshipura, Durga Prasad Roy, Sandip Pakvasa, Gaetano Lambiase, Aragam Prasanna and others from whom I have learnt much of what appears in this book.

I thank my colleagues at PRL for their support.

I thank my wife Srubabati and daughter Anushmita for among other things keeping me fed and watered in the trying times during the completion of the book.

I thank B. Ananthanaryan, Indian Institute of Science, Bangalore (Editorial Board member of SpringerBriefs in Physics) for his encouragement and support.

And last but not the least, I thank my Springer Editor Lisa Scalone for patiently guiding me till the completion of this book.

Ahmedabad, India   Subhendra Mohanty
May 2020

Contents

About the Author

Subhendra Mohanty obtained his Ph.D. from the University of Wisconsin–Madison and, after a few post-doctoral stints at CERN, ICTP and other institutions, has been at the Physical Research Laboratory (India), where he is currently a Senior Professor. He has also been a Visiting Professor at the Universitat Autonoma de Barcelona. Dr. Mohanty has given numerous courses on various aspects of particle physics and cosmology at his home institution and outside, being an early adopter of the genre of Astroparticle Physics. He has published more than a hundred research papers with his local and international collaborators and has trained numerous Ph.D. students and young researchers in the field.

Acronyms

BH	Black hole
CDM	Cold dark matter
CMB	Cosmic microwave background
CW	Coleman–Weinberg
DE	Dark energy
DM	Dark matter
EFT	Effective field theory
FOPT	First order phase transition
FRW	Friedmann–Robertson–Walker
GW	Gravitational waves
HDM	Hot dark matter
LSS	Large scale structure
NS	Neutron star
PBH	Primordial black hole
SIDM	Self interacting dark matter
SOPT	Second order phase transition
WD	White dwarf
WDM	Warm dark matter
WIMP	Weakly interacting massive particle

Introduction

1

Abstract

We give a survey of the recent experimental observations which started the multi messenger era of astronomical observations like observations of gravitational waves, gamma rays, neutrinos and cosmic rays. We also list some important unsolved issues raised by these observations.

1.1 Gravitational Wave Observation by LIGO

From the earliest time of observations of astronomical photons have been our messengers from the universe outside. The spectral range of photons which is currently used for imaging the universe goes the whole range from radio, infra-red, sub-millimeter, optical, UV, x-ray and gamma rays. In addition to this wide range of photons, extra-terrestrial signals in the form of PeV energy neutrinos at IceCube. The most anticipated signal in astronomy which was fulfilled is the first observation of gravitational waves from merger of binary black holes [1]. This opened a new window to the observation of the universe through gravitational waves.

The first gravitational direct observation of gravitational waves was from of black-hole binary merger event GW150914 by the Hanford and Livingstone gravitational wave detectors of Ligo [1]. This event also marked the first observation of black holes. The GW150914 even was from the merger of black holes of masses $36^{+5}_{-4}M_\odot$ and $29 \pm 4M_\odot$ from a distance of 410^{160}_{180} Mpc which formed a final black hole of mass $62 \pm 4M_\odot$ and radiated away $3.0 \pm 0.5M_\odot$ energy in the form of gravitational waves. In a typical black hole merger into a single black hole, the frequency and the amplitude rise till the merger (the "chirp signal") and there is a "ring down" of the final black hole which loses the energy in the perturbations of its horizon as gravitational waves, as shown in Fig. 1.1.

© Springer Nature Switzerland AG 2020
S. Mohanty, *Astroparticle Physics and Cosmology*,
Lecture Notes in Physics 975, https://doi.org/10.1007/978-3-030-56201-4_1

Fig. 1.1 The first detection
by Ligo of gravitational
waves from the black hole
binary merger event
GW150914. Plot taken from
[1]

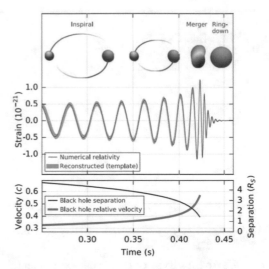

The first indirect but high precision observation of gravitational waves was the Hulse–Taylor binary binary pulsar [2]. The observation of the time period through many decades established and matching the observed time period loss with the prediction of the period loss by gravitational wave [3] the existence of gravitational waves was established. Binary pulsars in pulsar-neutron star binaries [4] and pulsar-white dwarf binaries [5] continue to provide the most stringent test of Einsteins gravity.

1.2   Multi-Messenger Signal at LIGO, VIRGO, IceCube and Fermi

The significant new development in multi-messenger astronomy was the observation of neutron-star binary mergers from the gravitational wave signal GW170817 by the two LIGO detectors the VIRGO detector [6]. By the triangulation of the gravitational wave directions by the three detectors the sours was identified to be in the NGC4993 at a distance of 40 Mpc. Gamma ray signals observed by the Fermi and INTEGRAL telescopes [7] after 1.74 ± 0.05 s of the end of the GW signal and lasting 2 s identified the event as the gamma-ray burst event GRB170817, see Fig. 1.2. After time delay of 15.9 s in the same source was observed by Swift in the ultraviolet [7], after 9 days the source was observed in x-rays by Chandra [7] and 16 days later the same source was observed by radio telescopes [9].

A different multi messenger observation was the observation of a 290 TeV neutrino by IceCube following the observations by Fermi and Swift of gamma rays (of upto 400 GeV) from the blazer TXS 0506+056 [10] which was in a flaring state. This constituted the first observation of simultaneous emission in gamma rays photons in the flaring and 100 TeV neutrinos from a source. It also is the

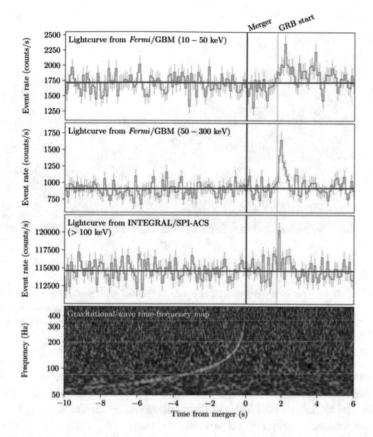

Fig. 1.2 Gravitational wave signal from the neutron star merger event GW170817 observed by Ligo and the gamma rays from the same event identified as GRB170817observed by Fermi and Integral after 1.7 s of the merger. Plot taken from [8]

first identification of a source of the IceCube extraterrestrial neutrinos, subsequent archival search by IceCube the same source had was could account for ν_μ.

1.3   IceCube Observation of PeV Neutrinos

IceCube [11] has observed more than 100 events with neutrino energy in the 0.1–10 PeV range. There are two types of signals. The upward going muon neutrinos which produce muons within a km of the detector are registered as $\nu_+ \bar{\nu}_\mu$ events and they give clear directionality of the source. Since the earth is needed as a shielding for the atmospheric muons, only half the sky is accessible to the muon neutrino events. The flux of the muon neutrinos is $\Phi_{\nu_\mu + \bar{\nu}_\mu} = \left(1.01^{+0.26}_{-0.23}\right) \times 10^{-18}\,\mathrm{Gev}^{-1}\,\mathrm{cm}^{-2}\,\mathrm{s}^{-1}\,\mathrm{sr}^{-1}$ (normalized at 100 TeV and a spectral index of 2.19 $\pm$

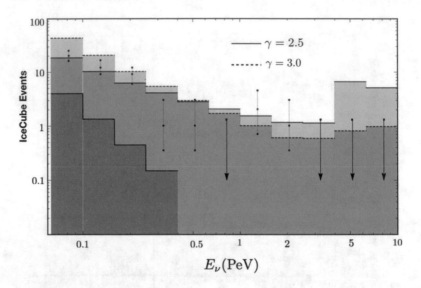

Fig. 1.3 Neutrino events in 6 years of IceCube data [11] fitted with a single power law neutrino spectrum. The neutrino spectrum is taken as a power law $d\Phi_\nu/dE_\nu \propto E_\nu^\gamma$, with gamma in the range $2.5 \leq \gamma \leq 3$. $\gamma = 2.5$ gives a better fit at the low energy while missing the upper bounds at the Glashow resonance region $E_\nu \simeq 6.3\,\text{PeV}$

0.1. The other type of signal is the one where the neutrinos produce e or τ showers inside the detector. These so called HESE (high energy starting events) events constitute the majority of the PeV neutrinos observed by IceCube.

As one can see from Fig. 1.3 a common power-law does not fit the entire spectrum of observations. It is possible that the PeV events are not part of a continuum but are the peak as expected from high energy neutrinos produced from PeV mass dark matter by either annihilation or decay [12]. Finding a PeV mass dark matter candidate which has the correct cosmological relic density whose lifetime is of the order of 10^{27} s required to explain IceCube flux provides a motivation for new physics like PeV scale Supersymmetry [13].

1.4   LSS and CMB

Measurements of temperature and the E-mode polarisation anisotropies of the cosmic microwave background (CMB) has established the idea of inflation as the most successful theory which can explain the CMB anisotropies. The specific particle physics or general relativity model of inflation which was responsible for the exponential expansion and generation of the scale invariant perturbation is not clear. Measurement of the B-mode polarization in the CMB (see Fig. 1.4) which is generated by the gravitational waves from the inflationary era will not only establish the general idea of inflation but can also go a long way in nailing down the

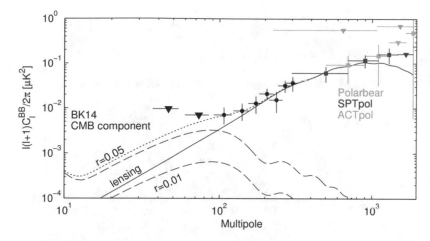

Fig. 1.4 B-mode polarisation measurement by BICEP2/Keck (BK14) [14], ACTPol, SPTPol and Polarbear. Inverted triangle denotes upper bound at 95% CL. Long dashed lines show B-mode polarisation predicted for different values of tensor to scalar ratio r which can arise in inflation model. Solid line is the B-mode from gravitational lensing of E-modes. Dotted line is the total B-mode from lensing plus primordial B with $r = 0.05$. Figure taken from Ref. [14]

specific inflation model -or will rule out many models of inflation. Ground based observations from the next generation terahertz telescopes [15] can measure the tensor to scalar ratio $r < 0.002$ (at 95% CL) and will be able to test some favorite models of inflation like Starobinsky inflation from R^2 gravity [16].

Galaxy surveys like the 2dF (2 degree Field) redshift survey and Sloan Digital Sky Survey (SDSS) have played an important role in establishing the standard ΛCDM cosmological model. The matter power spectrum which the theories predict can be tested with the galaxy distributions revealed by these surveys. The ongoing Dark Energy Survey (DES) and Extended Baryon Oscillation Spectroscopic Survey (eBOSS) and the upcoming Euclid satellite will map the matter power spectrum which will enable to unravel fundamental issues like properties of dark matter and dark energy and put bounds on neutrino masses and interactions [17].

1.5   Phase Transitions and Stochastic Gravitational Waves

As the universe cools the shape of the scalar potentials change sometime and develop new minima. If the universe evolves adiabatically from one minima to another then it is a Second order phase transition. There is no energy latent energy released in this process. There is the other possibility that the universe tunnels through an energy barrier from the higher energy local minima (false vacuum) to the lower energy minima (true vacuum) and the false vacuum. This is called a First order phase transition which is accompanied by release of the energy difference in the potential between the higher minima and the lower minima. First order PT

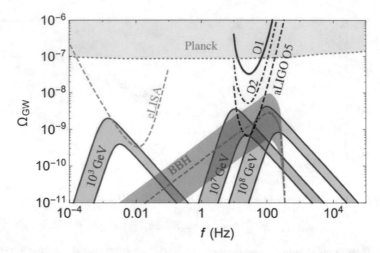

Fig. 1.5 The spectrum of stochastic gravitational waves from phase transitions at different temperatures. Figure taken from [21]

takes place through the process of bubble nucleation. The true vacuum bubble nucleates in the false vacuum background. The bubbles then grow and collide till the whole universe is in the true vacuum state. The growth and collision of the bubbles generates stochastic gravitational waves [18]. The gravitational waves redshift in te expanding universe and the frequency of the gravitational waves is related to the time (or temperature) of the phase transition. The standard model electroweak symmetry breaking is a second order PT. The QCD phase transition which occurs at a temperature of 1 GeV is strong First order PT and the gravitational waves from QCD phase transitions should be observable with pulsar timing arrays [19]. The peak frequency of the stochastic gravitational waves in this case is 10^{-9} Hz. It is proposed that observations of a group of 40 pulsars with 100 nano-second timing accuracy can reveal this stochastic gravitational wave signal [20].

In Fig. 1.5 taken from Ref. [21] the temperature of different phase transitions is shown along with the planned gravitational wave detector which will have the optimal sensitivity for detecting it. If there is a phase transition at the PeV scale it may be observed by a LIGO.

References

1. B.P. Abbott et al. [LIGO Scientific and Virgo Collaborations], Observation of Gravitational Waves from a Binary Black Hole Merger. Phys. Rev. Lett. **116**(6), 061102 (2016)
2. R.A. Hulse, J.H. Taylor, Astrophys. J. **195**, L51 (1975)
3. P.C. Peters, J. Mathews, Gravitational radiation from point masses in a Keplerian orbit. Phys. Rev. **131**, 435 (1963)
4. J. Antoniadis et al., A massive pulsar in a compact relativistic binary. Science **340**, 6131 (2013)

5. P.C.C. Freire et al., The relativistic pulsar-white dwarf binary PSR J1738+0333 II. The most stringent test of scalar-tensor gravity. Mon. Not. Roy. Astron. Soc. **423**, 3328 (2012)
6. B.P. Abbott et al. [LIGO Scientific and Virgo Collaborations], GW170817: Observation of gravitational waves from a binary neutron star inspiral. Phys. Rev. Lett. **119**(16), 161101 (2017)
7. B.P. Abbott et al., Multi-messenger observations of a binary neutron star merger. Astrophys. J. **848**(2), L12 (2017)
8. B.P. Abbott et al. [LIGO Scientific and Virgo and Fermi-GBM and INTEGRAL Collaborations], Gravitational waves and gamma-rays from a binary neutron star merger: GW170817 and GRB 170817A. Astrophys. J. **848**(2), L13 (2017)
9. G. Hallinan et al., A radio counterpart to a neutron star merger. Science **358**, 1579 (2017)
10. M.G. Aartsen et al., Multimessenger observations of a flaring blazar coincident with high-energy neutrino IceCube-170922A. Science **361**(6398), eaat1378 (2018)
11. C. Kopper [IceCube], Observation of astrophysical neutrinos in six years of IceCube data. PoS **ICRC2017**, 981 (2018)
12. A. Esmaili, P.D. Serpico, Are IceCube neutrinos unveiling PeV-scale decaying dark matter? JCAP **11**, 054 (2013)
13. J.D. Wells, PeV-scale supersymmetry. Phys. Rev. D **71**, 015013 (2005)
14. P.A.R. Ade et al., Improved constraints on cosmology and foregrounds from BICEP2 and Keck array cosmic microwave background data with inclusion of 95 GHz band. Phys. Rev. Lett. **116**(3) (2016). https://dash.harvard.edu/handle/1/30403683
15. K.N. Abazajian et al. [CMB-S4 Collaboration], *CMB-S4 Science Book*, 1st edn. arXiv:1610.02743 [astro-ph.CO]
16. A.A. Starobinsky, A new type of Isotropic cosmological models without singularity. Phys. Lett. **91B**, 99 (1980). [Adv. Ser. Astrophys. Cosmol. **3**, 130 (1987)]
17. L. Amendola, S. Appleby, D. Bacon, et al., Cosmology and fundamental physics with the Euclid satellite. LRR **16**, 6 (2013)
18. C. Caprini, D.G. Figueroa, Cosmological backgrounds of gravitational waves. Classical Quantum Gravity **35**(16), 163001 (2018)
19. C. Caprini, R. Durrer, X. Siemens, Detection of gravitational waves from the QCD phase transition with pulsar timing arrays. Phys. Rev. D **82**, 063511 (2010)
20. F.A. Jenet, G.B. Hobbs, K. Lee, R.N. Manchester, Detecting the stochastic gravitational wave background using pulsar timing. Astrophys. J. Lett. **625**, L123–L126 (2005)
21. P.S.B. Dev, A. Mazumdar, Probing the scale of new physics by advanced LIGO/VIRGO. Phys. Rev. D **93**(10), 104001 (2016)

Dark Matter

2

Abstract

We discuss the phenomenology of dark matter at cosmological and galactic scales. Direct detection experiments like Xenon-1T have ruled out a large swathe of the mass vs. DM-nucleon cross-section parameter space challenging the conventional WIMP paradigm of the $\sim 100\,\text{GeV}$ weakly interacting dark matter. New mechanisms for relic density like the freeze-in production of dark matter or the $3 \rightarrow 2$ annihilation process can give the correct relic density. These dark matter particles have very small couplings to standard model particles and lower ($\sim \text{MeV}$) mass therefore they can evade the stringent constraints from direct detection experiments. Dark matter can be self interacting thereby solving some problems with cold dark matter at galactic scales. Dark matter may be very light with mass $10^{-21}\,\text{eV}$ or they may be inelastic.

2.1 Introduction

The first evidence of the existence of dark matter was given by Fritz Zwicky, who noticed that velocities of galaxies in the Coma cluster were too large ($\sim 1000\,\text{km/s}$) for the galaxies to be gravitationally bound to the cluster unless the cluster mass was much larger than the mass of the luminous stars [1,2].

The evidence of dark matter in the galaxies came from the measurement of the circular velocity of gas in the galaxies as a function of distance from the center. These rotation curves which were plotted using optical [3,4] and 21-cm observations [5–8] from galaxies of many types, established the idea that the there was dark matter in the galaxies with mass to luminosity ratio $M/L \sim 10$.

The first elementary particle to be considered as dark matter was the neutrino . It was pointed out by Gershtein and Zeldovich [9] and Cowsik and McClelland [10] that since the neutrino temperature is close to the photon temperature the neutrino number density $n_\nu \sim 115\,\text{cm}^{-3}$, for the neutrinos to not overclose the universe

© Springer Nature Switzerland AG 2020

S. Mohanty, *Astroparticle Physics and Cosmology*,

Lecture Notes in Physics 975, https://doi.org/10.1007/978-3-030-56201-4_2

the total mass neutrino mass must be less than $93\,\mathrm{eV\,h^2}$. The neutrino as a hot dark matter (HDM are relativistic at the time of decoupling) was not favoured for structure formation as their free -streaming after decoupling would erase density perturbations smaller than $\sim 20(\mathrm{eV}/m_\nu)^{1/2}\,\mathrm{h^{-1}}$ Mpc [11–14].

Weinberg and Lee [15] and Hut [16] observed that if neutrinos are heavy and freeze-out when they are non-relativistic then the Cowsik–Mclleland bound did not apply anymore and they derived the cosmological neutrino bound for heavy neutrinos. Kolb and Olive [17] did the calculation for both Dirac and Majorana neutrinos. They showed that weakly interacting massive particles decouple when they are non-relativistic and can provide the closure density of the universe if they have masses in the $\sim 10\,\mathrm{GeV}$ scale. This was the foundation of the Wimp (Weakly interacting massive particle) dark matter paradigm: dark matter with mass $m \simeq (10\text{–}100)\,\mathrm{GeV}$ and annihilation cross section $\sigma v \simeq 3 \times 10^{-26}\,\mathrm{cm^3\,s^{-1}}$ has a relic density $\Omega_m \simeq 0.25$ which matches the observations [18].

In particle physics the ready candidates for Wimp dark matter were the neutralinos—the super-partners of neutral gauge and scalar bosons of the supersymmetric standard model [19]. The R-parity in SUSY ensures that the lightest SUSY particle is stable and a candidate for dark matter.

Goodman and Witten [20] proposed the method of direct detection of dark matter by coherent scattering of heavy nuclei. They classified the DM-nucleus interactions into spin-dependent and spin-independent couplings.

Major experimental efforts to observe the direct DM-nucleus scattering have ruled out a large parameter space for the allowed Wimp cross sections. This has spurred investigation into new models of dark matter which can give the correct relic density with smaller couplings to standard model particles and have lower masses. In this chapter we will study the galactic dark matter which reveals the small scale problem with cold dark matter. Then we will study relic density calculations for Wimps, SIDM, freeze-in dark matter and ultra-light dark matter. We then give a survey of the direct detection experiments and their implications for the dark matter mass and couplings parameter space.

2.2   Dark Matter in the Galaxies

Evidence of dark matter in galaxies comes from the rotation curves which are a plot of the circular velocity of gas and stars in the galaxy as a function of distance. If all the mass in the galaxy had been located in the central bulge where most of the luminous stars reside then the rotation speed away from the core would be $v(r) = \sqrt{GM/r}$. What is seen in observation of galaxies is that $v(r)$ rises linearly near the center and becomes flat away from the core. The flat rotation curves prove that the total enclosed mass grows linearly with distance and there is non-luminous mass than the stars and the gas in the galaxy.

The circular velocity of the gas is related to the density profile $\rho(r)$ in the galaxy as

$$v^2(r) = \frac{G}{r} \int_0^r 4\pi r'^2 dr' \, \rho(r') \tag{2.1}$$

The early observations of the rotation curves gave a flat rotation curve $v(r > r_c) \simeq constant$ away from the cores of the galaxies. To get a flat rotation curve the density profile must go as $\rho(r) \propto r^{-2}$ at $r \gg r_c$. Observations of the rotation curves at the central regions show that at $r < r_c$ the velocity rises linearly $v(r) \propto r$. To achieve this the density profile near the core must behave as $\rho(r) \simeq constant$.

To fit the rotation curves there are several possibilities for the dark matter distribution, which we will consider in turn.

2.2.1   Isothermal Distribution

If dark matter consists collisionless non-relativistic particles then we can assume that their phase space distribution function $f(\mathbf{x}, \mathbf{v}, t)$ does not change in time. This gives us the homogenous Boltzmann equation,

$$\frac{df}{dt} = 0 \Rightarrow \frac{\partial f}{\partial t} + \mathbf{v} \cdot \nabla_x f + \dot{\mathbf{v}} \cdot \nabla_v f = 0 \tag{2.2}$$

The acceleration is due to the gravitational potential of all the other particles. We denote the gravitational potential of all the particles as $\Phi(\mathbf{x}, t)$, then we have

$$\dot{\mathbf{v}} = -\nabla\Phi \tag{2.3}$$

and we can write the BE (2.2) for the steady state distribution ($\partial f/\partial t = 0$) as

$$\mathbf{v} \cdot \nabla_x f - \nabla\Phi \cdot \nabla_v f = 0. \tag{2.4}$$

One solution of this equation is

$$f(\mathcal{E}) = \frac{\rho_1}{(2\pi\sigma^2)^{3/2}} e^{-\mathcal{E}/\sigma^2}, \quad \mathcal{E} = \frac{\mathbf{v}^2}{2} + \Phi(\mathbf{x}) \tag{2.5}$$

The integral of (2.5) w.r.t velocity $\mathbf{v}$ gives the spatial distribution of the dark matter in the galaxy

$$\rho(r) = \int d^3\mathbf{v} \, f(\mathcal{E}) = \rho_1 e^{-\Phi(r)/\sigma^2} \tag{2.6}$$

Although the dark matter fluid is assumed to be collisionless the form of the distribution function is that of a particles in thermal equilibrium with an effective temperature $k_B T = m\sigma^2$, hence the name isothermal distribution. From (2.6) we get the relation between the dark matter density and the Newtonian potential for the isothermal distribution

$$\Phi(r) = -\sigma^2 \ln \rho(r) \tag{2.7}$$

This relation can be used to derive the form of Φ and $\rho(r)$ by using the Poisson equation

$$\nabla^2 \Phi(r) = -4\pi G \rho(r) \tag{2.8}$$

Substituting (2.7) in (2.8) we obtain

$$\frac{1}{r^2} \frac{d}{dr} \left(r^2 \frac{\partial \ln \rho}{\partial r} \right) = \frac{4\pi G}{\sigma^2} \rho \tag{2.9}$$

This has the solution

$$\rho(r) = \frac{\sigma^2}{2\pi G} \frac{1}{r^2} \tag{2.10}$$

Poisson's equation (2.8) relates the Newtonian potential to the density profile but in this case we can simply use the relation (2.7) to derive Newtonian potential of the dark matter distribution,

$$\Phi(r) = v_c^2 \ln(r) . \tag{2.11}$$

Using the density profile (2.10) we calculate the mass of dark matter within a radius r,

$$M(r) = 4\pi \int_0^r dr' r'^2 \rho(r') = \frac{2\sigma^2}{G} r , \tag{2.12}$$

If this mass dominates the galactic gravitational mass then the rotation velocity of the gas and stars in the galaxy under the influence of dark matter mass is

$$v_c(r) = \left(\frac{GM(r)}{r} \right)^{1/2} = \sqrt{2}\sigma , \tag{2.13}$$

So the isothermal distribution gives a constant rotation velocity which is a good approximation to the observations at large distances from the center. The density profile (2.10) is called the singular-isothermal profile as it diverges at $r \to 0$. We can regulate this distribution by using the pseudo-isothermal profile [21] defined as

$$\rho_{PI}(r) = \frac{\rho_c R_c^2}{r^2 + R_c^2} \tag{2.14}$$

Here ρ_c is the central density and R_c is the core radius such that $\rho(r \ll R_c) = \rho_c$. The rotation velocity derived from (2.14) has the property $v_c(r \ll R_c) = (4\pi G\rho R_c^2)^{1/2}$ and $v_c(r \gg R_c) = (4\pi G\rho_c R_c^2/3)^{1/2}r$.

The velocity distribution of dark matter in the isothermal distribution is

$$f(\mathcal{E}) = \frac{\rho_1}{(\pi v_c^2)^{3/2}} e^{-v^2/v_c^2} \tag{2.15}$$

plays an important role in the calculation of the event rates for direct detection experiments and in particular the annual variation of dark matter detection rates. For calculating the event rates of dark matter scattering nuclei in the direct detection experiments we normally use the isothermal velocity distribution formula (2.15) although we use a different $\rho(r)$ like the NFW or the Bukhert profiles which fit the rotation curve of Milky Way better. The correct procedure would be to calculate the velocity distribution from the empirical rotation velocity curve $v_c(r)$ from observations like the [22] survey. An analytic expression connecting the gravitational potential Φ with the phase space function $f(\mathcal{E})$ was derived by Eddington [23] which assumes spherical symmetry. A more accurate approach is to numerically calculate the velocity distribution which is consistent with the dark matter density distribution function [24,25]. Analytical expressions generalising the Eddington formula to non-spherical cases have been derived in [26].

2.2.2 NFW Profile

The Navarro Frenk–White distribution [27] is an analytic fit to the collisionless cold dark matter simulations and is given by

$$\rho_{\text{NFW}} = \frac{\rho_0}{\left(\frac{r}{R_s}\right)\left(1 + \left(\frac{r}{R_s}\right)^2\right)} \tag{2.16}$$

This has two parameters (ρ_s, R_s). At $r \ll R_s$, $\rho_{\text{NFW}} \simeq (\rho_0 R_s)r^{-1}$ This differs from the requirement from observations that the velocity curve at the core goes as $v_c \sim r$ which implies $\rho(r < r_c) = constant$. The core of the halos predicted by CDM simulations is very cuspy (there is a steep rise in the density while the rotation curve observations require a core with constant density. This is the *cuspy-core problem of CDM* at small scales [21]. To summarise, the density profile has a r dependence near the core $\rho(r < R - c) \sim r^\alpha$ and observations demand that $\alpha = 0$ while CDM simulations give $\alpha = -1$. The solution of this problem may be theta instead of collisionless cold dark matter we have self interacting dark matter as illustrated in Fig. 2.1.

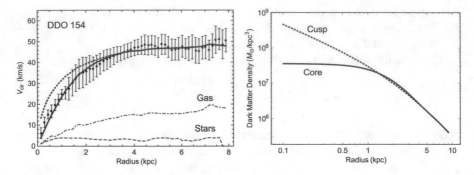

Fig. 2.1 Rotation curve of dwarf galaxy DDO 154 (black data points) compared to models with the cuspy NFW profile from CDM (dotted line) and cored profile from self-interacting dark matter (solid line). Figure taken from [28]

2.2.3   Burkert Profile

The Burkert profile [29] is a phenomenological fit to rotation curve data,

$$\rho_{\text{Burkert}}(r) = \frac{\rho r_0^3}{(r + r_0)(r^2 + r_0^2)} .\qquad(2.17)$$

It has two parameters (ρ_0, r_s), here ρ_0 is the central density r_0 is the core radius. The density goes as $\rho \propto r^0$ at $r < r_0$ and gives $v_c \propto r$ near the core. At $r \gg r_0$ it goes as $\rho \propto r^{-3}$ as the NFW profile.

2.2.4   The Local Density of Dark Matter

The local density of dark matter is obtained from the rotation curve and the vertical motion of the stars in the disc [30–32]. The recent determination of the dark matter density at the location of the Sun is [32],

$$\rho(R_\odot) = 0.46^{+0.07}_{-0.09}\,\text{GeV}\,\text{cm}^{-3} .\qquad(2.18)$$

The value of the local dark matter density is a critical input for determining the expected rates from the direct detection experiments and a change in this number shifts the exclusion region of dark matter—nucleon scattering cross section.

2.3   Dark Matter at Cosmological Scales

The density of dark matter at cosmological scales is inferred from its effect on the cosmic microwave temperature anisotropy spectrum. The change in CMB temperature anisotropy with baryon fraction Ω_b is shown in Fig. 2.2. The peaks and

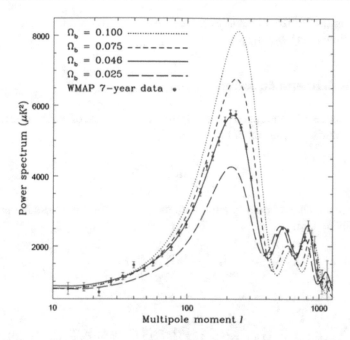

Fig. 2.2 The change in cosmic microwave anisotropy temperature with baryon fraction Ω_b. Comparing with WMAP-7 year data [33] we see that the baryon fraction is 4.6%. The location of the first peak shows $\Omega_{total} = 1.001 \pm 0.002$ proving the existence of dark matter spread at cosmological scales

valleys are due to plasma oscillations of baryon-photon fluid in the gravitational potential of dark matter. The location of the first peak at $l \sim 200$ shows that the universe is flat with $\Omega_c + \Omega_b + \Omega_\lambda = 1.001 \pm 0.002$ [10]. The height of the first peak that is due to baryon acoustic oscillations baryon fraction $\Omega_b h^2 = 0.2233 \pm 0.00015$. The cold dark matter fraction $\Omega_c h^2 = 0.1198 \pm 0.0012$.

2.3.1   Relic Density of Dark Matter in the Universe

Dark matter can be in thermal equilibrium in the early universe but if it remains in thermal equilibrium at temperatures much below its mass then its number density will decrease exponentially as $n_{eq} = g(mT/2\pi)^{3/2} \exp(-m/T)$. In order to survive as dark matter with appreciable density in the present universe the number changing interactions of dark matter particle to standard model particles $\chi\chi \rightarrow f\bar{f}$ must freeze-out. This means that at some temperature known as the freezout temperature the interaction rate $\Gamma = \langle \sigma v \rangle n_\chi$ must drop below the Hubble expansion rate H. The decoupling if number changing interactions does not occur at a sharp temperature boundary and a better calculation of the dark matter relic density can be done by considering the Boltzman equation which takes into account the interactions from the early time to the and the present epoch.

2.3.2 Boltzmann Equation

In the absence of interactions the number of particles of each species $a^3 n$ is conserved during expansion of the universe

$$\frac{1}{a^3}\frac{d}{dt}(a^3 n) = 0 \tag{2.19}$$

When there are interactions say annihilations $1+2 \leftrightarrow 3+4$ the Boltzmann equation has a collision term

$$\frac{1}{a^3}\frac{d}{dt}(a^3 n_1) = C(f) \tag{2.20}$$

The collision term is given by

$$C(f) = \sum_{s_i} \int d\Pi_1 \, d\Pi_2 \, d\Pi_3 \, d\Pi_4 \, (2\pi)^4 \delta^4 (P_1 + P_2 - P_3 - P_4)$$

$$\times \left\{ |M_{34 \to 12}|^2 f_3 f_4 (1 \pm f_1)(1 \pm f_2) - |M_{12 \to 34}|^2 f_1 f_2 (1 \pm f_3)(1 \pm f_4) \right\} \tag{2.21}$$

where the phase space factor

$$d\Pi_i = \frac{d^3 p_i}{(2\pi)^3 2 E_i}. \tag{2.22}$$

and the summation is over the spin states s_i, $(i = 1, 2, 3, 4)$ of all the particles. When CP or T is conserved then the amplitude squared of the forward and backward reactions are equal $|M_{12 \to 34}|^2 = |M_{34 \to 12}|^2 = |M|^2$. The rate of decrease in the number of a particle species (for example species 1) will be proportional to the amplitude squared $|M|^2$ and the distribution functions f_1 and f_2. The reverse reaction which increases n_1 would be proportional to f_3 and f_4. The probability of outgoing 3 and 4 would be Fermi-blocked or Bose enhanced by factors $(1 \pm f_3)(1 \pm f_4)$ (where the $+$ sign is for bosons and—sign for fermions).

For our application of relic density of cold dark matter calculation the distribution functions are of the dilute gas $f_i \ll 1$ and the Fermi-blocking and Bose-enhancement factors are just unity. The equilibrium distribution is

$$f_i^{eq}(E_i) = \frac{1}{e^{E_i/T} \pm 1} \simeq e^{-E_i/T} \tag{2.23}$$

The actual distribution functions f_i will deviate from the equilibrium distributions due to interactions. The deviation of the actual distribution functions from the equilibrium values is represented by a chemical potential μ_i,

$$f_i(E_i) = \frac{1}{e^{(E_i - \mu_i)/T} \pm 1} \simeq e^{\mu_i/T} e^{-E_i/T} \tag{2.24}$$

The ratio between the actual number densities n_i and their equilibrium value n^{eq} is therefore

$$\frac{n_i}{n_i^{eq}} = e^{\mu/T} \tag{2.25}$$

Under these conditions the factors of the distribution functions in the expression (2.21) for $C(f)$ reduce to the form

$$f_3 f_4 (1 \pm f_1)(1 \pm f_2) - f_1 f_2 (1 \pm f_3)(1 \pm f_4) = f_3 f_4 - f_1 f_2$$
$$= e^{(\mu_3 + \mu_4)/T} e^{-(E_3 + E_4)/T} - e^{(\mu_1 + \mu_2)/T} e^{-(E_1 + E_2)/T}$$
$$= e^{-(E_1 + E_2)/T} \left[\frac{n_3 n_4}{n_3^{eq} n_4^{eq}} - \frac{n_1 n_2}{n_1^{eq} n_2^{eq}} \right] \tag{2.26}$$

where in the final step we used the energy-momentum conservation relation $E_1 + E_2 = E_3 + E_4$. In the dilute particle limit the collision term (2.21) can be written as

$$C(f) = \sum_{s_i} \int d\Pi_1 \, d\Pi_2 \, d\Pi_3 \, d\Pi_4 \, (2\pi)^4 \delta^4 (P_1 + P_2 - P_3 - P_4) |M|^2$$
$$\times e^{-(E_1 + E_2)/T} \left[\frac{n_3 n_4}{n_3^{eq} n_4^{eq}} - \frac{n_1 n_2}{n_1^{eq} n_2^{eq}} \right] \tag{2.27}$$

The scattering cross-section for the interaction $1 + 2 \rightarrow 3 + 4$ is

$$\sigma_{12} = \int d\Pi_3 d\Pi_4 \frac{1}{v_{12}} \frac{1}{g_1 \, g_2} \frac{1}{4E_1 \, E_2} (2\pi)^4 \delta^4 (p_1 + p_2 - p_3 - p_4) \times \sum_{s_i = 3,4} |M|^2 \tag{2.28}$$

where v_{12} is the Moller velocity,

$$v_{12} = \frac{1}{E_1 E_2} \left((p_1 \cdot p_2)^2 - (m_1 m_2)^2 \right)^{1/2} \tag{2.29}$$

The thermal average of the scattering rate $\sigma_{12}v_{12}$ over the initial particle momentum distribution is therefore

$$
\begin{aligned}
\langle \sigma_{12}v_{12} \rangle &= \frac{1}{n_1 n_2} \sum_{s_i=1,2} \int \frac{g_1 \, d^3 p_1}{(2\pi)^3} \frac{g_2 \, d^3 p_2}{(2\pi)^3} \, \sigma_{12}v_{12} \, f_1 f_2 \\
&= \frac{1}{n_1^{eq} n_2^{eq}} \int d\Pi_1 \, d\Pi_2 \, d\Pi_3 \, d\Pi_4 \, (2\pi)^4 \delta^4(P_1 + P_2 - P_3 - P_4) \\
&\quad \times \sum_{s_i=1,2,3,4} |M|^2 \, e^{-(E_1+E_2)/T}
\end{aligned}
\tag{2.30}
$$

where we used the relation $f_i = e^{\mu_i/T} e^{-E_i/T} = (n_i/n_i^{eq}) e^{-E_i/T}$. Similarly we can derive the thermal averaged scattering rate of the $3 + 4 \rightarrow 1 + 2$ process to be

$$
\begin{aligned}
\langle \sigma_{34}v_{34} \rangle &= \frac{1}{n_3 n_4} \sum_{s_i=3,4} \int \frac{g_3 \, d^3 p_3}{(2\pi)^3} \frac{g_4 \, d^3 p_4}{(2\pi)^3} \, \sigma_{34}v_{34} \, f_3 f_4 \\
&= \frac{1}{n_3^{eq} n_3^{eq}} \int d\Pi_1 \, d\Pi_2 \, d\Pi_3 \, d\Pi_4 \, (2\pi)^4 \delta^4(P_1 + P_2 - P_3 - P_4) \\
&\quad \times \sum_{s_i=1,2,3,4} |M|^2 \, e^{-(E_1+E_2)/T}
\end{aligned}
\tag{2.31}
$$

Using (2.27), (2.30), and (2.31) we can express the collision term as

$$
C(f) = \langle \sigma_{34}v_{34} \rangle n_3 n_4 - \langle \sigma_{12}v_{12} \rangle n_1 n_2
\tag{2.32}
$$

From (2.30) and (2.31) we obtain the detailed balance relation between the particle numbers at equilibrium

$$
\langle \sigma_{12}v_{12} \rangle \, n_1^{eq} \, n_2^{eq} = \langle \sigma_{34}v_{34} \rangle \, n_3^{eq} \, n_4^{eq} .
\tag{2.33}
$$

Using the detailed balance relation (2.33) the collision term (2.32) can be also written as

$$
C(f) = \langle \sigma_{12}v_{12} \rangle \, n_1^{eq} \, n_2^{eq} \left(\frac{n_3 n_4}{n_3^{eq} n_4^{eq}} - \frac{n_1 n_2}{n_1^{eq} n_2^{eq}} \right)
\tag{2.34}
$$

There would be no change in the particle numbers if all particles are at their thermal equilibrium values.[1] The Boltzmann equation for the change in n_1 is thus

$$\frac{1}{a^3} \frac{d}{dt}(a^3 n_1) = -\langle \sigma_{12} v_{12} \rangle \left(n_1 n_2 - \left(\frac{n_1 n_2}{n_3 n_4}\right)_{eq} n_3 n_4 \right) \tag{2.35}$$

The interaction rate of particle species "1" is $\Gamma_1 \equiv \langle \sigma_{12} v_{12} \rangle n_2$ and we can also write (2.35) as

$$\frac{dn_1}{dt} + 3H n_1 = -\Gamma_1 n_1 \left(1 - \left(\frac{n_1 n_2}{n_3 n_4}\right)_{eq} \frac{n_3 n_4}{n_1 n_2} \right) \tag{2.36}$$

which shows that when $\Gamma_1 \ll H$ the change in the number of particles in the combing volume $(a^3 n_1)$ is negligible and the particle densities only change due to Hubble expansion. When the expansion rate $\Gamma_1 \gg H$ the collision term on the r.h.s of (2.36) is significant and the particle number density changes mainly due to interactions. The temperature T_D at which $H(T_d) = \Gamma_1(T_D)$ is called the decoupling temperature and at $T < T_D$ the particle number density mainly changes due to Hubble expansion, and particle "1" decouples from the thermal bath of other particles.

There are two kinds of decoupling. The chemical equilibrium is maintained by annihilation processes like

$$\chi \chi \leftrightarrow f \bar{f} \tag{2.37}$$

while the kinetic equilibrium is maintained by scattering processes like

$$\chi f \leftrightarrow \chi f \tag{2.38}$$

Since the cold dark matter density is much smaller than the thermal bath particles $n_\chi \ll n_f$ the chemical decoupling (conventionally called freeze-out in the case of cold dark matter) occurs much before kinetic decoupling. For neutrinos and hot-dark matter where $n_\chi \sim n_f$ the kinetic decoupling and chemical decoupling occur at same time.

2.3.3   Relic Density of Cold Dark Matter

In most applications we have annihilations of particle-antiparticle pairs into another set of lighter standard model particle-antiparticle pair. If the final state particles

---

[1]For non-relativistic particles $n_i^{eq} = g_i(m_i T/2\pi)^{3/2} e^{-m_i/T}$. For relativistic particles $n_i^{eq} = g_i(\zeta(3)/\pi^2)T^3$ (bosons) and $(3/4)g_i(\zeta(3)/\pi^2)T^3$ (fermions).

produced attain thermal equilibrium with the radiation bath at time scales much smaller than the interaction rate then $n_3 = n_3^{eq}$, $n_4 = n_4^{eq}$. The Boltzmann equation (2.35) for number density n of particle X in the process $X + \bar{X} \leftrightarrow f + \bar{f}$ simplifies to the form

$$\frac{1}{a^3} \frac{d}{dt} \left(a^3 n \right) = -\langle \sigma v \rangle \left(n^2 - n_{eq}^2 \right) \tag{2.39}$$

The total entropy $(a^3 s)$ of all species is a constant even in the presence of number changing interaction so we can write Eq. (2.39) as

$$(a^3 s) \frac{1}{a^3} \frac{d}{dt} \left(\frac{a^3 n}{a^3 s} \right) = -\langle \sigma v \rangle \left(n^2 - n_{eq}^2 \right) \tag{2.40}$$

Now we can define the $a(t)$ independent quantity $Y \equiv n/s$ which does not change in the absence of interactions. The change of Y due to annihilations can then be obtained from (2.40),

$$\frac{dY}{dt} = -\langle \sigma v \rangle s \left(Y^2 - Y_{eq}^2 \right) \tag{2.41}$$

It is more convenient to use $x \equiv m/T$ as the time variable. The Boltzmann equation in terms of x is

$$\frac{dx}{dT} \frac{dT}{dt} \frac{dY}{dx} = -\langle \sigma v \rangle s \left(Y^2 - Y_{eq}^2 \right) \tag{2.42}$$

Now $dx/dT = -m/T^2$ and dT/dt can be computed using the conservation of total entropy

$$\frac{1}{g_* T^3 a^3} \frac{d}{dt} (g_* T^3 a^3) = 3\frac{\dot{a}}{a} + 3\frac{\dot{T}}{T} + \frac{\dot{g_*}}{g_*} = 0 \tag{2.43}$$

where overdot represents time derivative. Writing

$$\dot{g_*} = \frac{dT}{dt} \frac{dg_*}{dT} \tag{2.44}$$

and substituting in (2.43) we get the form of dT/dt,

$$\frac{dT}{dt} = -HT \left(1 + \frac{1}{3} \frac{d \ln g_*}{d \ln T} \right)^{-1} \tag{2.45}$$

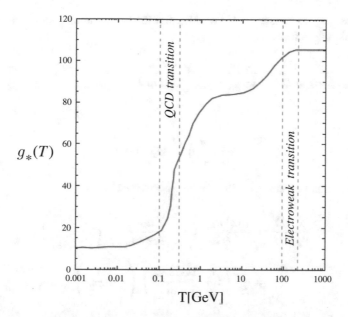

$g_*(T)$

Fig. 2.3 Total entropy degrees of freedom in the in the standard model as a function of temperature

Using (2.45) in Eq. (2.42) we get the Boltzman equation in the form

$$\frac{dY}{dx} = \frac{\langle \sigma v \rangle s}{Hx} \left(1 + \frac{1}{3}\frac{d \ln g_*}{d \ln T}\right) \left(Y_{eq}^2 - Y^2\right) \tag{2.46}$$

The entropy degrees of freedom $g_*(T)$ as a function of temperature is plotted in Fig. 2.3. There is a sharp drop during the quark hadron transition at which should be taken into account for the calculation of relic density. The Hubble expansion rate in the radiation era is

$$H = \left(\frac{8\pi G}{3}\right)^{1/2} \left(\frac{\pi^2}{30}\right)^{1/2} \sqrt{g_\rho} \left(\frac{m^2}{x^2}\right) \tag{2.47}$$

and the entropy density

$$s = \left(\frac{2\pi^2}{45}\right) g_* \frac{m^3}{x^3}. \tag{2.48}$$

Using these expressions in (2.46) we find that the Boltzmann equation can be put in the dimensionless form

$$\frac{dY}{dx} = \frac{\lambda}{x^2} \frac{g_*}{g_\rho^{1/2}} \left(1 + \frac{1}{3} \frac{d \ln g_*}{d \ln T}\right) \left(Y_{eq}^2 - Y^2\right) \tag{2.49}$$

with

$$\lambda = \left(\frac{\pi}{45}\right)^{1/2} M_P \, m \, \langle \sigma v \rangle \tag{2.50}$$

The Boltzman equation (2.49) can also be written as

$$\frac{x}{Y_{eq}} \frac{dY}{dx} = \frac{\langle \sigma v \rangle n_{eq}}{H} \left(1 + \frac{1}{3} \frac{d \ln g_*}{d \ln T}\right) \left(1 - \frac{Y^2}{Y_{eq}^2}\right)$$

$$= \frac{\Gamma}{H} \left(1 + \frac{1}{3} \frac{d \ln g_*}{d \ln T}\right) \left(1 - \frac{Y^2}{Y_{eq}^2}\right) \tag{2.51}$$

where $\Gamma = \langle \sigma v \rangle n_{eq}$ is the annihilation rate of the species. At high temperature when the annihilation rate is large i.e. $\Gamma \gg H$ then the comoving number Y changes rapidly and is driven to the equilibrium value $Y = Y_{eq} = 0.278(g/g_*)$ if the particles are relativistic at that temperature and $Y = Y_{eq} = 0.145(g/g_*)x^{3/2}e^{-x}$ if the particles are non-relativistic. If the annihilation rate stays in equilibrium with the heat bath then when the particles become non-relativistic ($x > 3$) the particles comoving number density Y_{eq} would go to zero exponentially with increasing x. The only way they can survive in substantial numbers is if the interaction rate drops below the Hubble expansion while x is not too large. The freeze-out temperature is defined as the temperature at which the rate of number changing interactions (like $\chi\chi \to f\bar{f}$) of cold dark matter decouples and is determined from the condition $\Gamma(T_f) = H(T_f)$. At temperatures $T < T_f$, $\Gamma < H$ and $dY/dx \to 0$ and the particle number changes mainly due dilution in the expanding universe. This implies that $Y(x_f) \simeq Y(x \to \infty)$.

Particle which survive annihilation by decoupling when they are already non-relativistic $x > 3$ are called cold relics. Prior to decoupling (when $x \ll x_f$ their comoving number density is $Y(x) = Y_{eq} = 0.145(g/g_*(x))x^{3/2}e^{-x}$. Since Y is changing exponentially with x at the time of freeze-out the simple criterion of of determining freeze-out temperature $H(t_f) = \Gamma(T_f)$ is not accurate and one must integrate the Boltzmann equation (2.49) to determine $Y(T_0)$ in the present epoch [34, 35]. An approximate analytical solution can be obtained as follows.

For non-relativistic initial states we can expand the cross section in powers of v,

$$(\sigma v) = \sigma_0(a + bv^2) \tag{2.52}$$

where for the s-wave annihilation a is non-zero while for p-wave annihilation $a = 0$ and $b \neq 0$. The thermal averaged cross section is then

$$
\langle \sigma v \rangle = \frac{x^{3/2}}{2\sqrt{\pi}} \int_0^\infty dv\, e^{-xv^2/4}\, \sigma_0 (a + bv^2)
$$

$$
= \sigma_0 \left(a + \frac{6b}{x} \right) \tag{2.53}
$$

We will consider the general case of both types of terms being present for the annihilation process $\chi\chi \to$ SM particles. We start with the Boltzmann equation (2.49) which we can write in the form

$$
\frac{dY}{dx} = \frac{\lambda_*}{x^2} \left(a + 6bx^{-1} \right) \left(Y_{eq}^2 - Y^2 \right) \tag{2.54}
$$

where

$$
\lambda_*(x) \equiv 0.26 \frac{g_*}{g_\rho^{1/2}} \left(1 + \frac{1}{3}\frac{d\ln g_*}{d\ln T} \right) \sigma_0 M_P\, m. \tag{2.55}
$$

The At high temperatures $Y(x) = Y_{eq}$ and $dY/dx \simeq 0$. At the *freeze-out temperature* (2.63) Y will start deviating from the equilibrium value Y_{eq} as the annihilation rate drops below the Hubble expansion.

After the freeze-out $Y(x)$ remains a constant while Y_{eq} goes to zero exponentially with x. After freeze-out when $Y(x) \gg Y_{eq}$ and we can drop Y_{eq} from the Boltzmann equation (2.54) to obtain the equation for $Y(x)$,

$$
\frac{dY}{dx} = -\lambda_*(a + 6bx^{-1})x^{-2}Y^2 \qquad x > x_f \tag{2.56}
$$

This equation can be integrated from $x = x_f$ to the present epoch $x = \infty$ to give

$$
\frac{1}{Y_\infty} - \frac{1}{Y_f} = \int_{x_f}^\infty dx\, \lambda_* \, (a + 6bx^{-1})x^{-2} \tag{2.57}
$$

If we ignore the variation of λ_* with x and evaluate the integral with $\lambda_*(x) = \lambda_*(x_f)$ (where the dominant contribution to the integration comes from) and use the approximation $1/Y_f \ll 1/Y_\infty$ then we get the present comoving density of cold relics as

$$
Y_\infty = \frac{1}{\lambda_*(x_f)} \frac{x_f}{\left(a + 3bx_f^{-1} \right)} \tag{2.58}
$$

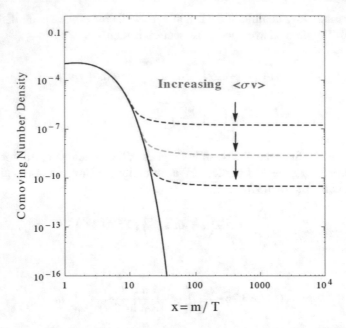

Fig. 2.4 Comoving relic density $Y(x)$ as a function of $x = m/T$ for different values of annihilation cross section. Larger cross-section results in later decoupling and lower relic density

The comoving relic density $Y(x)$ as a function of $x = m/T$ for different values of annihilation cross section is shown in Fig. 2.4. Larger cross-section results in later decoupling and lower relic density.

The number density of the cold relics in the present universe is

$$n_0 = s_0 Y_\infty \tag{2.59}$$

and the energy density of the cold relics in the present epoch is

$$\rho = m n_0 = m \, s_0 \frac{1}{\lambda_*(x_f)} \frac{x_f}{\left(a + 3bx_f^{-1}\right)}$$

$$= 3.78 s_0 \left[\frac{g_\rho^{1/2}}{g_*} \left(1 + \frac{1}{3}\frac{d \ln g_*}{d \ln T}\right)^{-1} \right]_{x=x_f} \frac{1}{\sigma_0 M_P} \frac{x_f}{\left(a + 3bx_f^{-1}\right)} \tag{2.60}$$

In order to get the numerical value for ρ we need to give an expression for x_f where $\Gamma(x_f) = H(x_f)$. The annihilation rate $n\langle \sigma v \rangle$ is

$$\Gamma(x) = n\langle \sigma v \rangle = 0.063 \, g \, m^3 e^{-x} x^{-3/2} (\sigma_0)(a + 6bx^{-1}) \tag{2.61}$$

where $g = 2$ for Majorana fermions, and $g = 1$ for scalar DM. The Hubble expansion rate is

$$H(x) = 1.66\, g_\rho^{1/2}\, m^2\, M_P^{-1} x^{-2} \tag{2.62}$$

The freeze-out temperature is obtained by solving $\Gamma(x_f) = H(x_f)$,

$$e^{x_f} = A x_f^{1/2}(a + 6bx_f^{-1}) \tag{2.63}$$

where

$$A = 0.038 g g_\rho^{-1/2} m M_P \sigma_0 \tag{2.64}$$

This equation can be solved iteratively for give the freeze-out x_f to be used in Eq. (2.60),

$$
\begin{aligned}
x_f &= \log(A) + \frac{1}{2}\log\left[\log(A)\right] + \log\left(a + 6b\left[\log(A)\right]^{-1}\right) \\
&= 23.5 + \log\left(\frac{\langle \sigma v \rangle}{2 \times 10^{-26}\,\text{cm}^3\,\text{s}^{-1}}\right) + \log\left(\frac{m}{100\,\text{GeV}}\right) \\
&\quad - 0.5\log\left(\frac{g_\rho}{100}\right) + \log(a + 0.27\,b)
\end{aligned}
\tag{2.65}
$$

The predicted relic density fraction of cold dark matter $\Omega_m \equiv \frac{\rho}{\rho_c}$ where $\rho_c = (3/8\pi G)H_0^2 = 1.05 \times 10^{-5} h^2\,\text{Gev/cm}^3$ is the critical density, turns out to be,

$$\Omega_m h^2 = 0.11\,\left(\frac{2 \times 10^{-26}\,\text{cm}^3 s^{-1}}{\langle \sigma v \rangle}\right)\left(\frac{10}{\bar{g}^{1/2}}\right)\left(\frac{x_f}{20}\right)\left(\frac{1}{a + 3bx_f^{-1}}\right) \tag{2.66}$$

where

$$\bar{g}^{1/2} \equiv \left[\frac{g_*}{g_\rho^{1/2}}\left(1 + \frac{1}{3}\frac{d\ln g_*}{d\ln T}\right)\right]_{x = x_f} \tag{2.67}$$

The cold dark matter density measured from the comic microwave anisotropy by PLANCK [10], $\Omega_{cdm} h^2 = 0.1188 h^2 \pm 0.001$. The fact that the weak interaction cross section gives the correct relic density supports the idea that dark matter particles are weakly interacting massive particles (WIMPS) and the coincidence of the weak interaction cross section giving the correct relic density is referred to as "the WIMP miracle". The relic density depends upon the annihilation cross section and on the mass of the dark matter through $\bar{g}(x_f)$.

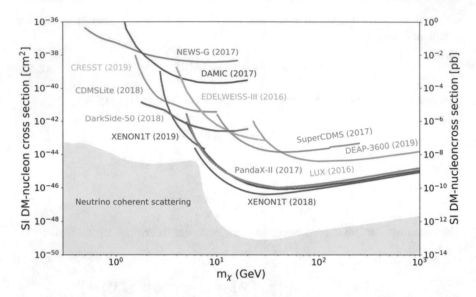

Fig. 2.5 Exclusion regions for WIMP-nucleon spin independent cross section versus WIMP mass at 90% CL from different experiments. The "neutrino floor" shaded region is the coherent scattering of solar neutrinos on nuclei expressed in equivalent Wimp-nucleon cross section. Plot taken from PDG-2018 [51]

2.3.4 Self Interacting Dark Matter

The WIMP miracle for producing the correct relic density works for 10–100 GeV mass dark matter which are severely constrained by non observation in direct detection experiments and CMB. Dark matter in the sub-GeV mass can have substantially larger couplings with standard model particles as can be seen from the direct detection exclusion plot Fig. 2.5. Due to the lower DM mass it will not impart sufficient energy to nucleons on collision and can evade detection.

Dark matter annihilation cross to standard model particles in the post recombination epoch will ionize Hydrogen which can be observed in the CMB. DM with annihilation cross section to SM particles $\langle \sigma v \rangle \simeq 3 \times 10^{-26} \, \text{cm}^3 \, \text{s}^{-1}$ is ruled out by Planck observations [10] for dark matter masses $m_\chi < 30 \, \text{GeV}$.

To obtain the required relic density sub-GeV dark matter, it must have large self interaction so that its number changes by the $3\chi \rightarrow 2\chi$ annihilation process [36,37] as we will discuss in this section.

The Boltzmann equation for the $1 + 2 + 3 \rightarrow 4 + 5$ annihilation process is given by

$$\frac{dn}{dt} + 3Hn = C_{3\rightarrow 2} \tag{2.68}$$

and the collision term is

$$C_{3 \to 2} = \frac{1}{s_i s_f} \int d\Pi_1 d\Pi_2 \cdots d\Pi_5 \, (2\pi)^4 \, \delta \left(\sum_{i=1}^5 p_i \right) |\mathcal{M}_{2 \to 3}|^2 \, (f_1 f_2 f_3 - f_4 f_5)$$

(2.69)

where the symmetry factors $s_i = 3!$ and $s_f = 2!$ for three identical particles in the initial state and two identical particles in the final state. We can express the distributions f_i in terms of the equilibrium distribution as follows

$$f_i = \left(\frac{n_i}{n_i^{eq}} \right) f_{eq} = \left(\frac{n_i}{n_i^{eq}} \right) e^{-\frac{E_i}{T}}$$

(2.70)

Using and (2.70) and the energy conservation $E_1 + E_2 + E_3 = E_4 + E_5$ we can write (2.69) in the form

$$C_{3 \to 2} = \left(\frac{1}{s_i s_f} \right) \frac{1}{n_1^{eq} n_2^{eq} n_3^{eq}} \int d\Pi_1 d\Pi_2 \cdots d\Pi_5 \, (2\pi)^4 \, \delta \left(\sum_{i=1}^5 p_i \right) |\mathcal{M}_{2 \to 3}|^2$$

$$\times e^{-(E_1 + E_2 + E_3)/T} \left(n_1 n_2 n_3 - \frac{n_1^{eq} n_2^{eq} n_3^{eq}}{n_4^{eq} n_5^{eq}} n_4 n_5 \right)$$

(2.71)

The collision term (2.71) can be written in terms of the thermal averaged $3 \to 2$ annihilation cross section.

The annihilation cross section is for the $3 \to 2$ process is

$$(\sigma v^2)_{3 \to 2} = \frac{1}{8 E_1 E_2 E_3} \int d\Pi_4 d\Pi_5 \, (2\pi)^4 \, \delta \left(\sum_{i=1}^5 p_i \right) |\mathcal{M}_{2 \to 3}|^2$$

(2.72)

and its thermal average over incoming particle distributions is

$$\langle \sigma v^2 \rangle_{3 \to 2} = \left(\frac{1}{s_i s_f} \right) \frac{1}{n_1^{eq} n_2^{eq} n_3^{eq}} \int \frac{d^3 p_1}{(2\pi)^3} \frac{d^3 p_2}{(2\pi)^3} \frac{d^3 p_3}{(2\pi)^3} e^{-(E_1 + E_2 + E_3)/T} (\sigma v^2)_{3 \to 2}$$

(2.73)

When we have identical particles in the initial and final states we can take $n_i^{eq} = n^{eq}$ and $n_i = n$ ($i = 1 \cdots 5$) and the collision term (2.71) can therefore be written in terms of the thermal averaged annihilation cross section as

$$C_{3 \to 2} = -\langle \sigma v^2 \rangle_{3 \to 2} \left(n^3 - n^{eq} n^2 \right).$$

(2.74)

For the $3 \to 2$ process the dimension of $[|\mathcal{M}_{2 \to 3}|^2] = E^{-1}$ and of cross section is $[\sigma v^2] = E^{-5}$. The annihilation rate is $\Gamma_{3 \to 2} = n^{eq2} \langle \sigma v^2 \rangle_{3 \to 2}$. The thermal freeze-out takes place at the freeze out temperature T_f when $\Gamma_{3 \to 2}(T_f) = H(T_f)$.

The Boltzmann equation (2.68) can be written in terms of $Y = n/s$ as a function of $x = m/T$,

$$\frac{dY}{dx} = -\frac{x \langle \sigma v^2 \rangle_{3 \to 2} s^2}{H(m)} \left(1 + \frac{1}{3} \frac{d \ln g_*}{d \ln T}\right) \left(Y^3 - Y^2 Y_{eq}\right) \tag{2.75}$$

Assuming the cross thermal averaged cross section to have a power law temperature dependence

$$\langle \sigma v^2 \rangle_{3 \to 2} = \sigma_0 \, x^{-n} \tag{2.76}$$

we have can write the Boltzmann equation as

$$\frac{dY}{dx} = -\kappa x^{-n-5} \left(1 + \frac{1}{3} \frac{d \ln g_*}{d \ln T}\right) \left(Y^3 - Y^2 Y_{eq}\right) \tag{2.77}$$

with

$$\kappa = 0.116 \frac{g_*^2}{g_\rho^{1/2}} M_P \, m^4 \sigma_0 \tag{2.78}$$

At freeze-out $Y \sim Y_{eq}$ and at $x \gg x_f$, the dark matter abundance $Y \gg Y_{eq}$. In this regime one can solve (2.77) analytically (neglecting the variation of g_* with T) as

$$\int_{Y_{eq}}^{Y_\infty} \frac{dY}{Y^3} = -\kappa \int_{x_f}^{\infty} dx x^{-(n+5)} \tag{2.79}$$

Performing the integrals we have the solution at $x = \infty$,

$$Y_\infty = \sqrt{\frac{n+4}{2\kappa}} x_f^{\frac{n+4}{2}}$$

$$= 4.14 \left(\frac{n}{2} + 2\right)^{1/2} \frac{g_\rho^{1/4}}{g_*} \frac{x_f^{(n+4)/2}}{\left(M_P \, m^4 \sigma_0\right)^{1/2}} \tag{2.80}$$

The freeze-out condition is determined from $\Gamma = n_{eq}^2 \langle \sigma v^2 \rangle = H$ which gives the equation for the decoupling temperature as

$$x_f = \frac{1}{2} \ln \left[0.0024 \frac{\sigma_0}{g_\rho^{1/2}} m^4 M_P\right] - \frac{1}{2}(n+1) \ln x_f \tag{2.81}$$

Solving for x_f iteratively we have the decoupling temperature,

$$x_f = 20 - \frac{1}{4}\ln\left(\frac{g_\rho}{10}\right) + \frac{3}{2}\ln\alpha - \frac{1}{2}\ln\left(\frac{m}{\text{GeV}}\right) \qquad (2.82)$$

where we have taken $\sigma_0 = \alpha^3/m^5$ and $n = 0$, The dark matter density in the present era is $\rho_{dm} = m s_0 Y_\infty$ and the self interacting dark matter relic density fraction turns out to be

$$\Omega_{dm}h^2 = 0.1 \left(\frac{10^{3/4}}{g_\rho^{-1/4} g_*}\right)\left(\frac{x_f}{20}\right)^2\left(\frac{4}{\alpha}\right)^{3/2}\left(\frac{m}{100\,\text{MeV}}\right)^{3/2} \qquad (2.83)$$

2.3.5   Cold Dark Matter Relic by Freeze-in

A new mechanism for obtaining a thermal relic density of dark matter with very small coupling to standard model particles is the freeze-in mechanism [38] where the dark matter was not in thermal equilibrium in the early universe and its number density grows only by production from the heat bath in the late epoch. For sub-MeV dark matter the freeze-in mechanism avoids the problem of extra particles creating problem for the Big-Bang nucleosynthesis by making the sub-meV dark matter dominate only after the BBN epoch of $T = 1\,\text{MeV}$ (and $t \sim 1\,\text{s}$).

Consider a coupling $\lambda\phi_1\phi_2 X$ of dark matter particle X and two standard model particles ϕ_1 and ϕ_2. If we take $m_{\phi_1} > m_{\phi_2} + m_X$ then at the epoch $m_x/T \geq 1$ the process $\phi_1 \to \phi_2 + X$ freezes-in a relic density of cold dark matter X (which were never present at thermal equilibrium in the early universe). The Boltzmann equation for the dark matter number density n_x is

$$\dot{n}_x + 3Hn_x = \sum_{s_i}\int d\Pi_x d\Pi_1 d\Pi_2 (2\pi)^4\delta^4(p_1 - p_2 - p_x)$$

$$\times\left[|M_{\phi_1\to\phi_2+X}|^2 f_1(1\pm f_2)(1\pm f_x) - |M_{\phi_2+X\to\phi_1}|^2 f_x f_2(1\pm f_1)\right]$$

$$(2.84)$$

As $f_x \ll 1$ prior to freeze-in the reverse reaction $\phi_2 + X \to \phi_1$ is negligible and one can drop the second term in the collision term of (2.84). The number density is small so that the Fermi-blocking/Bose enhancement factor $(1 \pm f_i)$ are unity. The particles ϕ_1 and ϕ_2 remain in equilibrium with the thermal bath and $f_1 = f_1^{eq} = e^{-E_1/T}$ and $f_2 = f_2^{eq} = e^{-E_2/T}$. The collision term in the r.h.s of (2.84) reduces to the form

$$C(f) = \sum_{s_i}\int d\Pi_x d\Pi_1 d\Pi_2 (2\pi)^4\delta^4(p_1 - p_2 - p_x)\left[|M_{\phi_1\to\phi_2+X}|^2 e^{-E_1/T}\right]$$

$$(2.85)$$

The collision term (2.85) can be written in terms of the decay width of $\phi_1 \rightarrow \phi_2 + X$ as follows. The decay width of ϕ_1 decay in the rest frame is

$$\Gamma_{\phi_1} = \int d\Pi_2 d\Pi_X \frac{1}{g_1} \frac{1}{2m_1} \sum_{s_i} |M_{\phi_1 \rightarrow \phi_2 + X}|^2 (2\pi)^4 \delta^4(p_1 - p_2 - p_x) \qquad (2.86)$$

Comparing (2.85) and (2.86) we see that the collisional term can be written in terms of the decay width as

$$C(f) = g_1 \int \frac{d^3 p_1}{(2\pi)^3} e^{-E_1/T} \frac{m_1}{E_1} \Gamma_{\phi_1} \qquad (2.87)$$

where m_1/E_1 is the Lorentz factor for the ϕ_1 decays with initial energy E_1. Changing the integration variable from p to E_1 we have $p^2 dp = (E_1 - m^2)^{1/2} E_1 dE_1$ and we can write the collision term as the integral over E_1 as

$$C(f) = \frac{g_1}{2\pi^2} \Gamma_{\phi_1} m_1 \int_{m_1}^{\infty} (E_1^2 - m_1^2)^{1/2} e^{-E_1/T} dE_1$$

$$= \frac{g_1}{2\pi^2} \Gamma_{\phi_1} m_1^2 T K_1(m_1/T) \qquad (2.88)$$

where K_1 is the Modified Bessel Function of the second kind.

We can write Boltzman equation (2.84) as

$$\frac{1}{a^3} \frac{d}{dt}(a^3 n) = \frac{g_1}{2\pi^2} \Gamma_{\phi_1} m_1^2 T K_1(m_1/T) \qquad (2.89)$$

The evolution equation for $Y = n/s$ is therefore

$$\frac{d}{dt} Y = \frac{g_1}{2\pi^2} \frac{1}{s} \Gamma_{\phi_1} m_1^2 T K_1(m_1/T) \qquad (2.90)$$

We change the evolution from t to $x = m_1/T$ using the relation (2.45)

$$\frac{d}{dt} = -\frac{m_1}{T^2} (-HT) \left(1 + \frac{1}{3} \frac{T}{g_*} \frac{dg_*}{dT}\right)^{-1} \frac{d}{dx} \qquad (2.91)$$

to get

$$\frac{dY}{dx} = \frac{g_1}{2\pi^2} \frac{1}{Hs} \Gamma_{\phi_1} \frac{m_1^2 T}{x} \left(1 + \frac{1}{3} \frac{T}{g_*} \frac{dg_*}{dT}\right) K_1(m_1/T)$$

$$= 0.069 \frac{g_1 M_P}{m_1^2} \Gamma_{\phi_1} \alpha(g) K_1(x) x^3 \qquad (2.92)$$

where

$$\alpha(x) \equiv \frac{1}{g_* g_\rho^{1/2}} \left(1 + \frac{1}{3} \frac{T}{g_*} \frac{dg_*}{dT} \right) \tag{2.93}$$

This can be integrated with the initial condition $Y(x = 0) \simeq 0$ to give

$$Y(x = \infty) = 0.069 g_1 \frac{M_P}{m_1^2} \Gamma_{\phi_1} \int_0^\infty \alpha(x) x^3 K_1(x) dx \tag{2.94}$$

The integrand $x^3 K_1(x)$ peaks at $x = 2.5$ and we can find an analytical approximation as[2]

$$Y(x = \infty) = 0.069 g_1 \frac{M_P}{m_1^2} \Gamma_{\phi_1} \bar{\alpha} \int_0^\infty x^3 K_1(x) dx$$

$$= 0.327 g_1 \frac{M_P}{m_1^2} \Gamma_{\phi_1} \tag{2.95}$$

where $\bar{\alpha} = \alpha(x = 2.5)$. The dark matter fraction in the present universe is therefore

$$\Omega h^2 = \frac{Y_\infty s_0 m}{1.05 \times 10^{-5} \, \text{Gev cm}^{-3}}$$

$$= 1.1 \times 10^{24} \frac{m}{m_1^2} \frac{\Gamma_{\phi_1}}{m_1^2} \left(\frac{\bar{\alpha}}{10^{-3}} \right) \tag{2.96}$$

which means that to obtain the required relic density of $\Omega h^2 = 0.11$ we must have $m\Gamma/m_1^2 \simeq 10^{-25}$. The couplings required between the dark matter and the thermal bath particles must be very small. Consider the following types of couplings between the dark matter χ and the thermal bath particles:

- Yukawa coupling $\lambda \phi_1 \bar{\psi}_2 \cdot \chi$. In this case the decay width is

$$\Gamma_{\phi_1 \to \psi_2 + \chi} = \frac{\lambda^2}{8\pi} m_1, \tag{2.97}$$

which implies that the relic density requires

$$\lambda \left(\frac{m}{m_1} \right)^{1/2} = 1.5 \times 10^{-12}. \tag{2.98}$$

---

[2] $\int_0^\infty x^3 K_1(x) dx = 3\pi/2.$

- Triple scalar coupling $\mu\,\phi_1\phi_2 X$. The decay width in this case is

$$\Gamma_{\phi_1\to\phi_2+X} = \frac{\mu^2}{8\pi m_1},\tag{2.99}$$

which implies that we require

$$\frac{\mu^2 m}{m_1^3} = 2.5\times 10^{-24}.\tag{2.100}$$

to obtain the required relic density of $\Omega h^2 = 0.11$ by the freeze-in.

2.3.6   Ultra-Light Dark Matter

Ultra-light particles of mass $m \sim 10^{-20}$ eV have been introduced as fuzzy dark matter for solving the small scale problems of CDM [39, 40]. The de Broglie wavelength of this particle is of the size of a dwarf galaxy

$$\lambda_{dB} = \frac{2\pi}{mv} = 2\,\text{kpc}\,\left(\frac{10^{-22}\,\text{eV}}{m}\right)\left(\frac{10\,\text{km/s}}{v}\right)\tag{2.101}$$

Prime candidates for ULDM are axion like goldstone bosons are described by the Lagrangian

$$\mathcal{L} = \sqrt{-g}\left[\frac{1}{2}g^{\mu\nu}\partial_\mu a\,\partial_\nu a + \mu^4\left(\cos\frac{a}{f}-1\right)\right]\tag{2.102}$$

which have a shift symmetry $a \to 2\pi f + a$. The mass of the ULDM particle is when it oscillates at the bottom of the potential is obtained by expanding the cosine term to quadratic order in a/f,

$$m = \frac{\mu^2}{f}\tag{2.103}$$

The equation of motion of the zero momentum modes in a FRW background is

$$\ddot{a} + 3H\dot{a} + m^2 \sin\frac{a}{f} = 0\tag{2.104}$$

where $H = \dot{R}/R$ is the Hubble expansion rate. When $H < m$ the solution of this equation is $a = $ constant. When H falls below m the dm field starts oscillating with frequency m and has the damped oscillations solution

$$a(t) = a_0\left(\frac{1}{R^{3/2}}\right)\sin mt\tag{2.105}$$

$$\Omega_a = \frac{m^{1/2} f^2 T_0^3}{\rho_c M_{pl}^{3/2}} \sim 0.1 \left(\frac{f}{10^{17}\,\text{GeV}}\right)\left(\frac{m}{10^{-22}\,\text{eV}}\right)^{1/2} \tag{2.106}$$

The fact that the mass required for galaxy formation is the same that provides the relic density is an interesting feature of this model [40].

2.4 Direct Detection of WIMP Dark Matter

The density of dark matter in the neighborhood of the sun is $\rho_\odot = (0.46 \pm 0.09)\,\text{GeV/cm}^3$ [32] and is a factor of 10^6 larger than the average cosmological density $\sim 0.57 \times 10^{-6}\,\text{GeV/cm}^3$ [10]. The flux of dark matter going through the earth is $10^5 (100\,\text{GeV}/m_\chi)\,\text{cm}^{-2}\,\text{s}^{-1}$. With a velocity of 220 km/s dark matter particles of mass 10–100 GeV can deposit 20–200 keV of recoil energy if they scatter off nuclei.

Direct detection experiments look for the nuclear recoil caused by weakly interacting massive particle (Wimp) dark matter in large (few ton mass) underground detectors [20, 41–43]. Consider a dark matter nucleus scattering $\chi(p) + N(k = 0) \to \chi(p-q) + N(q)$. The initial total energy is

$$E_i = \frac{\mathbf{p}^2}{2m_\chi} \tag{2.107}$$

and the energy after scattering is

$$E_f = \frac{|(\mathbf{p}-\mathbf{q})|^2}{2m_\chi} + \frac{|\mathbf{q}|^2}{2m_N}. \tag{2.108}$$

The recoil energy carried away by the nucleus is

$$E_R = \frac{|\mathbf{q}|^2}{2m_N}. \tag{2.109}$$

The conservation of energy gives

$$\frac{|\mathbf{p}||\mathbf{q}|\cos\theta}{m_\chi} = \frac{|\mathbf{q}|^2}{2\mu} \tag{2.110}$$

where $\cos\theta = \hat{\mathbf{p}}\cdot\hat{\mathbf{q}}$ and $\mu = m_\chi m_N/(m_\chi + m_N)$ is the wimp-nucleus reduced mass. The momentum transferred is

$$|\mathbf{q}| = \frac{2\mu\,|\mathbf{p}|}{m_\chi}\cos\theta = 2\mu\,v\cos\theta \tag{2.111}$$

where the DM velocity $v \sim 10^{-3}$ and the recoil energy is

$$E_R = \frac{2\mu^2 v^2}{m_N} \cos^2 \theta. \tag{2.112}$$

When $\mu = 10\text{--}100\,\text{GeV}$ the range of recoil momentum is $|\mathbf{q}|_{max} = 20\text{--}200\,\text{MeV}$ and recoil energy is $E_{Rmax} = 20\text{--}200\,\text{keV}$.

The differential event rate of a nuclear scattering (measured in differential rate units dru $=$ counts $\text{kg}^{-1}\,\text{day}^{-1}\,\text{keV}^{-1}$) in a given energy recoil energy range will depend upon the DM-nucleus cross section as

$$\frac{dR}{dE_R} = N_T \frac{\rho_\odot}{m_\chi} \int_{v_{min}}^{v_{esc}} v f(\mathbf{v}) \frac{d\sigma_{\chi N}}{dE_R} d^3\mathbf{v} \tag{2.113}$$

where N_T is the number of target nuclei per unit mass, $f(\mathbf{v})$ is the velocity distribution function of the dark matter in the galaxy, $v_{min} = \sqrt{m_N E_R / 2\mu^2}$ is the minimum DM velocity which can give a recoil energy E_R and $v_{esc} = 544\,\text{km/s}$ is the escape velocity of our Milky Way galaxy[44].

The total rate (counts $\text{kg}^{-1}\,\text{day}^{-1}$) is

$$R = \int_{E_{low}/Q}^{E_{high}/Q} dE_R \, \epsilon(QE_R) \frac{dR}{dE_R} \tag{2.114}$$

where E_{low} is the threshold of the detector and $E_{high} = \frac{2\mu^2 v_{esc}^2}{m_N}$ and $\epsilon(QE_R)$ is the energy dependent efficiency factor of the detector. Q is the quenching factor which accounts for of the loss of energy from the nuclear recoil into excitation of electrons in the atom or phonons and heat. The detector scintillation is calibrated by electron beams and the $Q(E_R)$ is the ration between the nuclear recoil energy and the equivalent electron energy which produces the same scintillation, $E_{ee} = Q(E_R)E_R$. The quenching factor $Q(E_R)$ depends upon the target nucleus and detection efficiency of the scintillation light signal by the detector.

2.4.1   Annual Modulation of the Signal

Over the random dark matter velocity with distribution function $f\mathbf{v}$ there is a time variation in the dark matter detection rate due to the motion of the sun in the galaxy and the earth around the sun [45]. The isothermal velocity distribution function (2.15) which is in the galactic reference frame must be boosted to the earths reference frame by making a Galilean transformation

$$f_{lab}(\mathbf{v}) = \frac{1}{N(v_0)} \exp\left(-\frac{(\mathbf{v} + \mathbf{v}_l)^2}{v_o^2}\right) \Theta(v_{esc} - |\mathbf{v} + \mathbf{v}_l|) \tag{2.115}$$

where

$$\mathbf{v}_l = v_\odot + \mathbf{v}_\oplus \tag{2.116}$$

The sun moves towards the constellation Cygnus with a speed of $v_\odot = 235 \, \text{km/s}$ and the earth moves around the sun with the average speed of $v_\oplus = 29.8 \, \text{km/s}$. The speed of lab frame in the is

$$v_l = v_\odot + v_\oplus \cos i \, \cos (\omega(t - t_0)) \tag{2.117}$$

where $\omega = 2\pi/T$, $T = 365.25 \, \text{d}$ and $t_0 = 152.5 \, \text{d}$ (2nd June) and i the inclination angle of the Earth's orbital plane with w.r.t the galactic plane is $60°$. The factor $N(v_0)$ is for normalization of f_{lab} over the velocity distribution and is given by

$$N(v_0) = \pi^{3/2} v_0^{3/2} \left(\text{erf} \left(\frac{v_{esc}}{v_0} \right) - \frac{2}{\sqrt{\pi}} \frac{v_{esc}}{v_0} \exp \left(-\frac{v_{esc}^2}{v_0^2} \right) \right) \tag{2.118}$$

where $v_0 = 200 \, \text{km/s}$ is the dispersion in the galactic dark matter velocity.

The effect of earths motion is to introduce a modulation in the rate of dark matter events with the time of the year with a modulation of 5–10% and with the maximum event rate predicted in June when the earth and sun move in the same direction and the minimum rate in December when their motions are in opposite directions. This annual modulation has been see by DAMA/LIBRA experiment [46] but not confirmed by other experiments like COSINE-100 [47] which also use the same type of NaI crystals detector as DAMA.

2.4.2 From Nucleon Cross Section to Nuclear Scattering Rates

For dark matter dark of mass 10–100 Gev is the transferred momentum is $1/|\mathbf{q}| \simeq$ 1–10 fm which implies that the nucleons in the nucleus are scattered coherently. For higher momentum transferred the dark matter starts scattering the individual nucleons and there is loss of coherence. This is accounted for by introducing a $|\mathbf{q}|$ dependent form factor $F(q)$ which is the Fourier transform of the nucleon density in the nucleus.

We can relate the cross section of Wimp-nucleon scattering with the Wimp-nucleus scattering as follows. The differential scattering cross section for the DM nucleon scattering $\chi(k) + n(p) \rightarrow \chi(k') + n(p')$ is

$$
\begin{aligned}
d\sigma_{\chi n} &= \frac{\overline{|M|^2}}{4 m_\chi m_n v} (2\pi)^4 \delta^4 (p + k - p' - k') \frac{d^3 p'}{(2\pi)^3 2 m_n} \frac{d^3 k'}{(2\pi)^3 2 m_\chi} \\
&= \frac{\overline{|M|^2}}{(4 m_\chi m_n)^2} \frac{1}{4\pi v^2} \delta \left(\cos\theta - \frac{|\mathbf{q}|}{2 \mu_{\chi n} v} \right) d|\mathbf{q}|^2 \, d\cos\theta
\end{aligned}
\tag{2.119}
$$

where $\overline{|M|^2}$ is the spin-averaged amplitude squared for the χ-nucleon scattering and $\mu_{\chi n}$ is the reduced mass of the χ-nucleon system. This can be integrated to give the total cross section of the Wimp-nucleon scattering

$$\sigma_{\chi n} = \frac{\mu_{\chi n}^2}{\pi} \frac{\overline{|M|^2}}{(4 m_\chi m_n)^2} \tag{2.120}$$

To obtain the differential cross section for Wimp-nucleus scattering we start from the expression (2.119) for the differential Wimp-nucleon cross section and multiply it with the nuclear form factor $|F(q)|^2$ and identify $d|\mathbf{q}^2| = 2 m_N d E_R$. The Wimp-nuclear cross section then given by

$$\frac{d\sigma_{\chi N}}{d E_R} = \frac{\overline{|M|^2}}{(4 m_\chi m_n)^2} \frac{m_N}{2\pi v^2} F(q)^2 A^2 \delta \left(\cos\theta - \frac{|\mathbf{q}|}{2\mu v} \right) d\cos\theta$$

$$= \frac{\sigma_{\chi n}}{\mu_{\chi n}^2} \frac{m_N}{2 v^2} F(q)^2 A^2 \tag{2.121}$$

where the normalization of the nuclear form factor is such that $F(q = 0) = 1$. When the target neutron and proton cross sections differ then we generalize (2.121) as

$$\frac{d\sigma_{\chi N}}{d E_R} = \frac{\sigma_{\chi n}}{\mu_{\chi n}^2} \frac{m_N}{2 v^2} F(q)^2 \left(f_p Z + f_n (A - Z) \right)^2$$

$$= \frac{2 m_N}{\pi v^2} F(q)^2 \left(f_p Z + f_n (A - Z) \right)^2 \tag{2.122}$$

where f_p and f_n represent the coupling of Wimp with protons and neutrons respectively. If in some specific Wimp model $f_p / f_n = -(A - Z)/Z$ then there is no spin-independent coupling of the nucleus to the Wimp.

The nuclear form factor which represents the distribution of nucleons in the nucleus is usually taken to be the Helms form factor [48].

$$F(q) = \frac{3 j_1(q r_n)}{q r_n} e^{-(qs)^2/2} \tag{2.123}$$

where j_1 is the spherical Bessel function of the first kind, the nuclear radius is $r_n = 1.14 \, A^{1/3}$ and the nuclear skin thickness is $s = 0.5$ fm.

Using (2.122), the scattering rate of Wimps with nuclei in the detectors (2.113) can be written in terms of Wimp nucleon spin-independent cross sections as

$$\frac{dR}{d E_R} = N_T \frac{\rho_\odot}{m_\chi} \frac{\sigma_{\chi n} m_N}{2 \mu_{\chi n}^2} F(q)^2 \left(f_p Z + f_n (A - Z) \right)^2 \int_{v_{min}}^{v_{esc}} \frac{f(\mathbf{v})}{v} d^3 \mathbf{v} \tag{2.124}$$

A quick check of the E_R dependence of the differential scattering rate (2.124) can be made by assuming a Maxwellian distribution for

$$f(v) = \frac{1}{\pi^{3/2}v_0^3}e^{-v^2/v_0^2}. \tag{2.125}$$

Evaluating the integral over v and assuming $v_{max} = \infty$ we see that that the scattering rate falls exponentially with E_R,

$$\frac{dR}{dE_R} \propto e^{-v_{min}^2/v_0^2} = \exp\left(-\frac{m_N E_R}{2\mu^2 v_0^2}\right) \tag{2.126}$$

To estimate the amplitude of annual modulation we can use the distribution function $f(v) \sim (v/v_l v_0\sqrt{\pi})\, e^{-(v_l-v)^2/v_0^2}$ and the speed of the lab frame is $v_l = v_0(1.05 + 0.007\cos\omega(t - t_0))$ where the galactic speed of the sun is $1.05v_0$. The velocity integral then gives

$$\int_{v_{min}}^{\infty} \frac{f(\mathbf{v})}{v}d^3\mathbf{v} = \frac{1}{2v_l}\left(\text{erf}\left(\frac{v_{min} + v_l}{v_0}\right) + \text{erf}\left(\frac{v_{min} - v_l}{v_0}\right)\right) \tag{2.127}$$

This factor gives a 7 % annual modulation of the Wimp signal which has been seen in DAMA/LIBRA [46] but not in COSINE-100 [47] experiment.

Comparing the experimental upper bound on the Wimp nucleus scattering rate with the theoretical prediction (2.124) leads to bounds on the Wimp-nucleon cross section. The bounds from different dark matter direct detection experiments (XENON1T, PandaX, COSINE, DarkSide, CRESST) on the Wimp-nucleon cross section (assuming $f_p = f_n$) vs the Wimp mass is shown in Fig. 2.5. For Wimp mass $m_\chi > 5\,\text{GeV}$ the most stringent bounds come from XENON-1T experiment [49] which uses 2 ton of liquid Xenon as target. Large nuclei like Xenon give the nest bounds due to the coherence A^2 enhancement of the Wimp-nucleus cross section. In the mass range $1.8\,\text{GeV} \le m_\chi \le 5\,\text{GeV}$ the best bound comes from DarkSide-50 experiment [50] which uses liquid Argon as target. For low mass Wimps lighter nuclei are more suitable as targets like Argon and Germanium are more suitable as a lower target mass will result in more energy transfer to the target.

2.4.3   From Quarks Couplings to Nuclear Couplings of Dark Matter

A fundamental particle physics theory with dark matter candidates will predict have DM-quark four field operators. To test the theory with experiment one must relate the quark level operators to nuclear operators. One starts with the fundamental theory where there is effective coupling between DM and quarks and gluons, and then taking the matrix elements of the nucleons between these QCD operators. Consider the trace anomaly of QCD which gives mass to the nucleons

$$m_n \langle n|n \rangle = \langle n|\theta^\mu{}_\mu|n \rangle$$

$$= \sum_q \langle n|m_q \bar{q}q|n \rangle + \sum_{Q=c,b,t} m_Q \langle n|\bar{Q}Q|n \rangle$$

$$+ \frac{\alpha_s}{8\pi}\left(\frac{2n_q}{3} - \frac{11}{3}N_c\right)\langle n|G^a_{\mu\nu}G^{a\mu\nu}m| \rangle \tag{2.128}$$

where the number of quark flavors $n_q = 6$ and for QCD $N_c = 3$, The contribution of the heavy quarks to the nucleon mass is from the heavy quark effective theory [52]

$$\sum_{Q=c,b,t} m_Q \langle n|\bar{Q}Q|n \rangle = -\frac{3\alpha_s}{12\pi}\langle n|G^a_{\mu\nu}G^{a\mu\nu}|n \rangle \tag{2.129}$$

Substituting (2.129) in (2.128) we have

$$m_n \langle n|n \rangle = \sum_{q=u,d,s} m_q \langle n|\bar{q}q|n \rangle + \frac{9\alpha_s}{8\pi}\langle n|G^a_{\mu\nu}G^{a\mu\nu}|n \rangle \tag{2.130}$$

The matrix element of the nucleon states of the various quark and gluon operators is defined as

$$\langle n(k')|m_q\bar{q}q|n(k) \rangle = m_n\, f^n_{T,q}\, \bar{u}_n(k')u_n(k), \tag{2.131}$$

If the DM couples to quarks with a scalar exchange, it will couple to gluons through a quark loop. For a scalar exchange quark term $g_s\bar{\chi}\chi\bar{q}q$ there will be a loop generated DM gluon coupling

$$-g_s\frac{\alpha_s}{12\pi}G^a_{\mu\nu}G^{a\mu\nu} \tag{2.132}$$

The matrix element of the nucleon state for the gluon operator is therefore

$$\langle n(k')|\alpha_s G^a_{\mu\nu}G^{a\mu\nu}|n(k) \rangle = -\frac{8\pi}{9}m_n\sum_{q=u,d,s}\left(1 - f^n_{T,q}\right)\bar{u}_n(k')u_n(k)$$

$$\equiv -\frac{8\pi}{9}m_n\, f^n_{T,G}\,\bar{u}_n(k')u_n(k). \tag{2.133}$$

The coefficients $f^n_{T,q}$ are obtained from experiments like pion nucleon scattering and lattice calculations and $f^n_{T,G} = \sum_{q=u,d,s}(1 - f^n_{T,q})$ are given in Table. 2.1.

Table 2.1 Matrix elements of the light quarks and gluon operator between nucleon states [43, 53, 54]

	$f_{T,G}^n$	$f_{T,u}^n$	$f_{T,d}^n$	$f_{T,s}^n$
Neutron	0.910 (20)	0.013 (3)	0.040 (10)	0.037 (17)
Proton	0.917 (19)	0.018 (5)	0.027 (7)	0.037 (17)

The numbers in the parenthesis are one sigma uncertainties in the last digits

Consider the operator

$$\mathcal{L}_q^S = g_q \bar{\chi} \chi \bar{q} q, \tag{2.134}$$

which can couple a fermionic dark matter to quarks via a heavy scalar exchange. The scalar couples to the quarks at the tree level but has no tree level coupling with gluons.

The couplings of the DM to protons and neutrons is then given by

$$\mathcal{L}_n^S = f_p \bar{\chi} \chi \bar{p} p + f_n \bar{\chi} \chi \bar{n} n \tag{2.135}$$

where the coupling of protons to DM can be written in terms of quarks and gluons fraction of the nucleons as

$$f_p = \sum_{q=u,d,s} g_q \frac{m_p}{m_q} f_{T,q}^p + \frac{2}{27} f_{T,G}^p \sum_{Q=c,b,t} g_\phi \frac{m_p}{m_Q}, \tag{2.136}$$

with a similar expression for neutrons DM coupling f_n. The coherent scattering cross section of nucleus with DM can be described by the effective Lagrangian

$$\mathcal{L}_N^S = \left(f_p Z + (A - Z) f_n \right) F(q) \bar{\chi} \chi \bar{n} n \tag{2.137}$$

We can write the nucleon spinor as

$$n_s(p) = \begin{pmatrix} \sqrt{p \cdot \sigma} \xi^s \\ \sqrt{p \cdot \bar{\sigma}} \xi^s \end{pmatrix}$$

and similarly for the DM χ. Here s is the spin index and ξ^s are two component spinors with the property $\sum_s \xi^{s\dagger} \xi^s = 1$. The spinor bilinear in the non-relativistic limit reduces to

$$\bar{n}_{s'}(p') n_s(p) \simeq 2 m_n \xi^{s'\dagger} \xi^s \tag{2.138}$$

The spin averaged amplitude squared for scattering is

$$\overline{|M|^2} = \frac{(4m_n m_\chi)^2}{(2s_\chi + 1)(2s_n + 1)} \left(f_p Z + (A - Z) f_n\right)^2 F^2(q) \sum_{r,r'ss'} |\xi^{s\dagger}\xi^{s'}|^2 |\xi^{r\dagger}\xi^{r'}|^2$$

$$(2.139)$$

The differential scattering cross section for the Wimp-nucleus scattering is then given by

$$\frac{d\sigma_{\chi N}}{dE_R} = \frac{2m_N}{\pi v^2} \overline{|M|^2}$$

$$= \frac{2m_N}{\pi v^2} \left(f_p Z + (A - Z) f_n\right)^2 F^2(q) \qquad (2.140)$$

Using expression (2.136) for the proton coupling f_p and a similarly for f_n we can relate the DM-nuclear scattering rates observed in experiments to quark-dark matter couplings g_q.

For Dirac dark matter the exchange of a gauge boson gives rise to a coupling

$$\mathcal{L}_q^V = b_q \, \bar{\chi}\gamma^\mu \chi \, \bar{q}\gamma^\mu q, \qquad (2.141)$$

Here when considering DM nucleon scatterings there is no contribution of the s-quark or gluons to the vector current via loops. therefore for DM-nucleon scatterings only the u, d valence quarks need to be considered. The Wimp-nucleon interaction therefore is $\mathcal{L}_n^V = b_n \, \bar{\chi}\gamma^\mu \chi \, \bar{n}\gamma^\mu n$ with $b_p = 2b_u + b_d$ and $b_n = 2b_d + b_u$ for protons and neutrons respectively. In the nucleus the scattering of DM due to vector couplings will also be coherent like in the case of scalar couplings and the DM-nucleus scattering matrix will be described by the interaction Lagrangian

$$\mathcal{L}_N^V = B_N \, F(q) \, \bar{\chi}\gamma^\mu \chi \, \bar{N}\gamma^\mu N \qquad (2.142)$$

where $F(q)$ is the nuclear form factor and B_N for a nucleus is

$$B_N = b_u(A + Z) + b_d(2A - Z) \qquad (2.143)$$

The DM-nuclear scattering cross section is

$$\frac{d\sigma_{\chi N}^V}{dE_R} = \frac{m_N}{128\pi v^2} B_N^2 \, F^2(E_R) \qquad (2.144)$$

where m_N is the mass of the nucleus and μ is the DM-nucleus reduced mass. Comparing this rate with experimental bounds will place limits on the quark-DM vector coupling b_q.

Vector couplings of Majorana dark matter is zero unless it is inelastic scattering where the incoming DM and outgoing DM are in different mass eigenstates [55].

2.4.4   Spin-Dependent Cross Section

Certain types of DM-quark interactions lead to a coupling between the quark and DM spins and in this case in the nuclear scattering the scattering rates depend on the spin of the nucleus [20, 41]. As an example of an operator that leads to spin-dependent scattering cross section consider the axial vector operator between DM and quarks

$$\mathcal{L}_q^A = d_q \, \bar{\chi}\gamma^\mu\gamma_5\chi \, \bar{q}\gamma_\mu\gamma_5 q \tag{2.145}$$

which can arise for example by the exchange of the standard model Z boson. The axial-bilinear $\bar{q}\gamma^\mu\gamma_5 q$ in the non-relativistic limit measures the spin- of the quarks. The quark spin sandwiched between nucleons operators can be written as

$$\langle n|\bar{q}\gamma^\mu\gamma_5 q|n\rangle = \Delta_q^n \, \bar{n}\gamma_\mu\gamma_5 n \tag{2.146}$$

where Δ_q^n the axial vector matrix elements between nucleon states are determined by polarized beam scattering experiments. Table 2.2 lists the values of Δ_q^n.

The effective DM-nucleon operator can therefore be written as

$$\mathcal{L}_n^A = d_q \, \Delta_q^n \, \bar{\chi}\gamma^\mu\gamma_5\chi \, \bar{n}\gamma_\mu\gamma_5 n \tag{2.147}$$

In the non-relativistic limit (2.147) reduces to

$$\mathcal{L}_{nr}^A = \sum_{q=u,d,s} d_q \, \Delta_q^n \, (4m_\chi m_n) \left(\xi^{s'\dagger}\sigma^i\xi^s\right)\left(\xi^{r'\dagger}\sigma_i\xi^r\right)$$

$$= (16 m_\chi m_n) \sum_{q=u,d,s} d_q \, \Delta_q^n \left(S_n^i\right)(S_{\chi i}) \tag{2.148}$$

where S_n^i is the spin of the nucleon and $S_{\chi i}$ is the spin of the dark matter. Only the light quarks make significant contribution to the spin of the nucleons. The axial vector interaction therefore gives rise to an interaction between the DM's spin and the nucleon's spin. For the scattering of DM with nuclei we need to express the spin

Table 2.2 Matrix elements of the light quark axial vector operators between nucleon states [43, 56]

	Δ_u^n	Δ_d^n	Δ_s^n
Neutron	−0.46 (4)	0.80 (3)	−0.12 (8)
Proton	0.80 (3)	−0.46 (4)	−0.12 (8)

The numbers in the parenthesis are one sigma uncertainties in the last digits of the protons and neutrons in terms of the nuclear spin. According to the Wigner–Eckart theorem any linear combination of the total proton and neutron spin vectors in the nucleon must be proportional to the nuclear spin J_N, therefore we can write

$$\sum_p \mathbf{S}_p \, \Delta_q^p + \sum_n \mathbf{S}_n \, \Delta_q^n = \lambda_q \mathbf{J}_N \tag{2.149}$$

The proportionality factor can be found by sandwiching the z-component of the spin operators in (2.149) between the highest spin state of the nucleus $|J_N, M_J = J_N\rangle$, to give

$$\lambda_q = \frac{\langle S_p \rangle}{J_N} \Delta_q^p + \frac{\langle S_n \rangle}{J_N} \Delta_q^n \tag{2.150}$$

where $\langle S_p \rangle \equiv \langle J_N, J_N | \sum_p S_{zp} | J_N, J_N \rangle$ and a similar definition for $\langle S_p \rangle$. By summing over the quarks we can define the quantities

$$a_p = \sum_q \frac{d_q \Delta_q^p}{\sqrt{2} G_F} \quad \text{and} \quad a_n = \sum_q \frac{d_q \Delta_q^p}{\sqrt{2} G_F} \tag{2.151}$$

and

$$\Lambda = \frac{1}{J_N} \left[a_p \langle S_p \rangle + a_n \langle S_p \rangle \right] \tag{2.152}$$

The matrix element of the nuclear scattering is then given by

$$M = 16 m_\chi m_N \sqrt{2} G_F \Lambda \, \mathbf{J}_N \cdot \mathbf{S}_\chi \tag{2.153}$$

and the spin-averaged amplitude squared is

$$\overline{|M|^2} = 64 m_\chi^2 m_N^2 G_F^2 \, \Lambda^2 \, J_N (J_N + 1) \tag{2.154}$$

The spin-dependent scattering differential cross section can then be written as

$$\frac{d\sigma_{\chi N}^{SD}}{dE_R} = \frac{16 m_N}{\pi v^2} \Lambda^2 G_F^2 J_N (J_N + 1) \frac{S(E_R)}{S(0)} \tag{2.155}$$

where $S(E_R)$ is the spin-dependent form factor of the nucleus.

Since the spin-dependent scattering does not have the coherence factor of A^2 enhancement the bounds obtained on the SD cross section are poorer by 5–6 orders of magnitude compared to the exclusion limits on SI cross section [42].

2.4.5   Inelastic Dark Matter

Direct detection experiments put severe constraint on scattering cross section which for a 50 GeV dark matter is $\sigma_{\chi-n} v < 10^{-46} \, \text{cm}^2$. This cross section is many orders of magnitude different from the annihilation cross section for getting the correct relic density $\langle \sigma v \rangle = 3 \times 10^{-26} \, \text{cm}^2 \, \text{s}^{-1} = 10^{-36} \, \text{cm}^2$. This poses a problem for constructing dark matter models.

One way to explain the non-observation of dark matter in direct detection experiments is the idea that dark matter is inelastic [55] i.e. there are two mass states χ_1 and χ_2 with masses $m_1 = m_\chi$ and $m_2 = m_\chi + \delta$. The interaction eigenstates are made off-diagonal i.e. the interaction vertex with mediator Z' is of the form $\chi_1 \chi_2 Z'$. The dark matter in the galaxy is in the lower mass eigenstate χ_1. In scattering with nuclei the dark matter has to get up-scattered to the higher mass state. Kinematically however, for the interaction $\chi_1 N \rightarrow \chi_2 N$ to take place, we must satisfy the condition

$$\delta < \frac{\beta^2 m_\chi m_N}{2 m_\chi m_N} \tag{2.156}$$

where $\beta \sim 220 \, \text{km/s}$ is the dark matter velocity. For $m = 100 \, \text{GeV}, m_N \simeq 100 \, \text{GeV}$ we get $\delta \sim 13.46 \, \text{keV}$. If in the model δ is larger than the r.h.s (which depends on the nuclear mass m_N) then the detection by scattering with that nucleus is evaded.

As an example consider a Dirac fermion $\psi = (\eta \; \bar{\xi})$. This can have axial and vector couplings to quarks, there is a Dirac mass m and a small Majorana mass δ. The Lagrangian describing the interaction and the masses is

$$\mathcal{L} = \bar{\psi} \gamma_\mu \left(g'_V + g'_A \gamma_5 \right) \psi \, \bar{q} \gamma^\mu \left(g_V + g_A \gamma_5 \right) q + m \left(\bar{\eta} \xi + \bar{\xi} \eta \right) + \frac{\delta}{2} \left(\eta \eta + \bar{\eta} \bar{\eta} \right) \tag{2.157}$$

The axial-axial coupling leads to spin dependent interaction, axial-vector coupling gives velocity suppressed cross section and the largest interaction cross section comes from the vector coupling which will be spin independent. The mass eigenstates and eigenvalues are

$$\chi_1 = \frac{i}{2} (\eta - \xi), \quad m_1 = m - \delta,$$

$$\chi_2 = \frac{1}{2} (\eta + \xi), \quad m_2 = m + \delta. \tag{2.158}$$

The vector- vector interaction is off-diagonal while the diagonal Interaction is suppressed by δ/m,

$$\bar{\psi} \gamma_\mu \psi = i \left(\bar{\chi}_1 \bar{\sigma}_\mu \chi_2 - \bar{\chi}_2 \bar{\sigma}_\mu \chi_1 \right) + \frac{\delta}{2m} \left(\bar{\chi}_1 \bar{\sigma}_\mu \chi_1 - \bar{\chi}_2 \bar{\sigma}_\mu \chi_2 \right) \tag{2.159}$$

This will give a spin independent $\chi_1 N \rightarrow \chi_2 N$ inelastic scattering which if kinematically allowed is unsuppressed compared to the $\chi_1 N \rightarrow \chi_1 N$ process.

A natural example of inelastic scattering comes from dark matter with electric or magnetic dipole moments [57]. For Majorana particles only the transition electric and magnetic moments are non-zero. The Majorana electric and magnetic dipole couplings to photons are the following (with $i \neq j$)

$$\mathcal{L} = -\frac{i}{2} \bar{\chi}_i \sigma_{\mu\nu} \left(\mu_{ij} + i\gamma_5 \mathcal{D}_{ij} \right) \chi_j \qquad (2.160)$$

For Dirac dark matter one can have $i = j$ couplings also.

The following relations hold for direct detection calculations of general inelastic dark matter. The differential cross section per unit recoil energy of the nucleus E_R for the process $\chi_1 N \rightarrow \chi_2 N$ remains the same for the elastic or inelastic cases (apart from the constant factors of the couplings which are different in the two cases). In the integration over velocity distribution function

$$f(v) = \frac{4v^2}{\sqrt{\pi} v_0^2} e^{-v^2/v_0^2} \qquad (2.161)$$

where $v_0 = 220$ km/s, the minimum velocity changes due to the mass difference $\delta = m_2 - m_1$ and is given by

$$v_{min} = \frac{1}{\sqrt{2m_N E_{R\,min}}} \left(\frac{m_N E_{R\,min}}{\mu_N} + \delta \right) \qquad (2.162)$$

This change in the v_{min} changes the event rates when we have inelastic scattering.

Dirac dark matter with direct dipole moments have long range couplings with baryons and this can change the large scale structures and the matter power spectrum due to the dark matter-baryon drag [58].

2.5 Dark Matter Signals in High Energy Photons and Neutrinos Observations

The density of dark matter in the neighbourhood of the sun is $(0.3$–$0.4)$ GeV/cm^3 [31] and is a factor of 10^6 larger than the average cosmological density 0.57×10^{-6} GeV/cm^3. Due to the enhanced concentration from gravitational clumping, dark matter again annihilates into standard model particles and the signal from this annihilation can be inferred from the observations of high energy gamma rays, positrons, antiprotons and neutrinos.

For a $\chi\chi \to X\bar{X}$ annihilation of DM into a standard model particles $(X, \bar{X})$ in the galaxy, the flux spectrum of photons or neutrinos on earth can be written as

$$\frac{d\Phi}{dE} = \frac{1}{2}\,\kappa\,\frac{\langle\sigma v\rangle}{4\pi m_\chi^2}\,\frac{dN_X}{dE}\,J \tag{2.163}$$

where dN_X/dE is the number of photons or neutrinos which we get from the primary SM $X\bar{X}$ particles. The astrophysical factor J is the integral along the direction of the incident photon or neutrino (line of sight) of the dark matter density squared

$$J = \int d\Omega \int_{l.o.s} \rho^2(r(l,\psi))\,dl\,, \qquad r(l,\psi) = \sqrt{R_\odot - 2lR_\odot\cos\psi + l^2}\,. \tag{2.164}$$

Here $\rho(r)$ is the density profile of the dark matter as a function of the radial distance r from the galactic center, ψ is the angle between the direction of the galactic center and the line of sight and $R_\odot = 8.5\,\text{kpc}$ is the distance of the sun from the galactic center. For the NFW profile (2.16), with $\rho_0 = 0.6\,\text{GeV cm}^{-3}$ and $R_s = 25.5\,\text{kpc}$ the value of the J factor is $J = 3.23 \times 10^{23}\,\text{GeV}^2\,\text{cm}^{-5}$.

In (2.163) the factor of $1/2$ is to correct for the double counting in the pairwise annihilation rate, κ is 1 for Majorana /real scalar DM and for Dirac/complex scalar DM $\kappa = 1/2$ as a particle can only annihilate with half of the remaining particles.

If the dark matter is a Majorana fermion then the annihilation takes place between two identical fermions while for Dirac dark matter annihilation is between a fermion and antiparticle which are not identical. The general dark matter annihilation into standard model fermions, $\chi\chi \to f\bar{f}$ cross sections can be expanded in powers of the relative velocity as $\langle\sigma v\rangle = a + bv^2$. For Majorana fermions there is a helicity suppression of the first term, $a \propto (m_f/m_\chi)^2$ so the leading order contribution to the annihilation cross section can be the second term $\sigma v = bv^2$ which is velocity suppressed. The velocity of DM in the galaxy, $220\,\text{km/s} \simeq 10^{-3}$ is smaller than the velocity of DM during thermal decoupling which for cold dark matter is $(3/x_f)^{1/2} \sim 0.3$. The cross annihilation cross section has a large velocity suppression from the value at decoupling depending for Majorana DM compared to Dirac fermion DM. The helicity suppression can be overcome by a photon emission in the final state in a $\chi\chi \to f\bar{f}\gamma$ process but with a suppression by α due to the photon vertex [59].

Photons from DM annihilation in the galaxy can be observed in Fermi [60] and CTA [61] and neutrinos can be observed at Borexino, Super-Kamiokande and IceCube [62]. Dark matter annihilation can also produces electro-positron pairs but the flux of positrons measured by AMS-02 [63] for instance cannot be calculated with a simple line of sight integral like (2.163) as $e^\pm$ undergo multiple scattering by the galactic magnetic field before arrival [64].

Dark matter may also decay into standard model particles with decay time τ greater than the age of the universe. If dark matter in the galaxy decays into SM particle X, in the process $\chi \rightarrow \chi' X$ then the flux of photons or neutrinos on the earth can be written as

$$\frac{d\Phi}{dE} = \frac{\Gamma_X}{4\pi m_\chi} \frac{dN_X}{dE} \left[\int d\Omega \int_{l.o.s} \rho(r(l,\psi)) \, dl \right] \tag{2.165}$$

where $\Gamma_X = 1/\tau$ is the decay width and the quantity in the square bracket is the astrophysical factor J_D which depends upon the dark matter profile. For the NFW profile, $J_D = 1.15 \times 10^{24} \, \text{GeV cm}^{-2}$.

2.6   Conclusion

Dark matter has a large experimental signal in large scale structures, CMB and galactic rotation curves but there is no hint of what the nature of the dark matter is. It could be as light as 10^{-20} eV fuzzy dark matter or it could be primordial black holes with mass 10^{19} gm. For large scale structures and CMB the cold dark matter fits all experimental requirements of seeding structure formation and providing the gravitational potential for Baryon Acoustic Oscillations, The dark matter in cosmology is required to be cold i.e. it should be non-relativistic when it decouples kinematically from the plasma. Ultra-light dark matter they were produced in a cold coherent state and they stay in the zero momentum coherent state, providing the energy density due to there oscillation but no pressure, therefore for the purpose of CMB an LSS they behave the same as any other CDM.

References

1. F. Zwicky, Die Rotverschiebung von extragalaktischen Nebeln. Helv. Phys. Acta **6**, 110–127 (1933). [arXiv:1011.1669v3]
2. F. Zwicky, On the masses of nebulae and of clusters of nebulae. Astrophys. J. **86**, 217–246 (1937)
3. V.C. Rubin, W.K. Ford Jr., Rotation of the Andromeda nebula from a spectroscopic survey of emission regions. Astrophys. J. **159**, 379 (1970)
4. V.C. Rubin, W.K Ford Jr., N. Thonnard, Extended rotation curves of high-luminosity spiral galaxies. Astrophys. J. **225**, L107–L111 (1978)
5. M.S. Roberts, A high-resolution 21-cm hydrogen-line survey of the Andromeda nebula. Astrophys. J. **144**, 639 (1966)
6. D.H. Rogstad, G.S. Shostak, Gross properties of five Scd galaxies as determined from 21-cm observations. Astrophys. J. **176**, 315 (1972)
7. H.S. Roberts, A.H. Rots, Comparison of rotation curves of different galaxy types. Astron. Astrophys. **26**, 483 (1973)
8. A. Bosma, The distribution and kinematics of neutral hydrogen in spiral galaxies of various morphological types. Ph.D. thesis, Groningen University,1978. http://ned.ipac.caltech.edu/level5/March05/Bosma/TOC.html
9. S. Gershtein, Y. Zeldovich, Rest mass of muonic neutrino and cosmology. JETP Lett. **4**, 120–122 (1966)

10. R. Cowsik, J. McClelland, An upper limit on the neutrino rest mass. Phys. Rev. Lett. **29**, 669–670 (1972)
11. P. Peebles, Primeval adiabatic perturbations: effect of massive neutrinos. Astrophys. J. **258**, 415–424 (1982)
12. J. Bond, G. Efstathiou, J. Silk, Massive neutrinos and the large scale structure of the universe. Phys. Rev. Lett. **45**, 1980–1984 (1980)
13. J. Bond, A. Szalay, The collisionless damping of density fluctuations in an expanding universe. Astrophys. J. **274**, 443–468 (1983)
14. S.D. White, C. Frenk, M. Davis, Clustering in a neutrino dominated universe. Astrophys. J. Lett. **274**, L1–L5 (1983). https://doi.org/10.1086/161425
15. B.W. Lee, S. Weinberg, Cosmological lower bound on heavy neutrino masses. Phys. Rev. Lett. **39**, 165–168 (1977)
16. P. Hut, Limits on masses and number of neutral weakly interacting particles. Phys. Lett. B **69**, 85 (1977)
17. E.W. Kolb, K.A. Olive, The Lee–Weinberg bound revisited. Phys. Rev. D **33**, 1202 (1986)
18. E.W. Kolb, M.S. Turner, The early universe. Front. Phys. **69**, 1–547 (1990)
19. J.R. Ellis, J. Hagelin, D.V. Nanopoulos, K.A. Olive, M. Srednicki, Supersymmetric relics from the big bang. Nucl. Phys. B **238**, 453–476 (1984). https://doi.org/10.1016/0550-3213(84)90461-9
20. M.W. Goodman, E. Witten, Phys. Rev. D **31**, 3059 (1985)
21. W. de Blok, The core-cusp problem. Adv. Astron. **2010**, 789293 (2010)
22. A.G.A. Brown et al., (Gaia collaboration), *Gaia* data release 2: summary of the contents and survey properties Gaia collaboration. Astron. Astrophys. **616**, A1 (2018)
23. J. Binney, S. Tremaine, *Galactic Dynamics*, 2nd edn. (Princeton University Press, Princeton, 2008)
24. P. Bhattacharjee, S. Chaudhury, S. Kundu, S. Majumdar, Sizing-up the WIMPs of Milky Way: deriving the velocity distribution of Galactic dark matter particles from the rotation curve data. Phys. Rev. D **87**, 083525 (2013)
25. M. Fornasa, A.M. Green, Self-consistent phase-space distribution function for the anisotropic dark matter halo of the Milky Way. Phys. Rev. D **89**(6), 063531 (2014)
26. T. Lacroix, M. Stref, J. Lavalle, Anatomy of Eddington-like inversion methods in the context of dark matter searches. JCAP **9**, 040 (2018)
27. J.F. Navarro, C.S. Frenk, S.D.M. White, A universal density profile from hierarchical clustering. Astrophys. J. **490**, 2, 493 (1997)
28. S. Tulin, H.B. Yu, Dark matter self-interactions and small scale structure. Phys. Rep. **730**, 1–57 (2018). https://doi.org/10.1016/j.physrep.2017.11.004
29. A. Burkert, The structure of dark matter halos in dwarf galaxies. Astrophys. J. **447**(1), L25–L28 (1995)
30. J. Read, The local dark matter density. J. Phys. G **41**, 063101 (2014)
31. R. Catena, P. Ullio, A novel determination of the local dark matter density. JCAP **1008**, 004 (2010)
32. S. Sivertsson, H. Silverwood, J.I. Read, G. Bertone, P. Steger, The local dark matter density from SDSS-SEGUE G-dwarfs. Mon. Not. R. Astron. Soc. **478**(2), 1677 (2018)
33. E. Komastu et al., Astrophys. J. Suppl. **192**(2), 18 (2011)
34. G. Steigman, B. Dasgupta, J.F. Beacom, Precise relic WIMP abundance and its impact on searches for dark matter annihilation. Phys. Rev. D **86**, 023506 (2012)
35. M. Drees, F. Hajkarim, E.R. Schmitz, The effects of QCD equation of state on the relic density of WIMP dark matter. JCAP **1506**(6), 025 (2015)
36. Y. Hochberg, E. Kuflik, T. Volansky, J.G. Wacker, Mechanism for thermal relic dark matter of strongly interacting massive particles. Phys. Rev. Lett. **113**, 171301 (2014)
37. H.M. Lee, lectures on physics beyond the standard model. arXiv:1907.12409 [hep-ph]
38. L.J. Hall, K. Jedamzik, J. March-Russell, S.M. West, Freeze-in production of FIMP dark matter. JHEP **1003**, 080 (2010). [arXiv:0911.1120 [hep-ph]]
39. W. Hu, R. Barkana, A. Gruzinov, Cold and fuzzy dark matter. Phys. Rev. Lett. **85**, 1158 (2000)

40. L. Hui, J.P. Ostriker, S. Tremaine, E. Witten, Ultralight scalars as cosmological dark matter. Phys. Rev. D **95**(4), 043541 (2017)
41. G. Jungman, M. Kamionkowski, K. Griest, Supersymmetric dark matter. Phys. Rep. **267**, 195 (1996)
42. M. Schumann, Direct detection of WIMP dark matter: concepts and status. arXiv:1903.03026 [astro-ph.CO]
43. T. Lin, TASI lectures on dark matter models and direct detection. arXiv:1904.07915 [hep-ph]
44. M.C. Smith et al., The RAVE survey: constraining the local galactic escape speed. Mon. Not. Roy. Astron. Soc. **379**, 755 (2007)
45. K. Freese, M. Lisanti, C. Savage, Colloquium: annual modulation of dark matter. Rev. Mod. Phys. **85**, 1561 (2013)
46. R. Bernabei et al., First model independent results from DAMA/LIBRA-phase2. Universe **4**(11), 116 (2018) Nucl. Phys. Atom. Energy **19**(4), 307 (2018)
47. G. Adhikari et al., [COSINE-100 Collaboration], Search for a dark matter-induced annual modulation signal in NaI(Tl) with the COSINE-100 Experiment. Phys. Rev. Lett. **123**(3), 031302 (2019)
48. J.D. Lewin, P.F. Smith, Review of mathematics, numerical factors, and corrections for dark matter experiments based on elastic nuclear recoil. Astropart. Phys. **6**, 87 (1996)
49. E. Aprile et al., Dark matter search results from a one ton-year exposure of XENON1T. Phys. Rev. Lett. **121** 111302 (2018)
50. P. Agnes et al., Low-mass dark matter search with the DarkSide-50 experiment. Phys. Rev. Lett. **121** 081307 (2018)
51. M. Tanabashi et al., [Particle Data Group], Review of particle physics. Phys. Rev. D **98**(3), 030001 (2018). https://doi.org/10.1103/PhysRevD.98.030001
52. M.A. Shifman, A.I. Vainshtein, V.I. Zakharov, Remarks on Higgs–Boson interactions with nucleons. Phys. Lett. B **78**, 443 (1978)
53. M. Hoferichter, P. Klos, J. Menéndez, A. Schwenk, Improved limits for Higgs-portal dark matter from LHC searches. Phys. Rev. Lett. **119**(18), 181803 (2017)
54. J. Ellis, N. Nagata, K.A. Olive, Uncertainties in WIMP dark matter scattering revisited. Eur. Phys. J. C **78**(7), 569 (2018)
55. D. Tucker-Smith, N. Weiner, Inelastic dark matter. Phys. Rev. D **64**, 043502 (2001)
56. R.J. Hill, M.P. Solon, Standard model anatomy of WIMP dark matter direct detection II: QCD analysis and hadronic matrix elements. Phys. Rev. D **91**, 043505 (2015)
57. E. Masso, S. Mohanty, S. Rao, Dipolar dark matter. Phys. Rev. D **80**, 036009 (2009)
58. K. Sigurdson, M. Doran, A. Kurylov, R.R. Caldwell, M. Kamionkowski, Dark-matter electric and magnetic dipole moments. Phys. Rev. D **70**, 083501 (2004)
59. T. Bringmann, L. Bergstrom, J. Edsjo, New gamma-ray contributions to supersymmetric dark matter annihilation. JHEP **1**, 049 (2008)
60. C. Karwin, S. Murgia, T.M.P. Tait, T.A. Porter, P. Tanedo, Dark matter interpretation of the *Fermi*-LAT observation toward the Galactic Center. Phys. Rev. D **95**(10), 103005 (2017)
61. A. Morselli [CTA Consortium], The dark matter programme of the Cherenkov Telescope Array. PoS **ICRC2017**, 921 (2018)
62. K. Murase, R. Laha, S. Ando, M. Ahlers, Testing the dark matter scenario for PeV neutrinos observed in IceCube. Phys. Rev. Lett. **115**(7), 071301 (2015)
63. M. Aguilar et al. [AMS], Electron and positron fluxes in primary cosmic rays measured with the alpha magnetic spectrometer on the International Space Station. Phys. Rev. Lett. **113**, 121102 (2014)
64. I.V. Moskalenko, A.W. Strong, Production and propagation of cosmic ray positrons and electrons. Astrophys. J. **493**, 694–707 (1998)

Perturbations of the FRW Universe and Formation of Large Scale Structures

3

Abstract

Observations of cosmic microwave background and large scale structures are the main tools of precision cosmology whereby theories of dark matter, inflation, neutrino mass are tested with data from Planck, SDSS and other such experiments. Large scale structures like galaxies and clusters of galaxies are formed when matter perturbations grow as pressure-less cold dark matter falls into the gravitational potential of the primordial perturbations which are believed to arise from a period of inflation. In this chapter we will work out the general relativistic theory of density perturbations which will be used for comparing observations with theories of particle physics and cosmology.

3.1   Introduction

Large scale galactic surveys like the Sloan Digital Sky Survey, Dark Energy Survey and Baryon Oscillations Spectroscopy Survey are increasing providing data about the distribution of matter in the universe and provide a scope for us to tests theories of dark matter, dark energy, neutrino mass etc. To test theory with observations from the survey we need to understand perturbations of the metric and matter (dark matter, baryons, photons, neutrinos) in the framework of general relativity.

Galaxies and clusters of galaxies arise from amplification the primordial perturbations generated during inflation. The inflationary perturbations leave the horizon during the course of inflation and they re-enter the horizon after reheating in the subsequent radiation and matter domination era.

© Springer Nature Switzerland AG 2020   49
S. Mohanty, *Astroparticle Physics and Cosmology*,
Lecture Notes in Physics 975, https://doi.org/10.1007/978-3-030-56201-4_3

3.2   Perturbations in an FRW Universe

In the relativistic treatment we consider the perturbations in the Einsteins equations, $\delta G^{\mu\nu} = 8\pi G \delta T^{\mu\nu}$ in the background FRW universe. The general relativistic treatment of density and metric perturbations has been covered in Refs. [1–7].

3.2.1   Robertson Walker Background

The zeroth order equation $G^{\mu\,(0)}_{\ \nu} = 8\pi G \bar{T}^{\mu}_{\nu}$ describes the background unperturbed Freidmann-Robertson-Walker universe. The flat FRW metric describes a isotropic and homogenous expanding universe with zero spatial curvature,

$$ds^2 = a(\eta)^2 \left(-d\eta^2 + \delta_{ij} dx^i dx^j \right) \tag{3.1}$$

where

$$\eta = \int^t \frac{dt'}{a(t')} \tag{3.2}$$

is the conformal time variable. For the background FRW, the non-zero Christoffel connections are

$$\Gamma^0_{00} = \mathcal{H}\,, \quad \Gamma^i_{0j} = \mathcal{H}\delta^i_j\,, \quad \Gamma^0_{ij} = \mathcal{H}\delta_{ij}\,. \tag{3.3}$$

where $\mathcal{H} = a'/a = aH$ and $'$ denotes derivative w.r.t conformal time η.

The Ricci tensors and scalar are

$$R_{00} = -3\mathcal{H}'\,, \quad R_{0i} = 0\,,$$
$$R_{ij} = \left[\mathcal{H}' + 2\mathcal{H}^2 \right] \delta_{ij}\,,$$
$$R = \frac{6}{a^2} \left(\mathcal{H}' + \mathcal{H}^2 \right)\,, \tag{3.4}$$

and the Einstein tensors for the FRW metric are

$$G_{00} = 3\mathcal{H}^2\,, \quad G_{oi} = 0\,,$$
$$G_{ij} = -\left[2\mathcal{H}' + \mathcal{H}^2 \right] \delta_{ij}\,. \tag{3.5}$$

Taking the stress tensor to be

$$T^{\mu}_{\ \nu} = \text{diagonal}(-\bar{\rho}, \bar{P}, \bar{P}, \bar{P}) \tag{3.6}$$

The Einsteins equations $G^\mu{}_\nu = (8\pi G/3)T^\mu{}_\nu$ for the FRW metric give the Friedmann eqautions

$$\mathcal{H}^2 = \frac{8\pi G}{3}a^2\bar\rho \,,$$

$$2\mathcal{H}' + \mathcal{H}^2 = -8\pi G a^2 \bar P \,. \tag{3.7}$$

The stress tensor conservation equation for the background gives the relation

$$\bar\rho' + 3\mathcal{H}\left(\bar\rho + \bar P\right) = 0 \tag{3.8}$$

For the equation of state $\bar P = \omega\bar\rho$ the stress tensor conservation equation (3.8) gives us the relation between the background density and the scale factor,

$$\bar\rho \propto a^{-3(1+\omega)} \,. \tag{3.9}$$

Using this in the first of the Friedmann equations (3.7) we obtain the relation

$$a(\eta) = \begin{cases} \eta^{\frac{2}{1+3\omega}}, & \omega \neq -1 \,, \eta \in (0,\infty) \,, \\ \frac{-1}{H\eta}, & \omega = -1 \,, \eta \in (-\infty, 0) \,, \end{cases}$$

which is the relation between the scale factor and conformal time for a general equation of state parameter.

If the equation of state changes with time then we have the relation

$$\omega' = \left(\frac{\bar\rho}{\bar p}\right)' = 3\mathcal{H}\left(1 + \omega\right)\left(\omega - c_s^2\right) \tag{3.10}$$

where we made use of (3.8) and where $c_s^2 \equiv \bar\rho'/\bar p'$. This implies that if $\omega' = 0$ then $\omega = c_s^2$.

3.2.2 Gauge Dependence of Perturbations in GR

While writing the perturbations of the metric and energy momentum we have to take into consideration the general coordinate transformation symmetry of general relativity(GR), $x^\mu \to \tilde x^\mu = x^\mu + \xi^\mu(x)$. This gauge transformation can be used to eliminate 4 of the 10 components of the metric tensor. The physical degrees of freedom in the metric are the gauge invariant combinations of the metric perturbations.

One can utilise the gauge freedom to choose the metric perturbations which will result in the simplest form for the equations. In this section we discuss the scalar degree of perturbations in the metric which give rise to structure formation.

The vector perturbation of the metric has no growing solution and plays no role in seeding large scale structure. The tensor perturbation or the primordial gravitational waves have an important role in the anisotropy of the CMB temperature and polarisation and will be considered in the latter sections.

General Relativity is invariant under the general coordinate transformations or the gauge transformations

$$x^\mu \to \tilde{x}^\mu = x^\mu + \xi^\mu , \tag{3.11}$$

where $\xi^\mu(x^\rho)$ are four independent functions of x^μ.

The components of the metric transform as tensors under the gauge transformation (3.11),

$$\tilde{g}_{\mu\nu}(\tilde{x}^\rho) = \frac{\partial x^\mu}{\partial \tilde{x}^\alpha} \frac{\partial x^\nu}{\partial \tilde{x}^\beta} g_{\mu\nu}(x^\rho) \tag{3.12}$$

We split the metric $g_{\mu\nu}(x^\rho)$ into a homogenous-background and perturbation parts in the x^μ coordinates

$$g_{\mu\nu}(x^\rho) = \bar{g}_{\mu\nu}(x^\rho) + \delta g_{\mu\nu}(x^\rho), \tag{3.13}$$

Since $g_{\mu\rho} g^{\rho\nu} = \delta^\nu_\mu$ and $\tilde{g}_{\mu\rho}\tilde{g}^{\rho\nu} = \delta^\nu_\mu$, the contravariant perturbations $\delta g^{\mu\nu} = -\bar{g}^{\mu\rho}\bar{g}^{\nu\sigma}\delta g_{\rho\sigma}$.

On substituting (3.13) in (3.12) and keeping terms upto linear order in $\delta g_{\mu\nu}$ or ξ^μ we obtain,

$$\tilde{g}_{\mu\nu}(\tilde{x}^\rho) = \bar{g}_{\mu\nu}(x^\rho) + \delta g_{\mu\nu}(x^\rho) - \bar{g}_{\mu\delta}(x^\rho)\frac{\partial\xi^\delta}{\partial\tilde{x}^\nu} - \bar{g}_{\delta\nu}(x^\rho)\frac{\partial\xi^\delta}{\partial\tilde{x}^\mu} \tag{3.14}$$

We can also write $\tilde{g}_{\mu\nu}(\tilde{x}^\rho)$ as a background plus linear perturbation

$$\tilde{g}_{\mu\nu}(\tilde{x}^\rho) = \bar{g}_{\mu\nu}(\tilde{x}^\rho) + \delta\tilde{g}_{\mu\nu}(\tilde{x}^\rho). \tag{3.15}$$

The background metric $\bar{g}_{\mu\nu}(\tilde{x}^\rho)$ is the homogenous and isotropic FRW metric in $\tilde{x}^\rho$ coordinates. Under gauge transformations the background metrics in the two coordinates can be related as

$$\bar{g}_{\mu\nu}(x^\rho) = \bar{g}_{\mu\nu}(\tilde{x}^\rho - \xi^\rho) \simeq \bar{g}_{\mu\nu}(\tilde{x}^\rho) - \frac{\partial\bar{g}_{\alpha\beta}}{\partial\tilde{x}^\rho}\xi^\rho. \tag{3.16}$$

Here $\bar{g}_{\mu\nu}(\tilde{x}^\rho)$ is the FRW metric in $\tilde{x}^\rho$ coordinates while $\bar{g}_{\mu\nu}(x^\rho)$ is the FRW metric in x^ρ coordinates.

Using (3.16) in (3.14) we obtain

$$\tilde{g}_{\mu\nu}(\tilde{x}^\rho) = \bar{g}_{\mu\nu}(\tilde{x}^\rho) + \delta g_{\mu\nu}(x^\rho) - \frac{\partial\bar{g}_{\mu\nu}}{\partial\tilde{x}^\rho}\xi^\rho(x^\rho) - \bar{g}_{\mu\delta}(x^\rho)\frac{\partial\xi^\delta}{\partial\tilde{x}^\nu} - \bar{g}_{\delta\nu}(x^\rho)\frac{\partial\xi^\delta}{\partial\tilde{x}^\mu} \tag{3.17}$$

Comparing (3.17) with (3.15) we obtain the transformation law of metric perturbation under gauge transformations,

$$\delta g_{\mu\nu}(x^\rho) \rightarrow \delta \tilde{g}_{\mu\nu}(\tilde{x}^\rho) = \delta g_{\mu\nu}(x^\rho) - \frac{\partial \bar{g}_{\mu\nu}(\tilde{x}^\rho)}{\partial \tilde{x}^\delta}\xi^\delta - \bar{g}_{\mu\delta}(x^\rho)\frac{\partial \xi^\delta}{\partial \tilde{x}^\nu} - \bar{g}_{\delta\nu}(x^\rho)\frac{\partial \xi^\delta}{\partial \tilde{x}^\mu}.$$

(3.18)

The four gauge degrees of freedom $\xi^\mu = (\xi^0, \xi^i)$ can be decomposed into scalar and vector components. The three-vector ξ^i can be written as a sum of a divergence-less vector $\xi^i_\perp$ and a divergence of a scalar ξ,

$$\xi^i = \xi^i_\perp + \bar{\gamma}^{ij}\partial_i \xi,$$

(3.19)

where $\bar{\gamma}^{ij}$ is the spatial part of the background metric. The transverse part $\xi^i_\perp$ has the property $\partial_i \xi^i_\perp = 0$ and it represents 2 vector degrees of freedom. The scalar parameters in the general gauge transformation are ξ^0 and ξ.

For the FRW metric $\bar{g}_{00} = -a^2(\eta)$, $\bar{g}_{ij} = a^2(\eta)\delta_{ij}$ and using (3.18) we see that the gauge transformation of the components of perturbed FRW metric $g_{\mu\nu}(x^\rho)$ are

$$\delta \tilde{g}_{00} = \delta g_{00} - 2a\left(a\xi^0\right)',$$

$$\delta \tilde{g}_{0i} = \delta g_{0i} + a^2\left[\xi'_{\perp I} + \partial_i\left(\xi' - \xi^0\right)\right],$$

$$\delta \tilde{g}_{ij} = \delta g_{ij} + a^2\left[2\frac{a'}{a}\delta_{ij}\xi^0 + 2\partial_i\partial_j\xi + \partial_i\xi_{\perp j} + \partial_j\xi_{\perp i}\right],$$

(3.20)

where primes denote derivative w,r,t conformal time η.

The most general linearly perturbed spatially flat FRW metric can be written as

$$ds^2 = -(1 + 2\Psi)a^2 d\eta^2 + 2a^2 B_i d\eta dx^i + a^2\left[(1 - 2\Phi)\delta_{ij} + 2E_{ij}\right] dx^i dx^j,$$

(3.21)

where Ψ, Φ are the scalar perturbations, B_i are the vector perturbations and E_{ij} are the tensor perturbations.

The vector B_i can be decomposed into a gradient of as scalar, say B, and divergence free vector, say S_i, as

$$B_i \equiv \partial_i B - S_i, \qquad \text{where} \quad \partial^i S_i = 0,$$

(3.22)

and similarly, any second rank tensor E_{ij} can be written in terms of a divergence free vector and a traceless and divergence free tensor as

$$E_{ij} \equiv 2\partial_i\partial_j E + 2\partial_{(i}F_{j)} + h_{ij},$$

(3.23)

where $\partial^i F_i = 0$ and $h^i_i = \partial^i h_{ij} = 0$.

The 10 degrees of freedom of the metric can be decomposed into 4 scalar d.o.f. Ψ, B, Φ, E; 2 vectors S_i, F_i with 2 d.o.f each; and 2 d.o.f. of tensor h_{ij}. We can choose ξ^0 and ξ appropriately to remove two of the scalar functions of Ψ, Φ, B and E. This is called the *gauge fixing* or *gauge choice* which corresponds to choosing a set of *gauge transformation* parameters to simplify the calculations.

Using the metric perturbation transformation laws (3.20), one may check that the tensor perturbations h_{ij} are invariant under gauge transformations $\xi^\mu = (\xi^0, a^{-2}(\xi^i_\perp + \delta^{ij}\partial_j\xi))$ while scalar perturbations Ψ, B, Φ and E transform according to (3.20) as

$$\Psi \rightarrow \tilde{\Psi} = \Psi - \frac{1}{a}\left(a\xi^0\right)',$$

$$B \rightarrow \tilde{B} = B + \xi^0 - \xi',$$

$$\Phi \rightarrow \tilde{\Phi} = \Phi + \frac{a'}{a}\xi^0,$$

$$E \rightarrow \tilde{E} = E - \xi, \tag{3.24}$$

Physical observables must be written in terms of gauge-invariant quantities [8].

One can then construct gauge invariant quantities which do not depend on choice of coordinate system. Two gauge-invariant quantities introduced by Bardeen [8] are

$$\Psi^{GI} \equiv \Psi - \frac{1}{a}[a(E' - B)]', \tag{3.25}$$

$$\Phi^{GI} \equiv \Phi - \frac{a'}{a}(E' - B) \tag{3.26}$$

and these represent physical observables.

Under gauge transformations the vector component of metric perturbations on using (3.20) transform as

$$S_i \rightarrow S_i + \xi'_{\perp i}, \quad F_i \rightarrow F_i + \xi_{\perp i}, \tag{3.27}$$

and the vector

$$V_i^{GI} = S_i - F_i' \tag{3.28}$$

is gauge invariant.

One can restrict the gauge degrees of freedom by making a gauge choice which to fix some components of the metric perturbation. In the *conformal Newtonian gauge* we fix $B = E = 0$. The scalar perturbation of the metric are Ψ and Φ. In the *conformal Newtonian gauge* the metric perturbations are represented by the line element

$$ds^2 = -a(\eta)^2 (1 + 2\Psi) d\eta^2 + a(\eta)^2 \left[(1 - 2\Phi) \delta_{ij} + h_{ij}\right] dx^i dx^j. \tag{3.29}$$

Gauge Invariant Stress Tensor Perturbations

The energy momentum tensor for a general fluid can be expressed as

$$T_{\mu\nu} = (\rho + P)\, u_\mu u_\nu + P g_{\mu\nu} + \left[q_\mu u_\nu + q_\nu u_\mu + \pi_{\mu\nu} \right] \tag{3.30}$$

where ρ is the energy density, P is the pressure, q_μ is the heat flux and $\pi_{\mu\nu}$ is the viscous shear stress. For dealing with perturbations in the baryon, photons, CDM fluid we can treat the stress tensor with the terms in the square brackets equal to zero. Neutrino free-steaming can give a non-zero $\pi_{\mu\nu}$.

For the metric (5.18) the four velocity can be written as

$$u^\mu = \frac{dx^\mu}{\sqrt{-ds^2}} = \left(\frac{1}{a}(1 + \Psi), \frac{1}{a}(v^i - B^i) \right) \tag{3.31}$$

where $v^i = \frac{dx^i}{d\eta} = a\frac{dx^i}{dt}$ is the peculiar velocity. The covariant components of the four velocity are given by

$$u_\mu = g_{\mu\nu} u^\nu = (-a(1 + \Psi), a v_i) \tag{3.32}$$

and $u_\mu u^\mu = -1$.

Taking $T_{\mu\nu} = (\rho + p)u_\mu u_\nu + P g_{\mu\nu}$ the ideal fluid form, with the background $\bar{T}_\nu^\mu = \text{diag}(-\bar\rho(\eta),\ \bar{P}(\eta)\delta_j^i)$ and the perturbations $\delta T^\mu{}_\nu$ can be evaluated in terms of the density, pressure and velocity perturbations

$$\delta\rho(\eta, \mathbf{x}) \equiv \rho(\eta, \mathbf{x}) - \bar\rho(\eta), \quad \delta P(\eta, \mathbf{x}) = P(\eta, \mathbf{x}) - \bar{P}(\eta), \quad \delta u_\mu = (-a\Psi, a v_i). \tag{3.33}$$

The first order perturbations of $T_{\mu\nu}$ are

$$\delta T_{\mu\nu} = (\delta\rho + \delta P)\, u_\mu u_\nu + (\bar\rho + \bar{P})(\delta u_\mu\, u_\nu + u_\mu \delta u_\nu) + \delta P\, \bar{g}_{\mu\nu} + \bar{P}\delta g_{\mu\nu} \tag{3.34}$$

The perturbations of T_ν^μ are therefore given by

$$\delta T^\mu{}_\nu = \delta(g^{\mu\alpha} T_{\alpha\nu}) = \delta(g^{\mu\alpha}) T_{\alpha\nu} + g^{\mu\alpha} \delta(T_{\alpha\nu}). \tag{3.35}$$

from which we can get the components,

$$\delta T^0{}_0 = -\delta\rho,$$
$$\delta T^0{}_i = (\bar\rho + \bar{P})\, v^i,$$
$$\delta T^i{}_0 = -(\bar\rho + \bar{P})\, (v^i - B^i),$$
$$\delta T^i{}_j = \delta P\, \delta_{ij} + \pi_{ij}. \tag{3.36}$$

The stress tensor perturbations transform in the same way as the metric perturbations (3.18),

$$\delta T_{\mu\nu}(x^\rho) \to \delta \tilde{T}_{\mu\nu}(\tilde{x}^\rho) = \delta g_{\mu\nu}(x^\rho) - \frac{\partial \bar{T}_{\mu\nu}(\tilde{x}^\rho)}{\partial \tilde{x}^\delta}\xi^\delta - \bar{T}_{\mu\delta}(x^\rho)\frac{\partial \xi^\delta}{\partial \tilde{x}^\nu} \tag{3.37}$$

$$- \bar{T}_{\delta\nu}(x^\rho)\frac{\partial \xi^\delta}{\partial \tilde{x}^\mu}.$$

The density and pressure perturbations $\delta\rho$ and δP transform under gauge transformations as

$$\delta\rho \to \delta\tilde{\rho} = \delta\rho - \bar{\rho}'\xi^0,$$

$$\delta P \to \delta\tilde{P} = \delta P - \bar{\rho}'\xi^0. \tag{3.38}$$

A 4−vector $u^\mu(x^\rho)$ like the four velocity transforms as

$$\tilde{u}_\alpha(\tilde{x}^\rho) = \frac{\partial x^\mu}{\partial \tilde{x}^\alpha}u_\mu(x^\rho),$$

and its perturbation $\delta u_\alpha(x^\rho) = u_\alpha(x^\rho) - \bar{u}_\alpha(x^\rho)$ transforms as

$$\delta u_\alpha(x^\rho) \to \delta \tilde{u}_\alpha(\tilde{x}^\rho) = \delta u_\alpha(x^\rho) - \frac{\partial \bar{u}_\alpha(\tilde{x}^\rho)}{\partial \tilde{x}^\rho}\xi^\rho - \bar{u}_\mu(x^\rho)\frac{\partial \xi^\mu}{\partial \tilde{x}^\alpha}. \tag{3.39}$$

The velocity perturbation $\delta u_\alpha = (0, v_i)$ can be decomposed into a gradient and a
divergence-less vector

$$v_i = \partial_i v + v_{\perp i} \quad \partial^i v_{\perp i} = 0, . \tag{3.40}$$

Under gauge transformations

$$v_i \to \tilde{v}_i = v_i + \partial_i \xi^0, \tag{3.41}$$

therefore the velocity potential v transforms like a scalar while $v_{\perp i}$ is gauge
invariant,

$$\tilde{v} = v + \xi^0, \quad \tilde{v}_{\perp i} = v_{\perp i}. \tag{3.42}$$

Any general scalar $\phi(x^\mu)$, does not change under the coordinate transformation

$$\tilde{\phi}(\tilde{x}^\mu) = \phi(x^\mu),$$

however its split into the background and perturbation depends on the coordinates and perturbation $\delta\phi(x^\mu) = \phi(x^\mu) - \bar{\phi}(x^\mu)$ transforms as

$$\delta\phi(x^\mu) \rightarrow \delta\tilde{\phi}(\tilde{x}^\mu) = \delta\phi(x^\mu) - \frac{\partial\bar{\phi}(\tilde{x}^\mu)}{\partial\tilde{x}^\rho}\xi^\rho. \tag{3.43}$$

Since the background scalar $\bar{\phi}(x^0)$ is homogenous, the scalar perturbations transform as

$$\delta\tilde{\phi}(\tilde{x}^\mu) = \delta\phi(x^\mu) - \bar{\phi}(\eta)'\xi^0. \tag{3.44}$$

Gauge invariant quantities which can be constructed from the stress tensor perturbations are the co-moving density and pressure perturbations

$$\delta\rho^{GI} = \bar{\rho}\Delta \equiv \delta\rho + \bar{\rho}'v,$$
$$\delta P^{GI} = \delta P + \bar{P}'v, \tag{3.45}$$

which are the density and pressure perturbations respectively in a coordinate system where the peculiar velocity is zero.

An important gauge invariant observable is the co-moving curvature perturbation [9]

$$\mathcal{R} = \Phi - \mathcal{H}v = \Phi - \frac{\mathcal{H}}{\bar{\rho} + \bar{P}}\delta q. \tag{3.46}$$

where δ_q is the scalar component of the momentum flux $\delta T^0{}_i \equiv \partial_i \delta q$.

Under gauge transformations $\Phi \rightarrow \Phi + \mathcal{H}\xi^0$ and $v \rightarrow v + \xi^0$, therefore $\mathcal{R}$ is gauge invariant. $\mathcal{R}$ is the Ricci-curvature in the space-like hyper surface of zero peculiar velocity. It has the property that for perturbations of length scales larger than the horizon size $\mathcal{R}$ is conserved. Primordial co-moving curvature perturbations generated during inflation which are super-horizon, do not change in magnitude during the subsequent evolution of the universe as long as they remain outside the horizon. Perturbations entering the horizon in the subsequent radiation era are amplified and they seed formation of large scale structure when the dark matter and baryons fall into the local Newtonian potentials of these perturbations. Super-horizon perturbations which entered the horizon at the time when the free electrons bind with the protons to form neutral hydrogen give rise to large angle anisotropy in the cosmic microwave background radiation which has been observed by COBE, WMAP and Planck experiments.

Another gauge invariant variable is the curvature on space-like hypersurface of constant density

$$\zeta \equiv \Phi - \frac{\mathcal{H}}{\bar{\rho}'}\delta\rho. \tag{3.47}$$

Under gauge transformation $\delta\rho \to \delta\rho - \bar{\rho}'\xi^0$ and $\Phi \to \Phi - \mathcal{H}\xi^0$ one can choose $\xi^0 = \delta\rho/\bar{\rho}'$ such that in that space-like hyper-surface $\delta\tilde{\rho} = \delta\rho - \bar{\rho}'\xi^0 = 0$. The gauge invariant variable ζ (3.47) is interpreted as the curvature on space-like hypersurfaces of constant ρ i.e where $\delta\rho = 0$,

$$\zeta = \Phi\Big|_{\delta\rho=0} \tag{3.48}$$

3.2.3 Perturbations of Robertson Walker Universe

We choose the conformal Newtonian gauge in which the perturbations of the FRW metric can be written as

$$ds^2 = a(\eta)^2 \left(-(1 + 2\Psi(\eta, \mathbf{x}))\, d\eta^2 + (1 - 2\Phi(\eta, \mathbf{x}))\, \delta_{ij} dx^i dx^j \right). \tag{3.49}$$

Einsteins equation to the first order in metric and energy-momentum perturbations is

$$G^{\mu\,(0)}_{\ \nu} + \delta G^{\mu}_{\ \nu} = 8\pi G \left(\bar{T}^{\mu}_{\nu} + \delta T^{\mu}_{\ \nu} \right) \tag{3.50}$$

The first order perturbations equations are

$$\delta G^{\mu}_{\ \nu} = 8\pi G\, \delta T^{\mu}_{\ \nu}. \tag{3.51}$$

The perturbations in the Einstein tensor for the metric (3.49) is calculated in the following steps. The perturbations in the Christoffer connection in terms of metric perturbations can be written as

$$\delta\Gamma^{\mu}_{\nu\lambda} = \frac{1}{2}\delta g^{\mu\alpha}\left[g_{\alpha\nu,\lambda} + g_{\alpha\lambda,\nu} - g_{\nu\lambda,\alpha}\right] + \frac{1}{2}g^{\mu\alpha}\left[\delta g_{\alpha\nu,\lambda} + \delta g_{\alpha\lambda,\nu} - \delta g_{\nu\lambda,\alpha}\right]. \tag{3.52}$$

For the metric (3.49) the non-zero components of (3.52) are

$$\delta\Gamma^0_{ij} = -\delta_{ij}\left[2\mathcal{H}\left(\Phi + \Psi\right) + \Phi'\right],$$

$$\delta\Gamma^0_{00} = \Psi', \quad \delta\Gamma^0_{0i} = \Psi_{,i}, \quad \delta\Gamma^i_{00} = \delta^{ij}\Psi_{,j},$$

$$\delta\Gamma^i_{j0} = -\delta^i_j\Phi', \quad \delta\Gamma^i_{jk} = -2\delta^i_{(j}\partial_{k)}\Phi + \delta_{jk}\delta^{il}\Phi_{,l}. \tag{3.53}$$

The perturbations of the Ricci tensor and Ricci scalar are

$$\delta R_{\mu\nu} = \delta\Gamma^{\alpha}_{\mu\nu,\alpha} - \delta\Gamma^{\alpha}_{\mu\alpha,\nu} + \delta\Gamma^{\alpha}_{\mu\nu}\Gamma^{\beta}_{\alpha\beta} + \Gamma^{\alpha}_{\mu\nu}\delta\Gamma^{\beta}_{\alpha\beta} - \delta\Gamma^{\alpha}_{\mu\beta}\Gamma^{\beta}_{\alpha\nu} - \Gamma^{\alpha}_{\mu\beta}\delta\Gamma^{\beta}_{\alpha\nu}$$

$$(3.54)$$

$$\delta R = \delta g^{\mu\alpha} R_{\alpha\mu} + g^{\mu\alpha}\delta R_{\alpha\mu} . \tag{3.55}$$

The perturbations in the Einstein tensor can be written as

$$\delta G_{\mu\nu} = \delta R_{\mu\nu} - \frac{1}{2}\delta g_{\mu\alpha} R + g_{\mu\alpha}\delta R ,$$

$$\delta G^{\mu}{}_{\nu} = \delta g^{\mu\alpha} G_{\alpha\nu} + g^{\mu\alpha}\delta G_{\alpha\nu} . \tag{3.56}$$

Putting them together we obtain the perturbations of Einstein tensor which are as follows,

$$\delta G^0{}_0 = 2a^{-2}\left[3\mathcal{H}\left(\mathcal{H}\Psi + \Phi'\right) - \nabla^2\Phi\right] ,$$

$$\delta G^0{}_i = -2a^{-1}\left(\Phi'_{,i} + \mathcal{H}\Psi_{,i}\right) ,$$

$$\delta G^i{}_j = 2a^{-2}\left[\left(\mathcal{H}^2 + 2\mathcal{H}'\right)\Psi + \mathcal{H}\Psi' + \Phi'' + 2\mathcal{H}\Phi'\right]\delta^i_j$$

$$+a^{-2}\left[\nabla^2\left(\Psi - \Phi\right)\delta^i_j - \nabla^i\left(\Psi_{,j} - \Phi_{,j}\right)\right] . \tag{3.57}$$

We now turn towards the calculation of the perturbations of the energy momentum tensor, $\delta T^{\mu}{}_{\nu}$.

Using Einsteins tensor perturbation (3.57) and stress tensor perturbation (3.36) we can write the components of the first order perturbation of Einsten's equation, $\delta G^{\mu}{}_{\nu} = 8\pi G\delta T^{\mu}{}_{\nu}$.

The 00 component of this equation is

$$3\mathcal{H}\left(\mathcal{H}\Psi + \Phi'\right) - \nabla^2\Phi = -4\pi Ga^2\bar{\rho}\,\delta \tag{3.58}$$

The $0i$ component is

$$\partial^i\left(-\Phi' - \mathcal{H}\Psi\right) = 4\pi Ga^2(\delta\rho + \delta P)\,v^i \tag{3.59}$$

Equation (3.59) can be formally integrated and written as

$$\left(\Phi' + \mathcal{H}\Psi\right) = -4\pi Ga^2\left(\bar{\rho} + \bar{P}\right)v \tag{3.60}$$

where v is the velocity potential, defined as the scalar field whose divergence is the velocity field

$$\nabla v \equiv \mathbf{v}. \tag{3.61}$$

This equation is useful in computing the gauge invariant perturbations as we shall see in the next section. The ij component with $i \neq j$ gives

$$- a^{-2} \nabla^i \left(\Psi_{,j} - \Phi_{,j} \right) = \pi_{ij} \tag{3.62}$$

which implies that in the absence of anisotropic stress component ($\pi_{ij} = 0$)

$$\Psi = \Phi. \tag{3.63}$$

Finally the ij component with $i = j$ leads to the equation

$$\Phi'' + 2\mathcal{H}\Phi' + \mathcal{H}\Psi' + \left(\mathcal{H}^2 + 2\mathcal{H}' \right) \Psi = 4\pi G a^2 \delta P \tag{3.64}$$

Using (7.17) Eq. (3.64) can be written as

$$\Phi'' + 3\mathcal{H}\Phi' + \left(\mathcal{H}^2 + 2\mathcal{H}' \right) \Phi = 4\pi G a^2 \delta P \tag{3.65}$$

3.2.4   Continuity and Euler's equation

In addition to Einsteins equations we also have the stress tensor conservation

$$\nabla_\mu T^\mu{}_\nu = \partial_\mu T^\mu{}_\nu + \Gamma^\mu_{\mu\alpha} T^\alpha{}_\nu - \Gamma^\alpha_{\mu\alpha} T^\mu{}_\alpha = 0 \tag{3.66}$$

The $\nu = 0$ component of this equation gives us

$$\partial_0 T^0{}_0 + \partial_i T^i{}_0 + \Gamma^\mu_{\mu 0} T^0{}_0 - \Gamma^0_{00} T^0{}_0 - \Gamma^i_{j0} T^i{}_i = 0 \tag{3.67}$$

where we have retained terms upto first order in either metric or stress tensor perturbations. Using the expressions (3.3) and (3.53) for the zeroth and first order Christoffel connections and (3.6) and (3.36) for the background and perturbations stress tensor respectively we have

$$\begin{aligned} & - \partial_0 (\bar{\rho} + \delta\rho) - (\bar{\rho} + \bar{P})\partial_i v^i - (\mathcal{H} + \Psi' + 3\mathcal{H} - 3\Phi')(\bar{\rho} + \delta\rho) \\ & + (\mathcal{H} + \Psi')(\bar{\rho} + \delta\rho) + (\mathcal{H} + \Psi')(\bar{\rho} + \delta\rho) - (\mathcal{H} - \Phi')\delta^i_j \left[(\bar{P} + \delta P)\delta^j_i \right] = 0 \end{aligned}$$

$$\tag{3.68}$$

The zeroth order equation gives the continuity equation for the background

$$\bar{\rho}' + 3\mathcal{H}(\bar{\rho} + \bar{P}) = 0 \tag{3.69}$$

and collecting the first order terms we obtain

$$\delta\rho' + 3\mathcal{H}(\delta\rho + \delta P) - 3\Phi'(\bar{\rho} + \bar{P}) + (\bar{\rho} + \bar{P})\partial_i v^i = 0 \tag{3.70}$$

which can also be written as

$$\delta' + 3\mathcal{H}\left(\frac{\delta P}{\delta\rho} - \frac{\bar{P}}{\bar{\rho}}\right) + \left(1 + \frac{\bar{P}}{\bar{\rho}}\right)(\nabla \cdot \mathbf{v} - 3\Phi') = 0 \tag{3.71}$$

This is the *relativistic continuity equation* for the density perturbation.

The $\nu = i$ component of (3.66) is

$$\partial_0 T^0{}_i + \partial_j T^j{}_i + \Gamma^\mu_{\mu j} T^j{}_i - \Gamma^0_{0i} T^j{}_0 - \Gamma^j_{0i} T^0{}_j - \Gamma^0_{kj} T^k{}_j = 0 \tag{3.72}$$

which gives us

$$\left((\bar{\rho} + \bar{P})v_i\right)' + \partial_i \delta P - 4\mathcal{H}(\bar{\rho} + \bar{P})v_i - (\bar{\rho} + \bar{P})\partial_i \Psi = 0 \tag{3.73}$$

Using the zeroth order continuity equation (3.70), this reduces to the form

$$\mathbf{v}' + 3\mathcal{H}\left(\frac{1}{3} - \frac{\bar{P}'}{\bar{\rho}'}\right)\mathbf{v} = -\frac{1}{(\bar{\rho} + \bar{P})}\nabla\delta P - \nabla\Psi . \tag{3.74}$$

This is the *Euler equation* for acceleration of the velocity perturbation.

Summary

We summarise the result of this section by collecting the equations which we will use for calculating the power spectrum for large scale structures and CMB anisotropy.

These are the Einsteins equations for the first order metric perturbations

$$\nabla^2\Phi = 4\pi G a^2 \bar{\rho}\,\delta + 3\mathcal{H}\left(\mathcal{H}\Psi + \Phi'\right), \tag{3.75}$$

$$\left(\Phi' + \mathcal{H}\Psi\right) = -4\pi G a^2 (\bar{\rho} + \bar{P})\,v, \tag{3.76}$$

$$\Phi'' + 2\mathcal{H}\Phi' \quad +\mathcal{H}\Psi' + \left(\mathcal{H}^2 + 2\mathcal{H}'\right)\Psi = 4\pi G a^2 \delta P. \tag{3.77}$$

$$\nabla^i\left(\Psi_{,j} - \Phi_{,j}\right) = -a^2\pi_{ij} \tag{3.78}$$

(continued)

In the absence of anisotropic stress ($\pi_{ij} = 0$) we have $\Phi = \Psi$.

Equations (3.75) and (3.76) can be combined to give the relativistic Poisson equation

$$\nabla^2 \Phi = 4\pi G a^2 \bar{\rho} \Delta , \tag{3.79}$$

where

$$\Delta \equiv \delta - 3\mathcal{H} \left(1 + \frac{\bar{P}}{\bar{\rho}} \right) v . \tag{3.80}$$

is a gauge invariant quantity. The velocity potential therefore serves as a source for gravitational potential.

By combining the Einstein's equations (3.75), (3.76), and (3.77) and assuming adiabatic perturbations and zero-anisotropic stress we obtain an equation for the gravitational curvature $\Phi = \Psi$,

$$\Phi'' + 3(1 + c_s^2)\mathcal{H}\Phi' + \left[-c_s^2 \nabla^2 + 3\mathcal{H}^2 c_s^2 + 2\mathcal{H}' + \mathcal{H}^2 \right) \Phi = 0 \tag{3.81}$$

Using the relation of the background metric $\mathcal{H}' = (-1/2)(1 + \omega)\mathcal{H}^2$ this equations reduces to the form

$$\Phi'' + 3(1 + \omega)\mathcal{H}\Phi' - c_s^2 \nabla^2 \Phi = 0 \tag{3.82}$$

Here ω and c_s are the average equation of state and sound speed for all matter,

$$\omega = \frac{\sum_i \bar{P}_i}{\sum_i \bar{\rho}_i} , \quad \text{and} \quad c_s^2 = \frac{\sum_i \bar{P}_i'}{\sum_i \bar{\rho}_i'} . \tag{3.83}$$

In addition to the Einsteins equations we also have the continuity equation

$$\delta_i' + 3\mathcal{H} \left(\frac{\delta P_i}{\delta \rho_i} - \frac{\bar{P}_i}{\bar{\rho}_i} \right) = - \left(1 + \frac{\bar{P}_i}{\bar{\rho}_i} \right) (\nabla \cdot \mathbf{v}_i - 3\Phi') \tag{3.84}$$

and the Euler's equation

$$\mathbf{v_i}' + 3\mathcal{H} \left(\frac{1}{3} - \frac{\bar{P}_i'}{\bar{\rho}_i'} \right) \mathbf{v} = -\frac{1}{(\bar{\rho}_i + \bar{P}_i)} \nabla \delta P_i - \nabla \Psi . \tag{3.85}$$

obeyed by each species i. These equations are used for calculating the sub-horizon perturbations of dark matter, baryons and radiation.

3.2.5 Adiabatic and Iso-Curvature Perturbations

Perturbations are called adiabatic, if one can make a common shift of the local time coordinate for all species (labelled by i) and choose a local coordinate frame with no perturbations, It amounts to a common local shift in the time coordinate $\eta(\mathbf{x})$ for all species,

$$\delta \eta(\mathbf{x}) = \frac{\delta \rho_i(\eta, \mathbf{x})}{\bar{\rho}_i{}'} = -\frac{\delta \rho_i(\eta, \mathbf{x})}{3\mathcal{H}(\bar{\rho}_i + \bar{P}_i)}. \tag{3.86}$$

Therefore

$$\frac{\delta_i}{1 + \omega_i} = \frac{\delta j}{1 + \omega_j} \quad \text{for all species } i, j. \tag{3.87}$$

Where

$$\delta_i \equiv \frac{\rho_i(t, \mathbf{x}) - \bar{\rho}_i(t)}{\bar{\rho}_i}, \qquad \omega_i \equiv \frac{P_i}{\rho_i} \tag{3.88}$$

and we have used the conservation equation for each species $\bar{\rho}_i{}' + 3\mathcal{H}(\bar{\rho}_i + \bar{P}_i) = 0$.

In adiabatic perturbations of the different components (baryons, CDM, radiation) are thus related

$$\delta_c = \delta_b = \frac{3}{4}\delta_r \tag{3.89}$$

and there is no relative fluctuations in the number density of particles of different species, say non-relativistic matter and radiation,

$$\delta\left(\frac{n_m}{n_r}\right) = 0 \Rightarrow \frac{\delta n_m}{n_m} = \frac{\delta n_r}{n_r}. \tag{3.90}$$

In the case of adiabatic perturbations

$$\frac{\delta P_i}{\delta \rho_i} = \frac{P_i'}{\rho_i'} = c_{si}^2 \tag{3.91}$$

where c_{si} is the speed of pressure wave propagation of species i. The total pressure perturbation in adiabatic perturbations can be written in terms of the total density perturbation and an average speed of sound,

$$\delta P = \sum_i \delta P_i = \sum_i c_{si}^2 \, \delta \rho_i$$

$$= c_s^2 \sum_i \delta \rho_i = c_s^2 \delta \rho \tag{3.92}$$

where

$$c_s^2 \equiv \frac{\sum_i c_{si}^2 \, \bar{\rho}_i{}'}{\sum_i \bar{\rho}_i{}'} \tag{3.93}$$

In case the perturbations do not conserve entropy they are called non-adiabatic or isocurvature. One can also define a relative entropy perturbation between different species- say matter and radiation

$$S_{mr} = \frac{\delta_m}{\bar{\rho}_m + \bar{P}_m} - \frac{\delta_r}{\bar{\rho}_r + \bar{P}_r} \, . \tag{3.94}$$

Non-adiabatic perturbations are generated for instance when a heavy particle decays into relativistic particles.

The non-adiabatic perturbation in a single species can be defined as

$$\Gamma = \frac{\delta P}{P'} - \frac{\delta \rho}{\rho'} \, . \tag{3.95}$$

The non-adiabatic part of the pressure perturbations are defined as

$$\delta P_{nad} \equiv P' \Gamma \equiv \delta P - \frac{P'}{\rho'} \delta \rho \, , \tag{3.96}$$

From the CMB measurements it is seen that the isocurvature component of the primordial density perturbations is small [10]. In the rest of the section we shall assume adiabatic perturbations as the dominant perturbation mode which causes the formation of large scale structure and CMB anisotropy.

3.2.6　Conservation of Super-Horizon Curvature Perturbation

The co-moving curvature perturbation defined as

$$\mathcal{R} = \Phi - \mathcal{H} v \tag{3.97}$$

is a gauge invariant quantity. $\mathcal{R}$ is the Ricci 3-curvature in the local co-moving frame with zero peculiar velocity. The co-moving curvature does not change in time once modes leave the horizon. This is an important property since the perturbations generated during inflation exit the horizon where they remain frozen till they re-enter the horizon during radiation or matter era. So the density perturbations in the present era can be related to the primordial co-moving curvature perturbation generated during inflation.

The Poisson's equation (3.79) in Fourier space is

$$-k^2 \Phi_k = 4\pi G a^2 \bar{\rho} \delta_k - 3\mathcal{H} \left(1 + \frac{\bar{P}}{\bar{\rho}} \right) v_k . \tag{3.98}$$

where Φ_k is related to $\Phi(t, \mathbf{x})$ as

$$\Phi(\eta, \mathbf{x}) = \int \frac{d^3 k}{(2\pi)^3} e^{-i \mathbf{k} \cdot \mathbf{x}} \Phi_k(\eta) \tag{3.99}$$

and similarly for the other perturbation variables. This introduces the co-moving wavenumber k which is related to the physical wavelength of the perturbations $\Phi(\eta, \mathbf{x})$ as $\lambda(\eta) = 2\pi a(\eta) k^{-1}$. The size of the horizon is $R_h = H^{-1} = a\mathcal{H}^{-1}$. Modes are super-horizon if $\lambda \gg R_h$ or $k \ll \mathcal{H}$. For the super-horizon modes we have from (3.98)

$$-k^2 \Phi = \frac{3}{2} \mathcal{H}^2 \left[\delta - \frac{3\mathcal{H}}{\bar{\rho}} \left(\bar{\rho} + \bar{P} \right) v \right] \tag{3.100}$$

where we have used the Friedmann equation $4\pi G a^2 \bar{\rho} = (3/2)\mathcal{H}^2$. This implies that for super-horizon modes

$$v = \frac{\delta\rho}{3\mathcal{H}(\bar{\rho} + \bar{P})} , \quad \text{for} \quad k \ll \mathcal{H} . \tag{3.101}$$

Using this we see that for super-horizon modes the co-moving curvature perturbations (3.97) can be written as

$$\mathcal{R} = \Phi + \frac{\delta\rho}{3(\bar{\rho} + \bar{P})} \tag{3.102}$$

The continuity equation (3.71) for the super-horizon modes (where we can set $\nabla \cdot \mathbf{v} \simeq 0$) reduces to

$$\delta\rho' + 3\mathcal{H} \left(\delta\rho + \delta P \right) = 3 \left(\bar{\rho} + \bar{P} \right) \Phi' \tag{3.103}$$

We take the time derivative of Eq. (3.102) and use Eq. (3.103) to substitute for Φ' to obtain

$$3 \left(\bar{\rho} + \bar{P} \right) \mathcal{R}' = 3\mathcal{H} \left(\delta\rho + \delta P \right) + \frac{\bar{\rho}' + \bar{P}'}{\bar{\rho} + \bar{P}} \delta\rho . \tag{3.104}$$

Using $\bar{\rho}' = -3\mathcal{H}(\bar{\rho} + \bar{P})$ we can write this equation in the form

$$\frac{(\bar{\rho} + \bar{P})}{\mathcal{H}}\mathcal{R}' = \left(\delta P - \frac{\bar{P}'}{\bar{\rho}'}\delta\rho\right) \tag{3.105}$$

The term in the r.h.s is the non-adiabatic pressure perturbation (3.96) which vanishes if we have adiabatic perturbations, which implies that for adiabatic perturbations for super-horizon scale perturbations (for which $k \ll \mathcal{H}$) we have $\mathcal{R}' \simeq 0$ and the co-moving curvature $\mathcal{R}$ is conserved.

The other gauge independent variable the three curvature at constant energy density hypersurface which was introduced in (3.47),

$$\zeta = \Phi + \mathcal{H}\frac{\delta\rho}{\bar{\rho}'} = \Phi - \frac{\delta\rho}{3(\bar{\rho} + \bar{P})} \tag{3.106}$$

is related to $\mathcal{R}$ as

$$\zeta = \mathcal{R} - \frac{2\bar{\rho}}{3(\bar{\rho} + \bar{P})}\left(\frac{k^2}{\mathcal{H}^2}\right)\Psi \tag{3.107}$$

For super-horizon perturbations ($k^2 \ll \mathcal{H}^2$) and $\zeta \simeq \mathcal{R}$. Therefore $\zeta' = 0$ and ζ is conserved for adiabatic perturbations at super-horizon scales.

3.2.7   Time Evolution of Cosmological Perturbations

We now have the technology to study the large scale structure formation in the universe. We will derive the power spectrum of the density perturbations of baryons and dark matter and check if the predictions match the observations of the large scale surveys of galaxies and clusters of galaxies.

We will assume that the primordial perturbations (generated by Inflation or a Bouncing Universe) were adiabatic. We will now study the evolution of these initial perturbations as they enter the horizon at different epochs. The *particle horizon* is the distance photons have travelled from the beginning of the universe till any given time,

$$R_h = a(t)\int dx = a(t)\int_0^t \frac{dt'}{a(t')} = a\int_0^a \frac{da}{Ha^2} = a\int_0^a d\ln a\left(\frac{1}{aH}\right) \simeq \frac{1}{H} \tag{3.108}$$

The particle horizon represents the sphere of causal contact around any given point. The co-moving particle horizon is

$$\chi = \int_0^t \frac{dt'}{a(t')} = \int_0^a \frac{da}{Ha^2} = \int_0^a d\ln a \left(\frac{1}{aH}\right) \simeq \frac{1}{aH} = \frac{1}{\mathcal{H}} \qquad (3.109)$$

Perturbations enter the horizon when the physical wavelength $\lambda(\eta)$ equals the horizon $R_h = H^{-1} = a(\eta)\mathcal{H}^{-1}$.

The initial perturbations are generated during inflation and these exit the horizon of the inflationary universe $R_{inf} = H_{inf}^{-1}$. These super-horizon perturbations re-enter the horizon during the subsequent radiation and matter era. The physical wavelength of the perturbations in the present epoch $\lambda(\eta_0) = a_0 2\pi/k$ and k is measured in units of Mpc^{-1}. For example if perturbations entered the horizon at matter radiation equality z_{eq} given by the relation

$$z_{eq} = \frac{\Omega_m}{\Omega_r} \simeq 3700 \qquad (3.110)$$

where we have taken the present matter density $\Omega_m h^2 = 0.12$ and the radiation density (including three massless neutrinos) to be $\Omega_r h^2 = 3.24 \times 10^{-5}$.

Then the co-moving size of the horizon at matter-radiation equality is

$$\chi_{eq} = \int_0^{t_{eq}} \frac{dt}{a(t)} = \int_0^{a_{eq}} \frac{da}{a^2 H(a)} = \int_{z_{eq}}^\infty \frac{dz}{H_0 \left[(\Omega_m^2 (z+1)^3 + \Omega_r (z+1)^4\right]^{1/2}}$$
$$= 0.0258 H_0^{-1} = 115.14 \, \text{Mpc}. \qquad (3.111)$$

The co-moving wavenumber of the perturbations which entered the horizon at z_{eq} is $k_{eq} = \chi_{eq}^{-1} \simeq 0.01 \text{Mpc}^{-1}$ which implies that perturbations which entered the horizon at matter-radiation equality have a size $\lambda(a_0) = 2\pi a_0 k_{eq}^{-1} \simeq 600$ Mpc in the present era.

The theory of inflation (which we study in the next chapter) predicts that during inflation the perturbations in density and gravitational potential have an amplitude which has been measured from the large angle cosmic microwave anisotropy to be of magnitude $\mathcal{R} \sim 10^{-5}$ and nearly scale invariant i.e the amplitude is independent of the wavelength of the perturbations. The super-horizon co-moving curvature perturbation $\mathcal{R}$ remain frozen (as we have seen in Sect. 3.2.6). After they re-enter the horizon they grow at rate which depends on the era in which they enter the horizon. The growing sub-horizon perturbation then subsequently seed the formation of galaxies and galactic clusters. We now use the first order perturbations of Einstein's equation (3.2.4) to compute the evolution of gravitational potential and matter density perturbations which are initially super-horizon and then enter the horizon during radiation and matter domination era.

Evolution of Super-Horizon Perturbations

Equation (3.82) in Fourier space for the super-horizon modes $-k^2 \Phi_k \ll \mathcal{H} \Phi'$ reduces to

$$\Phi_k'' + 3(1 + \omega)\mathcal{H}\Phi_k' = 0. \tag{3.112}$$

The non decaying solutions for Φ_k in both radiation era ($\omega = 1/3.a \sim \eta, \mathcal{H} = 1/\eta$) and in matter era ($\omega = 0, a \sim \eta^2, \mathcal{H} = 2/\eta$) is a constant,

$$\Phi_k(\eta) = \text{constant}, \quad k \ll \mathcal{H} \tag{3.113}$$

Therefore for the super-horizon modes $\Phi_k = $ constant during both radiation and matter domination era. However when the equation of state changes during the transition from radiation to matter era Φ_k does change in this transition period. To determine the change of Φ in the transition from radiation to matter era, we can make use of the fact that $\mathcal{R}$ remains a constant through all epochs till the modes re-enter the horizon.

Using Eq. (3.76) we can write (3.97) (for the case where anisotropic stress is zero and $\Phi = \Psi$) as

$$\mathcal{R} = \Phi + \mathcal{H} \frac{(\Phi' + \mathcal{H}\Phi)}{4\pi G a^2 (\bar{\rho} + \bar{P})} \tag{3.114}$$

For adiabatic perturbations we can write (3.114) using the background equation $4\pi G a^2 \rho = (3/2)\mathcal{H}^2$ to obtain

$$\mathcal{R} = \Phi + \frac{2\Phi}{3(1 + \omega)} \left[1 + \frac{\Phi'}{\mathcal{H}\Phi} \right] \tag{3.115}$$

We see that during an epoch where the equation of state is $\omega = $ constant, the super-horizon the co-moving curvature is given by

$$\mathcal{R} = \Phi \left[\frac{5 + 3\omega}{3(1 + \omega)} \right] \tag{3.116}$$

as $\Phi = constant$.

In the radiation era $\omega = 1/3$ and we have

$$\mathcal{R} = \frac{3}{2}\Phi_{RD} \tag{3.117}$$

while in the matter era $\omega = 0$ and the gravitational potential is given by

$$\mathcal{R} = \frac{5}{3}\Phi_{MD} \tag{3.118}$$

Since $\mathcal{R}$ is identical in both epochs, we see that the relation between Φ in the radiation and matter is

$$\Phi_{MD} = \frac{9}{10}\Phi_{RD}. \tag{3.119}$$

So the super-horizon gravitational potential perturbations change by a factor $9/10$ in the transition from the radiation domination to matter domination.

Matter perturbations can be related to the gravitational potential using Eq. (3.75)

$$-k^2\Phi_k = 4\pi Ga^2\bar{\rho}\delta_k + 3\mathcal{H}^2\Phi_k + 3\mathcal{H}\Phi'_k \tag{3.120}$$

For the super-horizon modes $k^2\Phi_k$ term can be dropped. Also for $\Phi'_k \simeq 0$ for the super-horizon modes. Then using the relation $4\pi Ga^2\bar{\rho} = (3/2)\mathcal{H}^2$ we see that for super-horizon modes

$$\delta_k = -2\Phi_k = \text{constant}, \quad k \ll \mathcal{H} \tag{3.121}$$

Here δ_k is the total matter perturbation. The perturbations of the individual components (cold dark matter, baryons, radiation) will be in the ratio

$$\delta_c = \delta_b = \frac{3}{4}\delta_\gamma = \frac{3}{4}\delta_{v_i} \tag{3.122}$$

so as to maintain adiabatic perturbations. The δ_k in (3.121) is the total matter density fluctuation

$$\delta_k = (\delta_r\Omega_r + \delta_c\Omega_c + \delta_b\Omega_b) = -2\Phi_k \tag{3.123}$$

where we have defined the fractional density of each component $\Omega_i \equiv \bar{\rho}_i/\bar{\rho}$. In the radiation domination era $\Omega_r \simeq 1 \gg \Omega_c, \Omega_b$ and we have for the super-horizon modes

$$\delta_r = \frac{4}{3}\delta_c = \frac{4}{3}\delta_b = -2\Phi_{RD}, \quad k \ll \mathcal{H}. \tag{3.124}$$

In the matter era if we take $\Omega_c \simeq 1 \gg \Omega_r, \Omega_b$ we will have

$$\delta_c = \delta_b = \frac{3}{4}\delta_r = -2\Phi_{MD}, \quad k \ll \mathcal{H}. \tag{3.125}$$

Super-horizon matter perturbations can be related to R as

$$\mathcal{R}_k = \left[\frac{5+3\omega}{3(1+\omega)}\right]\Phi_k = -\frac{1}{2}\left[\frac{5+3\omega}{3(1+\omega)}\right]\delta_k \tag{3.126}$$

which implies that in the radiation era

$$\delta_r = -2\Phi_{RD} = -\frac{4}{3}\mathcal{R} \tag{3.127}$$

and in the matter era

$$\delta_m = \delta_c\Omega_c + \delta_b\Omega_b = -2\Phi_{MD} = -\frac{6}{5}\mathcal{R}. \tag{3.128}$$

We now turn to perturbations of Δ (3.80) which sources the gravitational potential through Poisson's equation (3.79),

$$-k^2\Phi_k = 4\pi Ga^2\bar{\rho}\Delta_k \tag{3.129}$$

Using the Friedman equation for the background $4\pi Ga^2\bar{\rho} = (3/2)\mathcal{H}^2$ we can write (3.129) as

$$-k^2\Phi_k = \frac{3}{2}\mathcal{H}^2\Delta_k \tag{3.130}$$

In the radiation era $a \sim \eta$, $H = 1/\eta$ and Δ_k grows as

$$\Delta_k = -\frac{2}{3}(k\eta)^2\Phi \propto a^2, \quad k\eta \ll 1. \tag{3.131}$$

In the matter era $a \sim \eta^2$, $H = 2/\eta$ and Δ_m grows as

$$\Delta_k = -\frac{1}{6}(k\eta)^2\Phi \propto a, \quad k\eta \ll 1. \tag{3.132}$$

So we see that at super-horizon scales δ_k remains a constant (and only changes when there is a transition from the radiation to matter era) while super-horizon Δ_k grows as $\Delta_k \propto a^2$ during radiation era and $\Delta_k \propto a$ during matter era. Just like Φ_k and δ_k, at radiation to matter era transition Δ_k reduces by a factor $9/10$.

Sub-Horizon Perturbations
The super-horizon perturbations of gravitational potential, matter and radiation are governed solely by Einsteins equations and on whether the perturbations are adiabatic or iso-curvature. After the perturbations enter the horizon their subsequent evolution depends on the background evolution (radiation era or matter era) and also of the equation of state of fluid itself. Each component falls into the gravitational potential sourced by dark matter and their fall is opposed by the pressure. In addition

the non-gravitational interactions play a role, for example the baryons are tightly coupled to the photons till the decoupling era $z_{dec} \simeq 1100$ and do not fall into the gravitational potential well of the cold dark matter. For this reason baryons by themselves without cold dark matter cannot account for the large scale structure of the universe.

The evolution of the sub-horizon gravitational potential Φ_k in radiation and matter era are as follows.

Sub-Horizon Gravitational Potential
We start with the evolution of Φ which is governed by Eq. (3.82). In the *radiation era* we have $c_s^2 = \omega = 1/3$ and $\mathcal{H} = 1/\eta$,

$$\Phi_k'' + \frac{1}{\eta}\Phi_k' + \frac{k^2}{3}\Phi_k = 0. \tag{3.133}$$

This equation has the solution

$$\Phi_k(\eta) = -2\mathcal{R}_k\left(\frac{\sin x - x\cos x}{x^3}\right), \quad x \equiv \frac{k\eta}{\sqrt{3}} \tag{3.134}$$

where we have imposed the initial conditions

$$\Phi_k(x = 0) = -\frac{2}{3}\mathcal{R}_k, \quad \text{and} \quad \Phi_k'(x = 0) = 0 \tag{3.135}$$

to match the solution for the super-horizon modes at $k\eta/\sqrt{3} \to 0$.

This solution has the limits, for super-horizon modes $x \ll 1$, $\Phi = -(2/3)\mathcal{R}$ while for sub-horizon modes $x \gg 1$ the solution is

$$\Phi_k(\eta) = -6\mathcal{R}_k\frac{1}{(k\eta)^2}\cos\left(\frac{k\eta}{\sqrt{3}}\right), \quad k > k_{eq}. \tag{3.136}$$

We see that Φ_k after entering the horizon has a damped oscillatory behaviour with damping factor $\propto \eta^{-2} = a^{-2}$.

In the *matter domination era*, we have from (3.82) with $\omega = c_s^2 = 0$ and $\mathcal{H} = 2/\eta$,

$$\Phi_k'' + \frac{6}{\eta}\Phi_k' = 0 \tag{3.137}$$

The non-decaying solution of this equation is a constant

$$\Phi_k(\eta) = \frac{3}{5}\mathcal{R}, \quad k < k_{eq},\tag{3.138}$$

where the amplitude is chosen to match the super-horizon. The modes which enter the horizon before matter radiation equality ($k > k_{eq}$) are damped as $\Phi_k \propto a^{-2}$ in the radiation era and become constant at t_{eq}. On the other hand modes which enter the horizon after radiation matter equality ($k < k_{eq}$) are constant in both matter and radiation era and only change by a factor $9/10$ at radiation matter transition.

Sub-Horizon Density Perturbations

The gauge invariant matter density perturbations $\Delta(k, \eta) \equiv \delta(k, \eta) - 3\mathcal{H}\left(1 + \frac{\bar{P}}{\bar{\rho}}\right)v$ in the sub-horizon regime ($k \gg \mathcal{H}$) simplifies to $\Delta(k, \eta) \simeq \delta(k, \eta)$. The evolution of the sub-horizon δ_m depends on the equation of state and speed of sound of the species m.

The equations which govern the perturbations of dark matter, baryons and radiation components are the continuity and Eulers equations (3.84) and (3.85) respectively. The continuity equation for each component of matter (denoted by subscript m) is

$$\delta'_m + 3\mathcal{H}\left(c_{s\,m}^2 - \omega_m\right)\delta_m = -(1 + \omega_m)\left(\nabla \cdot \mathbf{v}_m - 3\Phi'\right)\tag{3.139}$$

where we have assumed the adiabatic relation $\delta P_m/\delta\rho_m = \bar{P}'_m/\bar{\rho}'_m = c_{s\,m}^2$ and the equation of state relation $\omega_m = \bar{P}_m/\bar{\rho}_m$. For adiabatic perturbations $c_s^2 \simeq \omega_m$ and we can drop the second term on the l.h.s of (3.139).

The Euler's equation (3.85) obeyed by each component can be written in terms of the velocity potential, turns out to be

$$\mathbf{v}'_m + \mathcal{H}(1 - 3\omega_m)\mathbf{v}_m = -\frac{c_{s\,m}^2}{(1 + \omega_m)}\nabla\delta_m - \nabla\Phi.\tag{3.140}$$

We can combine (3.139) and (3.140) to obtain a second order equation for δ_m,

$$\delta''_m + (1 - 3\omega_m)\mathcal{H}\delta'_m - c_{s\,m}^2\nabla^2\delta_m = (1 + \omega_m)\nabla^2\Phi.\tag{3.141}$$

We can simplify this equation by using the Poisson equations (3.75) which for sub-horizon perturbations reduces to

$$\nabla^2\Phi = 4\pi G a^2\bar{\rho}\delta\tag{3.142}$$

where δ is the total matter density perturbation of all species which source the gravitational potential

$$\delta = \frac{1}{\bar{\rho}} \left(\delta_\gamma \bar{\rho}_\gamma + \delta_b \bar{\rho}_b + \delta_c \bar{\rho}_c \right) . \tag{3.143}$$

Substituting (3.142) in (3.141) we obtain the Jean's equation for the evolution of matter perturbation δ_m in the expanding universe,

$$\delta_m'' + (1 - 3\omega_m)\mathcal{H}\delta_m' + c_{sm}^2 k^2 \delta_m = (1 + \omega_m) 4\pi G a^2 \bar{\rho}\delta . \tag{3.144}$$

The solutions for the growth of perturbations of δ_c, δ_b and δ_r in radiation and matter era depends on the gravitational potential. For the perturbation modes

$$c_{sm}^2 k^2 \delta_m > (1 + \omega_m) 4\pi G a^2 \bar{\rho}\delta \tag{3.145}$$

the pressure term dominates over the gravitational potential and we will have a damped oscillatory solution for δ_m with the damping provided by the Hubble expansion. On the other hand for the long-wavelength modes where the gravitational term dominates we will have growth of perturbations as a power law or logarithmically in η and a. Once the perturbations become large enough that the gravitational potential term dominates over the Hubble friction term, the perturbations grow exponentially with time.

Cold Dark Matter
Cold dark matter (CDM) particles are non-relativistic when they decouple from the other particles in the heat bath and they are collision-less, so we have $\omega_m = 0$ and $c_{sm}^2 = 0$. The Euler equation for the CDM velocity perturbations is

$$\mathbf{v}_c' + \mathcal{H}\mathbf{v}_c = -\nabla\Phi . \tag{3.146}$$

In the absence of gravitational and pressure perturbations the velocity perturbation goes as $v_c' = -\mathcal{H}v_c \Rightarrow v_c \propto 1/a$. The peculiar velocity of collision-less non-relativistic matter gets dampened due to the Hubble expansion.

For CDM the Jean's equation (3.144) is of the form

$$\rho_c'' + \mathcal{H}\rho_c' = \frac{3}{2}\mathcal{H}^2 \left(\delta_r \Omega_r + \delta_c \Omega_c + \delta_b \Omega_b \right) \tag{3.147}$$

where we have used the Friedmann equation $4\pi G a^2 \bar{\rho} = \frac{3}{2}\mathcal{H}^2$ and defined the fractional density of each component $\Omega_m \equiv \bar{\rho}_m/\bar{\rho}$.

In the *radiation dominated era*, $\Omega_r \sim 1 \gg \Omega_c, \Omega_b$. The perturbations of radiation density δ_r oscillate at sub-horizon scales and the time average $\langle \delta_r \rangle \simeq 0$. As a result the local gravitational potential is sub-dominant compared to the Hubble friction term and we can drop the r.h.s of (3.147) to get the equation for the CDM perturbation in the radiation era

$$\delta_c'' + \frac{1}{\eta}\delta_c' = 0 \tag{3.148}$$

where we have used $a \sim \eta$ and $\mathcal{H} = 1/\eta$ for the radiation era. The general solution of this equation is

$$\delta_c(\eta) = A \ln \eta + B . \tag{3.149}$$

Therefore in the radiation era the growing mode of CDM perturbation grows as $\delta_c \simeq \ln a$ with the scale-factor a.

In the *matter domination era*, the dominant contribution to the gravitational potential comes from the CDM perturbations. Assuming that baryons are subdominant ($\Omega_b \ll \Omega_c$) we can take $\Omega_c \sim 1$ and Eq. (3.147) with ($a \sim \eta^2$, $\mathcal{H} = 2/\eta$ for the matter era) reduces to the form

$$\delta_c'' + \frac{2}{\eta}\delta_c' - \frac{6}{\eta^2}\delta_c = 0 . \tag{3.150}$$

This has the general solution

$$\delta_c(\eta) = A\eta^2 + B\eta^{-3} . \tag{3.151}$$

In the matter era the growing mode of the CDM perturbations therefore grows with the scale factor as $\delta_c \sim a$.

Radiation
Perturbations of radiation density obey equation (3.144) with $\omega_m = c_{sm}^2 = 1/3$, which in the radiation era reduces to,

$$\delta_r'' + \frac{1}{3}k^2\delta_r = \frac{16\pi}{3}Ga^2\bar{\rho}\left(\delta_r\Omega_r + \delta_c\Omega_c + \delta_b\Omega_b\right)$$

$$= \frac{2}{\eta^2}\delta_r \tag{3.152}$$

where we have taken $(16\pi/3)Ga^2\bar{\rho} = 2\mathcal{H}^2 = 2/\eta^2$ and assumed $\Omega_r \simeq 1 \gg \Omega_c, \Omega_b$. This has the solution

$$\delta_r(\eta, k) = -\frac{4}{3}\mathcal{R}\cos\left(\frac{k\eta}{\sqrt{3}}\right) \tag{3.153}$$

which matches the super-horizon solution $\delta_r(k\eta \to 0) = -(4/3)\mathcal{R}$. In the radiation era sub-horizon perturbation of radiation density have oscillatory solutions with $\langle\delta_r\rangle = 0$. This was assumed in the previous section while evaluating δ_c in the radiation era.

In the matter era, the CDM perturbations in the r.h.s of (3.144) are the source of potential well. Taking $\Omega_c \simeq 1 \gg \Omega_r, \Omega_b$ we see that (3.144) for radiation perturbation takes the form

$$\delta_r'' + \frac{1}{3}k^2\delta_r = 2\mathcal{H}^2\delta_c \tag{3.154}$$

In the matter era $\mathcal{H} = 2/\eta$ and $\delta_c \sim \eta^2$ which means that the r.h.s $2\mathcal{H}^2\delta_c \to C$ (constant). The solution of (3.154) is

$$\delta_r = \frac{3C}{k^2} + A\cos\left(\frac{k\eta}{\sqrt{3}}\right) + B\sin\left(\frac{k\eta}{\sqrt{3}}\right) \tag{3.155}$$

and sub-horizon radiation perturbations δ_r do not grow in the radiation era either.

Baryons
Nuclei and electrons which are bound together (prior to the decoupling era) by Coulomb interaction and are together referred to as the baryon fluid. The baryon fluid is tightly coupled to the photon fluid and obeys the equations and untill the time of decoupling at $z_{dec} \simeq 1100$. After the decoupling time which is well into the matter era (radiation matter quality occurs at $z_{eq} \sim 3400$, the equation governing baryon density perturbations is from (3.144) with $(c_{s\,m}^2 = \omega_m = 0)$,

$$\delta_b'' + \mathcal{H}\delta_b' = \frac{3}{2}\mathcal{H}^2\left(\delta_c\Omega_c + \delta_b\Omega_b\right) \tag{3.156}$$

The equation for CDM perturbations is the same as both baryons and CDM fall into the potential well created by the total matter perturbations

$$\delta_b'' + \mathcal{H}\delta_b' = \frac{3}{2}\mathcal{H}^2\left(\delta_c\Omega_c + \delta_b\Omega_b\right) \tag{3.157}$$

Defining $S_{bc} = \delta_b - \delta_c$ we see by subtracting (3.156) and (3.157) that S_{bc} obeys the equation

$$S''_{bc} + \frac{2}{\eta}S'_{bc} = 0 \tag{3.158}$$

which has the general solution

$$S_{bc} = A + B\eta^{-1} \tag{3.159}$$

The decaying solution $S_{bc} \sim \eta^{-1}$ shows that after decoupling from photons baryons perturbations approach the CDM perturbations asymptotically,

$$\delta_b \sim \delta_c \propto \eta^2 \propto a, \quad z \ll z_{dec}. \tag{3.160}$$

So we see that sub-horizon dark matter perturbations δ_c grows as a from the time of radiation matter equality η_{eq} which is at $z_{eq} \sim 3400$. The sub-horizon baryon perturbations δ_b oscillates with the radiation till decoupling era $z_{dec} \sim 1100$ and grows linearly with a only after decoupling. For this reason baryons by themselves are not sufficient to form large-scale structures since they don't have enough time to grow from the primordial value $\delta_b \sim \Delta T/T \sim 10^{-5}$ at z_{dec} to $\delta_c \sim 1$ needed for structure formation by $z \sim 10$.

Matter Perturbations in the Λ Dominated Universe
In the cosmological constant or dark energy Λ dominated epoch the evolution of matter perturbations obeys the equation

$$\delta''_m + \mathcal{H}\delta'_m - 4\pi G a^2 \bar{\rho} \left(\Omega_m \delta_m + \Omega_\Lambda \delta_\Lambda\right)\delta_m = 0 \tag{3.161}$$

The cosmological constant does not cluster therefore $\delta_\lambda = 0$. In the Λ dominated universe the second term $\mathcal{H}\delta'_m \simeq \mathcal{H}^2 \delta_m = (8\pi G/3)a^2 \rho_\Lambda \delta_m \gg 4\pi G a^2 \bar{\rho}_m \delta_m$. So the third term in (3.161) is much smaller compared to the second term and can be dropped, and the equation reduces to

$$\delta''_m + \mathcal{H}\delta'_m = 0. \tag{3.162}$$

In the Λ dominated universe $a \sim -1/\eta$, $\eta \in (-\infty, 0)$ and $\mathcal{H} = -1/\eta$ and we have

$$\delta''_m - \frac{1}{\eta}\delta'_m = 0. \tag{3.163}$$

This has the solutions

$$\delta_m = A\eta^2 = Aa^{-2} \quad \text{and} \quad \delta_m = \text{constant} \tag{3.164}$$

There is no growing mode solution and the growth of perturbation stops when the Λ domination starts.

3.2.8 Baryon Photon Coupling

When perturbations are of super-horizon size they are not affected by causal phenomenon like scattering. When these perturbations enter the horizon then the scattering with the particles in the heat bath becomes important specially for the photons, baryons and electrons. The main interaction is the Compton scattering $\gamma + e^- \leftrightarrow \gamma + e^-$. At energies ($\sim eV$) when the electrons are non-relativistic near the last scattering surface this scattering has the Thomson cross section cross section,

$$\sigma_T = \frac{8\pi\alpha^2}{3m_e^2} = 6.65 \times 10^{-25}\,\text{cm}^2 \tag{3.165}$$

where $\alpha = e^2/(4\pi) = 1/137$. The co-moving mean free path of the photons depends upon the free electron density n_e as

$$\lambda_c = \frac{1}{n_e \sigma_T a} \tag{3.166}$$

The electron number density is $n_e = (1 - Y_{He})x_e n_b$. Prior to the decoupling when the electron fraction $x_e \sim 1$ and taking the Helium fraction to be the value predicted from BBN $Y_{He} \simeq 0.25$ we have $n_e \simeq 10^4 cm^{-3}$ at $z \simeq 1200$. The co-moving mean free path is then $\lambda_c = 2.5\,\textbf{Mpc}$. The photons and electrons are therefore behave as tightly coupled and fluid. The protons are also coupled to this fluid due to electrostatic attraction and at length scales $\lambda \gg 2.5\text{Mpc}$ the baryons and photons are tightly coupled and have the same velocity $\mathbf{v}_\gamma \simeq \mathbf{v}_b$.

In the matter era CDM perturbations which has long decoupled from radiation grow as $\delta_c \sim a$. The growing CDM perturbations maintain a constant gravitational potential potential and the baryons fall into the potential well dragging the photons with them. As the density of the baryon photon fluid grows the photon pressure prevents further collapse and the fluid starts accelerating out the potential wells. The oscillations of the baryon-photon fluid is seen [11] in large scale structures at $k \sim 115\,h^{-1}\,\text{Mpc}$.

Summary
Primordial perturbations are created during inflation which become super-horizon during the course of inflation. These primordial perturbations then re-enter the horizon during the matter and radiation era. The co-moving curvature $\mathcal{R}(0)$ associated with the primordial metric and matter density perturbations is conserved if the perturbations are adiabatic.

Super-Horizon Perturbations: For the super-horizon perturbations the gravitational potential is a constant in both radiation and matter era,

$$\Phi_{RD} = \frac{2}{3}\mathcal{R}, \quad \Phi_{MD} = \frac{3}{5}\mathcal{R}, \quad k \ll \mathcal{H}. \tag{3.167}$$

but reduces by a factor of $9/10$ when there is a transition from the radiation to matter era.

Gravitational potential is sourced by the co-moving density perturbation Δ_m

$$-k^2\Phi = 4\pi G a^2 \bar{\rho}\Delta_m = \text{constant} \tag{3.168}$$

In the radiation era $a^2\bar{\rho} \propto a^2$ and the co-moving matter density $\Delta_m \propto a^2$. In the matter era, $a^2\bar{\rho} \propto a^{-1}$ therefore $\Delta_m \propto a$.

In the super-horizon regime the perturbations of CDM, baryons are proportional owing to adiabatic nature of perturbation and are constant in time. In the radiation era

$$\delta_r = \frac{4}{3}\delta_c = \frac{4}{3}\delta_b = -2\Phi_{RD} = -\frac{4}{3}\mathcal{R}, \tag{3.169}$$

while in the matter era

$$\delta_c = \delta_b = \frac{3}{4}\delta_r = -2\Phi_{MD} = -\frac{6}{5}\mathcal{R}. \tag{3.170}$$

Sub-Horizon Perturbations: The perturbations after horizon entry are amplified as matter falls into the Newtonian potential Φ sourced by Δ_m. At sub-horizon scales $\Delta_m \simeq \delta_m$.

In the radiation era, sub-horizon Φ has an oscillatory solution

$$\Phi_k(\eta) = -\frac{6R}{(k\eta)^2} \cos\frac{k\eta}{\sqrt{3}} \tag{3.171}$$

while in the matter era, $\Phi = \frac{3}{5}\mathcal{R}$ is constant at sub-horizon scales.

Sub-horizon perturbations of radiation δ_r are oscillatory in both radiation and matter era. The non-relativistic cold dark matter perturbations δ_c grow as $\delta_c \propto \ln a$

during radiation era and $\delta_c \propto a$ in the matter era. Baryons are tightly coupled to radiation and δ_b follows the oscillatory behaviour of δ_r till the decoupling epoch $z_{dec} \simeq 1089$. After decoupling δ_b grows as $\delta_b \propto a$ as the baryons fall into the potential well created by CDM. The oscillations of the baryon photon fluid called the Baryon Acoustic Oscillations can be observed in the large scale structures and as peaks and troughs in the CMB anisotropy spectrum.

3.2.9 Transfer Function

Primordial perturbations are generated during inflation and their length scale grows exponentially $\lambda_{phy} = a\lambda_{co-moving} = 2\pi e^{Ht}/k_{inf}$ while the inflation horizon $R_h = H_{inf}^{-1}$ is almost a constant, therefore the perturbations exit the horizon during inflation. After inflation there is a re-heating phase which begins the radiation era which is then followed by a matter era with the transition occurring at $z_{eq} \simeq 3400$. In the radiation era $\lambda_{phys} \propto t^{1/2}$ while the horizon grows as $R_h = 2t$. Therefore the modes which left the horizon during inflation re-enter the horizon during the radiation era. Similarly in the matter era $\lambda_{phys} \propto t^{2/3}$ while the horizon grows as $R_h = (3/2)t$ and super-horizon modes re-enter the horizon.

The same analysis can be done in terms of co-moving horizon (3.109). The co-moving length scale of the perturbations generated during inflation is $\lambda_{co-moving} = 2\pi/k_{inf}$. The co-moving horizon during inflation is $\chi_p = 1/(\mathcal{H})$. During inflation $a \sim -1/\eta H$ and $\chi_p = -\eta$ with $\eta \in (-\infty, 0)$ and the co-moving horizon therefore shrinks while $\lambda_{co-moving}$ is a constant, so the perturbations become super-horizon in the course of inflation. In the subsequent radiation era the co-moving horizon grows as $\chi_p \propto \eta$ and in the matter era $\chi_p = \eta/2$ with $\eta \in (0, \infty)$. Therefore the super-horizon perturbations generated during inflation re-enter the horizon in the subsequent radiation and matter era.

The amplitude of the primordial power spectrum generated during inflation is $\Delta_m(t_i, k) \simeq 10^{-5}$. The amplitude of these modes in the present epoch depends upon when the modes entered the horizon. The condition for horizon entry is that the physical wavenumber $k_{phy} = k/a = H$ or $k = \mathcal{H}$. Therefore the time of entry of a given mode depends upon the co-moving wavenumber k. The amplification factor from the initial time (end of inflation) to the present therefore depends upon the wavenumber. The relation between the initial amplitudes and its present value depends upon a function of k and is called the transfer function,

$$\Delta_m(\mathbf{k}, \eta_0) = T(k)\,\Delta_m(\mathbf{k}, \eta_i) \tag{3.172}$$

The matter-radiation equality occurs at $z_{eq} \simeq 3400$ and the modes that enter at z_{eq} have co-moving wavenumber $k_{eq} = \mathcal{H}_{eq} = a_{eq}H_{eq} = 0.01\,\mathrm{Mpc}^{-1}$. Modes with shorter wavelength ($k > k_{eq}$) entered the horizon before matter radiation equality. These modes grew as $\Delta_m \propto a^2$ when they were super-horizon, were almost constant after horizon entry in the radiation era and then grew as $\Delta_m \simeq \delta_c \propto a$ in the matter era. The total amplification of modes with $k > k_{eq}$ depends upon the horizon entry time (or scale factor) a_{ent},

$$T(k) = \left(\frac{a_{\text{ent}}}{a_i}\right)^2 \left(\frac{a_0}{a_{eq}}\right), \qquad k > k_{eq}.$$

$$= \left(\frac{a_{\text{ent}}}{a_{eq}}\right)^2 \left[\left(\frac{a_{eq}}{a_i}\right)^2 \left(\frac{a_0}{a_{eq}}\right)\right] \tag{3.173}$$

where a_i is the scale factor at reheating. The quantity in the square brackets is the same constant for all the modes independent of when they re-entered the horizon in the radiation era. We can relate the a_{ent} to the wavenumber k as

$$k = a_{\text{ent}} H_{\text{ent}} = \mathcal{H}_{ent} = \eta_{ent} \propto \frac{1}{a_{ent}} \tag{3.174}$$

Similarly $k_{eq} \propto 1/a_{eq}$. Therefore the transfer function (3.173) can be written as

$$T(k) = \text{constant} \left(\frac{k_{eq}}{k}\right)^2, \qquad k > k_{eq}. \tag{3.175}$$

Now consider the growth of perturbation modes Δ_m with $k < k_{eq}$ which entered the horizon in the matter era. These modes grow as $\Delta_m \propto a^2$ in the radiation era (when they are super-horizon) and then grow as $\Delta_m \propto a$ in the matter era (while they were still super-horizon). After horizon entry the matter density is dominated by CDM and the sub-horizon $\Delta_m \simeq \delta_c \propto a$. The total amplification of the modes which entered the horizon in the matter era is

$$T(k) = \left[\left(\frac{a_{eq}}{a_i}\right)^2 \left(\frac{a_0}{a_{eq}}\right)\right] = \text{constant}, \qquad k < keq. \tag{3.176}$$

independent of k. The k independence of the transfer function of modes which entered the horizon in the matter era is because they grew $\propto a$ both before and after horizon entry so the time of horizon entry or equivalently the k value during horizon entry does not change the growth factor.

The transfer function for cold dark matter is therefore,

$$T(k) = \text{constant} \times \begin{cases} 1, & k < k_{eq}, \\ k^{-2}, & k > k_{eq}. \end{cases} \tag{3.177}$$

The amplification of density perturbations of different species is encapsulated in the transfer function $T(k)$ which is different for cold dark matter, warm dark matter, baryons and photons.

3.2.10 Power Spectrum

What is measured in experiments is the two point correlation functions of the galaxies $\xi(|\mathbf{x} - \mathbf{y}|, t)$ which is the likelihood of finding two galaxies separated by distance $|\mathbf{x} - \mathbf{y}|$ at some redshift $t(z)$. This is related to the statistical average of the matter density contrast in position space as

$$\xi(|\mathbf{x} - \mathbf{y}|, t) = \langle \Delta_m(\mathbf{x}, t)\Delta_m(\mathbf{y}, t)\rangle \qquad (3.178)$$

assuming a isotropic and homogenous background. The two point correlation can be written in terms of the Fourier modes $\Delta_m(\mathbf{k}, t)$ as

$$\langle \Delta_m(\mathbf{k}, t)\Delta_m(\mathbf{k}', t)\rangle = \int d^3\mathbf{x}\, d^3\mathbf{y}\, e^{i(\mathbf{k}\cdot\mathbf{x} + \mathbf{k}'\cdot\mathbf{y})}\langle \Delta_m(\mathbf{x}, t)\Delta_m(\mathbf{y}, t)\rangle$$

$$= (2\pi)^3\delta_D^3(\mathbf{k} + \mathbf{k}')\int d^3\mathbf{r}\, e^{i\mathbf{k}\cdot\mathbf{r}}\xi(r, t) \qquad (3.179)$$

where $\mathbf{r} = \mathbf{x} - \mathbf{y}$.

The Fourier transform of the two point correlation function is called the Power spectrum

$$P(k, t) = \int d^3\mathbf{r}\, e^{i\mathbf{k}\cdot\mathbf{r}}\xi(r, t) \qquad (3.180)$$

The quantity $\Delta_m(\mathbf{k}, t)$ considered a random variable with probability distribution assumed as a Gaussian,

$$\text{Prob}\,(\Delta_m(\mathbf{k}, t)) = \frac{1}{\sqrt{2\pi P(k, t)}}\exp\left(\frac{-\Delta_m(\mathbf{k}, t)^2}{2P(k, t)}\right) \qquad (3.181)$$

The variance of the distribution is the power spectrum $\tilde{P}(k, t)$. This ensures that the two point correlation is the power spectrum

$$\langle \Delta_m(\mathbf{k}, t)\Delta_m(\mathbf{k}', t)\rangle = (2\pi)^3\delta_D^3(\mathbf{k} + \mathbf{k}')P(k, t). \qquad (3.182)$$

The power spectrum in the present epoch is therefore related to the initial power spectrum through the transfer function as

$$P(k, t_0) = T(k)^2 P(k, t_i) \qquad (3.183)$$

For CDM perturbations with transfer function given by (3.177), an initial power spectrum of the form $P(k, t_i) \propto k^n$ will evolve in the present universe to

$$P(k, t_0) = \text{constant} \times \begin{cases} k^n, & k < k_{eq}, \\ k^{n-4}, & k > k_{eq}. \end{cases} \tag{3.184}$$

The two point correlation is related to the power spectrum as

$$\langle \Delta_m(\mathbf{x}, t) \Delta_m(\mathbf{y}, t) \rangle = \int \frac{dk}{k} \frac{4\pi k^3}{(2\pi)^3} P(k) \, e^{i\mathbf{k}\cdot(\mathbf{x}-\mathbf{y})}. \tag{3.185}$$

The dimensionless quantity

$$\mathcal{P}(k, t) \equiv \frac{4\pi k^3}{(2\pi)^3} P(k, t) \tag{3.186}$$

is also used to represent the power spectrum. The initial power spectrum $P(k, t_i) \sim k^n$ corresponds to a $\mathcal{P}(k, t_i) \sim k^{n+3}$.

The gravitational potential perturbations are related to the density perturbations through the Poisson equation

$$-k^2 \Phi_k = 4\pi G \bar{\rho} a^2 \Delta_m(k) \tag{3.187}$$

We can define the dimensionless power spectrum of the Newtonian potential as

$$\mathcal{P}_\Phi(k) = \frac{k^3}{2\pi^2} \langle \Phi_k \Phi_k \rangle \tag{3.188}$$

Since $\Phi_k \propto k^{-2} \Delta_m(k)$, the power spectrum

$$\mathcal{P}_\Phi \propto k^{-4} \tilde{P}(k) \tag{3.189}$$

Therefore if the initial matter power spectrum $P(k, t_i) \sim k^n$ then the initial power spectrum of the gravitational perturbation will have a k dependence,

$$\mathcal{P}_\Phi \propto k^{n-1} \tag{3.190}$$

When $n = 1$ there will be equal power in gravitational perturbations at all length scale. This is the scale invariant Harrison-Zeldovich spectrum. Inflation models predicts an almost scale invariant curvature power spectrum, The value of n from CMB and Large scale observations is $n = 0.9670 \pm 0.0037$ at (68% CL) [10] which supports the prediction of infaltion.

In Fig. 3.1 we show the $P(k)$ for ΛCDM numerically generated using the CLASS code [12]. We see that $P(k) \propto k$ for $k < k_{eq}$ and $P(k) \propto k^{-3}$ for $k > k_{eq}$.

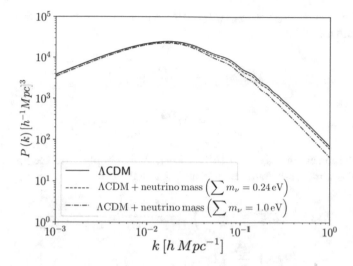

Fig. 3.1 Photons from different directions from the surface of last scattering have different temperatures owing to the local gravitational potential at the LSS (the SW effect), the changing gravitational potential in the course of the photon trajectory (he ISW effect) and the Doppler shift of temperature due to velocity of baryons at the LSS

The turnover occurs at $k = k_{eq} \sim 0.01$ Mpc^{-1}. Also shown is the power spectrum of ΛCDM with massive neutrinos. We see that there is a damping of power at large k with increasing neutrino masses due to their free-streaming after decoupling.

3.2.11 Free Streaming Scale of Warm Dark Matter

Warm dark matter decoupled much before matter radiation equality and they become non-relativistc much before matter radiation. The WDM travels at speed $\simeq c$ after decoupling till the temperature reaches its mass. After that its velocity falls as $v \propto 1/a^2$ and it does not disperse much when it becomes non-relativistic. The distance travelled by WDM between the time of decoupling and the time it becomes non-relativistic is the free-streaming length which defines the size of the smallest structures that can form from WDM. The co-moving free-steaming distance is

$$\lambda_{FS}^{com} = \int_{t_{dec}}^{t_{NR}} \frac{dt}{a(t)} \qquad (3.191)$$

In the radiation era $a(t) = a_{dec}(t/t_{dec})^{1/2}$ which implies that the free-streaming length from (3.191) is

$$\lambda_{FS}^{com} = \frac{t_{dec}^{1/2}}{a_{dec}} 2 \left(t_{NR}^{1/2} - t_{dec}^{1/2} \right)$$

$$\simeq \frac{t_{NR}^{1/2}}{a_{NR}} 2 t_{NR}^{1/2} = \frac{2}{a_{NR}} t_{NR} \qquad (3.192)$$

In the radiation era the $t_{NR} = 1/2H(T = m)$. Which implies that the present length scale of the warm-dark matter free-steaming length is

$$\lambda_{FS} = a_0 \lambda_{FS}^{com} = \left(\frac{a_0}{a_{NR}} \right) \frac{1}{H(T = m)} \simeq \frac{m}{T_0} \frac{M_P}{1.66 \sqrt{g_*} m^2}$$

$$= 99.5 \left(\frac{\text{keV}}{m} \right) \text{kpc} . \qquad (3.193)$$

To form galaxies confined within a size $100 \, \text{kpc}$ the mass of the warm-dark matter must be $m \geq 1 \text{keV}$. The smallest sized galaxies that can form from WDM have mass

$$M_{fs} = \frac{4\pi}{3} \rho_m \left(\frac{\lambda_{FS}}{2} \right)^3$$

$$\simeq 10^7 \left(\frac{\text{keV}}{m} \right)^3 M_\odot . \qquad (3.194)$$

Dwarf galaxies have masses in the range $(10^5 - 10^7) M_\odot$ and ulra-dwarf galaxies have masses $10^4 M_\odot$ so WDM with mass of a few keV can account for the observed galaxies. For a thorough analysis and results from simulation of galaxy formation with WDM see [13].

3.2.12 Window Function and Mass Distribution

The matter distribution in the universe is very inhomogeneous at small scales due to the fact that at late stages of the universe the peaks density perturbations have become large and the linear theory we have discussed no longer applies. To compare the observations with the linear theory one must smoothen the observed galaxy correlation function $\xi(\mathbf{x}, \mathbf{x}'; t)$ over a length scale of $\sim$Mpc to compare the observations with the linear theory. This smoothing is done with a *window function* or *filter function* $W(|\mathbf{x} - \mathbf{x}'|, R)$ defined with respect to a length scale R which smoothens the distribution on length scale smaller than R. The smoothened matter perturbation is related to the theoretical matter perturbation as

$$\Delta_m(\mathbf{x}, R) = \int d^3x' \, W(|\mathbf{x} - \mathbf{x}'|, R) \, \Delta_m(\mathbf{x}') \tag{3.195}$$

This relation is simpler in the Fourier space

$$\Delta_m(k, R) = \tilde{W}(k, R) \, \Delta_m(k) \,. \tag{3.196}$$

An example of a filter function is the Gaussian filter

$$W(|\mathbf{x} - \mathbf{x}'|, R) = \frac{1}{(2\pi)^3/2R^3} \exp\left(-\frac{|\mathbf{x} - \mathbf{x}'|^2}{2R^2}\right) \tag{3.197}$$

which in Fourier space is

$$\tilde{W}(kR) = \exp\left(-\frac{k^2R^2}{2}\right) \,. \tag{3.198}$$

Another filter function is the *top hat* filter defined as $W(|\mathbf{x}-\mathbf{x}'|, R) = 1$ for $|\mathbf{x}-\mathbf{x}'| \leq R$ and $W(|\mathbf{x} - \mathbf{x}'|, R) = 0$ for $|\mathbf{x} - \mathbf{x}'| > R$. In Fourier space the top hat filter is

$$\tilde{W}(kR) = \frac{3}{(kR)^3} \left[\sin kR - kR \cos kR\right] \,. \tag{3.199}$$

There is also a sharp k filter which is used for cutting of high k modes. This is defined as $\tilde{W}(kR) = 1$ for $kR \leq 1$ and $\tilde{W}(kR) = 0$ for $kR > 1$.

The mass enclosed in a sphere of radius R is a random variable with mean which depends upon the window function,

$$\bar{M}(R) = \int d^3x \, W(\mathbf{x}, R)\bar{\rho}$$

$$= \gamma \frac{4\pi}{3} R^3 \bar{\rho} \tag{3.200}$$

where $\bar{\rho}$ is background density and $\gamma = 1$ for the top-hat window function, $\gamma = 3\sqrt{\pi/2}$ for the Gaussian and $\gamma = 9\pi/2$ for the k-filter.

The mass in a radius R around a point $\mathbf{x}$ is related to the smoothed density perturbations as

$$M(\mathbf{x}, R) = \bar{M}(R)\left[1 + \Delta_m(\mathbf{x}, R)\right] \tag{3.201}$$

which means that the variance of the distribution of $M(R)$ is related to the power spectrum of $\Delta_m(\mathbf{x})$,

$$
\begin{aligned}
\sigma^2(M(R)) &= \langle \Delta_m(\mathbf{x}', R) \Delta_m(\mathbf{x}'', R) \rangle \\
&= \int d^3x' d^3x'' W(|\mathbf{x} - \mathbf{x}'|, R) W(|\mathbf{x} - \mathbf{x}''|, R) \langle \Delta_m(\mathbf{x}') \Delta_m(\mathbf{x}'') \rangle
\end{aligned}
\tag{3.202}
$$

The two point function of Δ_m is the power spectrum

$$
\langle \Delta_m(\mathbf{x}') \Delta_m(\mathbf{x}'') \rangle = \frac{1}{(2\pi)^3} \int d^3k\, e^{-i\mathbf{k}\cdot(\mathbf{x}'-\mathbf{x}'')} P(k)
\tag{3.203}
$$

Substituting (3.203) in (3.202) we obtain the relation between the variance of the mass distribution within a radius R and the power spectrum of the density perturbation

$$
\sigma^2(M(R)) \int \frac{d^3k}{(2\pi)^3} \tilde{W}^2(kR) P(k) .
\tag{3.204}
$$

From large scale survey of galaxies experimentalists report the σ_8 which is the variance of mass distribution at $R = 8h^{-1}$ Mpc. The Planck measurement [10] of lensing of CMB by galaxy clusters combined with BAO and CMB anisotropy data gives the value of σ_8 as

$$
\sigma_8 = 0.811 \pm 0.006 ,
\tag{3.205}
$$

which serves as a normalisation for the matter power spectrum $P(k)$ at $k = 0.125\,\mathrm{h\,Mpc}^{-1}$.

Galactic surveys observe the luminous matter in the galaxies and the distribution of luminous galaxies matter is only a fraction of the total matter,

$$
\delta_g = b\, \Delta_m
\tag{3.206}
$$

and the observed power spectrum of the galactic distribution at different length scales is related to the power spectrum of CDM and baryonic matter as

$$
P_{\delta_g}(k) = b^2 P(k) .
\tag{3.207}
$$

Using the Planck data of galactic cluster measurements using the Sunyaev–Zeldovich effect (up-scattering of CMB photons by the hot ($T \sim$ keV) electrons in the clusters) Ref. [14] puts the non-luminous fraction of mass as

$$
(1 - b) = 0.71 \pm 0.10
\tag{3.208}
$$

and the σ_8 parameter is

$$\sigma_8 \left(\frac{\Omega_m}{0.33}\right)^{0.25} = 0.765 \pm 0.035 . \tag{3.209}$$

The SZ effect gives the electron temperature which is related by the virial theorem to the total mass in the cluster. This is combined with the X-ray observations which provide the luminous mass and the ratio of the luminous mass to the virial mass is the bias parameter b in (3.208).

3.3   Problems with ΛCDM

The ΛCDM model where the density of the universe consists of $\Omega_\lambda \simeq 0.7$ dark energy, $\Omega_b \simeq 0.05$ baryons and $\Omega_c \simeq 0.25$ cold dark matter fraction, fits the CMB anisotropy and large scale structures quite well. There are however problems at explaining the observations at smaller scales (<1 Mpc) with the cold dark matter scenario [15]. These are as follows:

- *Missing Satellites Problem:* The CDM free streaming mass is extremely small $10^{-6} M_\odot$ so structures formation is hierarchial where the smallest structures form first. These small clumps of dark matter then merge to form dark matter halos. Galaxies form within these halos. The Milky Way has luminous mass of $10^{11} M_\odot$ is inside a dark matter halo of mass $2 \times 10^{12} M_\odot$ as inferred from the rotation curves of 21 cm Hydrogen.

    Simulations predict that there should be ~ 1000 dwarf galaxies of mass $M \sim 10^8 M_\odot$ surrounding Milky Way sized galaxies ($10^{12} M_\odot$) while observations show that there are ~ 50 dwarf galaxies in the Local Group within 2 Mpc of the Milky way [16–19]. This is the missing satellite problem of CDM.

    However recent estimates of number of dwarf galaxies around the Milky Way revises the expected number downwards to a few hundreds and correcting for the detection efficiency of galactic surveys for observing faint galaxies there may be no missing satellite problem [20, 21].

- *Too big to fail:* Simulations with collisionless and dissipationless CDM predict that Milky Way sized haloes of dark matter should have many sub-haloes of the size $(0.2 - 4) \times 10^{10} M_\odot$ with maximum circular velocities in the range $(30 - 70)$ km/s. Observations of the circular velocities of the dwarf spheroidal galaxies in the Local Group of the Milky Way show them to be in the range $12 < V_{max} < 25$ km s^{-1} while CDM simulations predict at least 10 subhaloes with $V_{max} > 25$ km s^{-1}. Massive sub-haloes are more likely host galaxies and cannot fail to form galaxies and thereby evade observations. This lack observed bright satellites around the Milky Way is called the "too big to fail problem" [22, 23].

- *Cuspy core of galaxies:* As we discussed in Chap. 2, observations of galactic rotation curve require that density profile at the core should be constant but the NFW profile which is a fit to CDM simulations is very cuspy at the core where it goes as $\rho \propto r^{-1}$. The solution to these may be in astrophysics of baryons in the gas and stars in the bulge or it may be that some new form of dark matter is required to address this issue.

References

1. S. Dodelson, *Modern Cosmology* (Academic Press, San Diego, 2003)
2. V. Mukhanov, *Physical Foundations of Cosmology* (Cambridge University, Cambridge, 2005)
3. D.H. Lyth, A.R. Liddle, *The Primordial Density Perturbation: Cosmology, Inflation and the Origin of Structure* (Cambridge University, Cambridge, 2009)
4. O. Piattella, *Lecture Notes in Cosmology* (Springer, Berlin, 2018)
5. A. Lewis, S. Bridle, Cosmological parameters from CMB and other data: a Monte Carlo approach. Phys. Rev. D **66**, 103511 (2002)
6. J. Lesgourgues, G. Mangano, G. Miele, S. Pastor, *Neutrino Cosmology.* (Cambridge University, Cambridge, 2013)
7. C.P. Ma, E. Bertschinger, Cosmological perturbation theory in the synchronous and conformal Newtonian gauges. Astrophys. J. **455**, 7 (1995)
8. J.M. Bardeen, Gauge invariant cosmological perturbations. Phys. Rev. D **22**, 1882–1905 (1980)
9. J.M. Bardeen, P. J. Steinhardt, M.S. Turner, Spontaneous creation of almost scale—free density perturbations in an inflationary universe. Phys. Rev. D **28**, 679 (1983). [Mon. Not. Roy. Astron. Soc. **489**(1), 401–419 (2019)]
10. N. Aghanim et al. [Planck], Planck 2018 results. VI. Cosmological parameters. [arXiv:1807.06209 [astro-ph.CO]]
11. D.J. Eisenstein et al. [SDSS Collaboration], Detection of the Baryon acoustic peak in the large-scale correlation function of SDSS luminous red galaxies. Astrophys. J. **633**, 560 (2005)
12. D. Blas, J. Lesgourgues, T. Tram, The Cosmic Linear Anisotropy Solving System (CLASS) II: approximation schemes. J. Cosmol. Astropart. Phys. **1107**, 034 (2011)
13. P. Bode, J.P. Ostriker, N. Turok, Halo formation in warm dark matter models. Astrophys. J. **556**, 93–107 (2001)
14. Í. Zubeldia, A. Challinor, Cosmological constraints from Planck galaxy clusters with CMB lensing mass bias calibration. Mon. Not. Roy. Astron. Soc. **489**(1), 401–419 (2019)
15. J.S. Bullock, M. Boylan-Kolchin, Small-Scale Challenges to the ΛCDM Paradigm. Ann. Rev. Astron. Astrophys. **55**, 343–387 (2017)
16. G. Kauffmann, S.D.M. White, B. Guiderdoni, The formation and evolution of galaxies within merging dark matter haloes. MNRAS, **264**, 201 (1993)
17. A.A. Klypin, A.V. Kravtsov, O. Valenzuela, F. Prada, Where are the missing Galactic satellites?. Astrophys. J. **522**, 82–92 (1999)
18. B. Moore, S. Ghigna, F. Governato, G. Lake, T.R. Quinn, J. Stadel, P. Tozzi, Dark matter substructure within galactic halos. Astrophys. J. Lett. **524**, L19–L22 (1999)
19. A. Drlica-Wagner et al. [DES], Eight ultra-faint galaxy candidates discovered in year two of the dark energy survey. Astrophys. J. **813**(2), 109 (2015)
20. S.Y. Kim, A.H.G. Peter, J.R. Hargis, Missing satellites problem: completeness corrections to the number of satellite galaxies in the milky way are consistent with cold dark matter predictions. Phys. Rev. Lett. **121**(21), 211302 (2018)

21. T. Abbott et al. [DES], Dark energy survey year 1 results: cosmological constraints from galaxy clustering and weak lensing. Phys. Rev. D **98**(4), 043526 (2018)
22. M. Boylan-Kolchin, J.S. Bullock, M. Kaplinghat, Too big to fail? The puzzling darkness of massive Milky Way subhaloes. MNRAS, **415**, L40 (2011)
23. M. Boylan-Kolchin, J.S. Bullock, M. Kaplinghat, The Milky Way's bright satellites as an apparent failure of LCDM. MNRAS, **422**, 1203 (2012)

Cosmic Microwave Background Anisotropy

4

Abstract

CMB anisotropy observations starting from observations of super-horizon anisotropy by COBE with precision observations by COBE, WMAP, Planck and Earth based telescopes like DASI, SPT, Bicep/KECK give precise measurements of the composition of the universe like dark matter, dark energy, neutrino masses and primordial gravitational waves.

4.1 Introduction

An isotropic cosmic microwave background radiation with temperature estimated at $3.5° \pm 1° K$ was discovered accidentally by Penzias and Wilson [1] in 1965. It was then immediately recognised by Dicke et al. [2] as the relic radiation expected from a hot big bang expanding universe. The existence of a background temperature was in fact predicted by Alpher and Herman [3] and Gamow [4] based on the theory that in the early universe there was nucleosynthesis of the light elements. Earlier analysis of diatomic molecular data from astronomical observations had placed a temperature of $2.3°K$ on the interstellar molecules like cyanogen (CN) [5] but the cosmological origin of this temperature was not recognised. Subsequent measurements of the microwave temperature with CN observations [6] and with balloon based experiments [7] improved the accuracy of the measurement.

The isotropic temperature is determined with remarkable precision $T_0 = 2.7258 \pm 0.00057K$ with the data from FIRAS instrument onboard the COBE satellite [8,9]. The FIRAS instrument also confirmed the blackbody spectrum of the radiation with an upper bound on spectral distortions, the photon chemical potential $|\mu| < 9 \times 10^{-5}$ and spectral distortion by Compton scattering $|y| < 1.2 \times 10^{-6}$ on the isotropic component of CMBR [9].

© Springer Nature Switzerland AG 2020

S. Mohanty, *Astroparticle Physics and Cosmology*,

Lecture Notes in Physics 975, https://doi.org/10.1007/978-3-030-56201-4_4

An importance advance in cosmology was the measurement of the anisotropy at large angles of the CMBR with the DMR instrument aboard COBE [10]. A temperature anisotropy of $\Delta T = 16 \pm 4\mu K$ was measured with COBE-DMR for angular separations $\theta > 7°$ and this established the theory of inflation which predicted the existence of such anisotropies at length scales larger than the size of the horizon at the surface of last scattering. COBE-DMR had an angular resolution of $7°$ which means it could measure multipole moments $l < 26$, ($\theta_{res} = 180°/l_{res}$). Subsequent satellite based experiments have measure anisotropies accurately at smaller angular resolutions and probed larger multipoles of angular spectrum. WMAP [11] had a resolution of $\theta_{res} = 0.23°$ which implied that it could probe $l < 783$. Planck [12] had a resolution of $5'$ and measured the angular anisotropy upto $l = 2160$.

In addition to the temperature anisotropy the CMB photons are expected to be polarised due to Thomson scattering near the last scattering surface. This polarisation was first observed by the Antarctica based DASI instrument [13] which measured E-mode polarisation and put upper bounds on the B-mode polarisation. B-mode polarisation which arises from lensing of E-mode polarisation was first measured at the South Pole Telescope [14]. Measurement of primordial B-mode polarisation caused by the gravitational waves produced during inflation are still undetected. The measurement of the primordial B-mode polarisation in the CMB remains the holy grail of the next generation CMB experiments [15]. The General Relativistic treatment of CMB anisotropy has been covered in many textbooks and review articles, for example see [16–22].

4.2 Boltzmann Equation for Photons

The probability of finding a particle in a volume h^3 of phase space is described by the distribution function

$$dP = f(\eta, x^i, p_j)\frac{d^3x d^3p}{h^3} = f(\eta, x^i, p_j)\frac{d^3x d^3p}{(2\pi\hbar)^3}, \qquad (4.1)$$

and we use the units $\hbar = c = k_B = 1$ in this book. Here x^i are the position coordinates and p_i the three-momenta. The phase space volume element

$$d^3x d^3p \equiv dx^1 dx^2 dx^3 dp_1 dp_2 dp_3 \qquad (4.2)$$

is invariant under general coordinate transformations.

To define number, energy, pressure densities we have to integrate the phase space distribution over the momentum. The invariant volume is then split into two separate covariant volume elements

$$N = \int dx^1 dx^2 dx^3 (-g)^{1/2} \int \frac{dp_1 dp_2 dp_3}{(2\pi\hbar)^3}(-g)^{-1/2} f(\eta, x^i, p_j), \qquad (4.3)$$

where g is the determinant of metric $g_{\mu\nu}$. The energy momentum tensor of a gas of particles can be defined in terms of its distribution function as

$$T^\mu{}_\nu = \int \frac{dp_1 dp_2 dp_3}{(2\pi)^3} (-g)^{-1/2} \frac{p^\mu p_\nu}{p^0} f(x^i, p_j, \eta), \tag{4.4}$$

The four momentum is defined as $p^\mu = \frac{dx^\mu}{d\lambda}$ where λ is some parameter which is normalised such that

$$g_{\mu\nu} p^\mu p^\nu = g^{\mu\nu} p_\mu p_\nu = -m^2 \tag{4.5}$$

where m is the invariant mass of the particle.

The distribution function is a function of temperature and is given by

$$f(t, x^i, p_i) = \frac{g_i}{\exp(E/T) \pm 1} \tag{4.6}$$

where E is the energy and the plus sign is for fermions while the minus sign is for bosons. In a coordinates system where the photon momentum is p^μ and the observer has a general four velocity u^μ the energy of the particle measured by the observer will be

$$E = g_{\mu\nu} p^\mu u^\mu = g^{\mu\nu} p_\mu u_\mu \tag{4.7}$$

If the particle momentum is defined with respect to the rest frame of the observer, then

$$E = \sqrt{-g^{00} p_0^2} = \sqrt{g^{ij} p_i p_j + m^2} \equiv \sqrt{p^2 + m^2} \tag{4.8}$$

where we have defined the magnitude of the three momentum as $p = \sqrt{g^{ij} p_i p_j}$. A photon is characterised by the energy $E = p$ and the direction $\hat{p}_i = p_i / \sqrt{\delta^{ij} p_i p_j}$. The temperature $T(\eta, x^i, \hat{p}_i)$ depends upon the direction of the photons observed $\hat{p}_i$.

We calculate the energy density $T^0{}_0$ of a particle of mass m using (4.4). The integration volume $dp_1 dp_2 dp_3 (-g)^{-1/2} = a^3 p^2 dp (-g)^{-1/2} d\Omega = a^{-1} p^2 dp \, d\Omega$ and $p_0 = aE$ therefore

$$\rho = T^0{}_0 = \frac{1}{(2\pi)^3} \int d\Omega \, p^2 \, dp \, E \, f(p, T) \tag{4.9}$$

The distribution function of photons in the unperturbed FRW background is

$$f^0(p) = 2\left[\exp\left(\frac{p}{\bar{T}(\eta)]} \right) - 1 \right]^{-1}. \tag{4.10}$$

and the energy density is

$$\rho_\gamma = 2 \int \frac{p^2 dp d\Omega}{(2\pi)^3} \frac{p}{e^p/T - 1}$$

$$= \frac{\pi^2}{15} T^4, \tag{4.11}$$

The present photon density as a fraction of the critical density is

$$\Omega_{\gamma 0} = \frac{\rho_{\gamma 0}}{3H_0^2/(8\pi G)} = 3.47 \times 10^{-5} h^{-2} \tag{4.12}$$

where we have taken $H_0 = 100h \,\mathrm{km\,s^{-1}\,Mpc^{-1}}$.

In the absence of interactions the particle number in a phase space volume does not change. A change in the phase space number occurs only through interactions through interactions. Creation/annihilation processes can change the number of particles in space-time volume d^3x and number conserving scatterings can change the number density in momentum space volume $d^3 p$. The Boltzmann equation is the change in distribution function $f(\eta, x^i, p_i)$ because of flow and interactions and can be formally written as

$$\frac{\partial f}{\partial \lambda} + \frac{dx^i}{d\lambda} \frac{\partial f}{\partial x^i} + \frac{dp_i}{d\lambda} \frac{\partial f}{\partial p_i} = \left(\frac{\partial f}{\partial \lambda} \right)_C \tag{4.13}$$

where λ is an affine parameter which parameterises the trajectory of a particle and $\frac{dx^\mu}{d\lambda} = p^\mu$ the four momentum. The change in momentum occurring due to non-gravitational interactions are treated through the collision term. in the r.h.s of BE, while the change in momentum due to gravity are treated as a geodesic flow in curved space and we use the geodesic equation to substitute for the gravitational acceleration $dp_i/d\lambda$ on the l.h.s of (4.13).

When perturbations are of super-horizon size they are not affected by causal phenomenon like scattering. After they enter the horizon photons are get scattered by electrons and the Boltzmann equation has to include Compton scattering through the collisional term.

For the change in photon temperature due to super-horizon gravitational perturbations it is sufficient to solve the homogenous Boltzmann equation

$$\frac{\partial f}{\partial \lambda} + \frac{dx^i}{d\lambda} \frac{\partial f}{\partial x^i} + \frac{dE}{d\lambda} \frac{\partial f}{\partial E} + \frac{dp_i}{d\lambda} \frac{\partial f}{\partial p_i} = 0 \tag{4.14}$$

governs the change in temperature of a non-interacting photon gas as it propagates in the gravitational potential of the perturbed FRW metric.

4.2.1 Photon Geodesic in Perturbed FRW Metric

We will now derive the temperature anisotropy measured by an observed in photons moving along the geodesics from the last scattering surface. We take the metric to be perturbed FRW in the conformal Newtonian gauge

$$ds^2 = a(\eta)^2 \left(-(1 + 2\Psi(\eta, \mathbf{x})) \, d\eta^2 + (1 - 2\Phi(\eta, \mathbf{x})) \, \delta_{ij} dx^i dx^j \right). \tag{4.15}$$

The four-momentum p_μ obeys the relation $g^{\mu\nu} p_\mu p_\nu = 0$ which for the metric (4.15) takes the form

$$a^{-2} (1 - 2\Psi) (p^0)^2 - a^{-2} (1 + 2\Phi) (p_i p_j \delta^{ij}) = 0 \tag{4.16}$$

We defined (4.8) the magnitude of the three momentum as $p \equiv \sqrt{g^{ij} p_i p_j}$ and we can write the components of p_μ as

$$p_\mu = (p_0, p_i) = \left(ap (1 + \Psi), ap (1 - \Phi) \, \hat{p}_i \right), \quad \hat{p}_i \equiv \frac{p_i}{\sum_i p_i^2}. \tag{4.17}$$

such that the relation $g_{\mu\nu} p^\mu p^\nu = 0$ is satisfied to the linear order in perturbations Φ and Ψ. The four momentum $p^\mu = g^{\mu\nu} p_\nu$ is

$$p^\mu = (p^0, p^i) = \left(-\frac{p}{a} (1 - \Psi), \frac{p}{a} (1 + \Phi) \, \hat{p}^i \right), \quad \hat{p}^i \equiv \hat{p}^i. \tag{4.18}$$

Here

$$\hat{p}^i \equiv \delta^{ij} \hat{p}_i \quad \text{and} \quad \delta_{ij} \hat{p}^i \hat{p}^j = \delta^{ij} \hat{p}_i \hat{p}_j = 1 \tag{4.19}$$

Note that $\hat{p}_i$ represents the unit vector in the direction of p_i, but $\hat{p}^i$ is not the unit vector in the direction of p^i. The upper index $\hat{p}^i$ is just $\hat{p}_i$ raised with δ^{ij}.

The trajectory of the photon in a perturbed orbit can be determined by solving for the magnitude $p(\eta)$ and direction $\hat{p}_i(\eta)$ of photon momentum as a function of time.

Going back to the homogenous BE (4.14) we need the geodesic equation for $dp_\alpha/d\lambda$. This is obtained by starting from the standard geodesic equation

$$\frac{dp^\mu}{d\lambda} + \Gamma^\mu_{\alpha\beta} p^\alpha p^\beta = 0 \tag{4.20}$$

and lowering the index by contracting by $g_{\mu\nu}$. The covariant metric can be taken through $d/d\lambda = u^{\delta}\nabla_{\delta}$ as $\nabla_{\delta}g_{\mu\nu} = 0$. Doing this and using the definition of $\Gamma^{\mu}_{\alpha\beta}$ we obtain the equation for the lower index four-momenta p_{ν} given by

$$\frac{dp_{\nu}}{d\lambda} = \frac{1}{2}\frac{\partial g_{\alpha\beta}}{\partial x^{\nu}}p^{\alpha}p^{\beta} \tag{4.21}$$

We can change the variable of the derivative by using $d/d\lambda = p^0\frac{d}{d\eta}$ and write this as

$$\frac{dp_{\nu}}{d\lambda} = \frac{1}{2}\frac{\partial g_{\alpha\beta}}{\partial x^{\nu}}\frac{p^{\alpha}p^{\beta}}{p^0} \tag{4.22}$$

The $\nu = 0$ component of (4.22) is therefore

$$\frac{dp_0}{d\eta} = -2p^0\frac{\partial(a^2\Psi)}{\partial\eta} - \frac{p^ip^j}{p^0}\delta_{ij}\frac{\partial(a^2\Phi)}{\partial\eta} \tag{4.23}$$

We can write this in terms of p by replacing $p_0 = ap(1+\Psi)$ and $p^0 = -a^{-1}p(1-\Psi)$ and obtain

$$\frac{1}{p}\frac{dp}{d\eta} = -\mathcal{H} - \frac{d\Psi}{d\eta} + \frac{\partial}{\partial\eta}(\Psi+\Phi) . \tag{4.24}$$

In the zeroth order in perturbations this equation is

$$\frac{1}{p}\frac{dp}{d\eta} = -\frac{1}{a}\frac{da}{d\eta} . \tag{4.25}$$

which implies that in the unperturbed FRW universe the magnitude of the photon momentum and energy red-shift as

$$E = p \propto \frac{1}{a} . \tag{4.26}$$

The energy of the photon redshifts $\propto 1/a$.

Next we solve the Boltzmann equation at first order in the perturbation variables.

4.2.2   Boltzmann Equation for Photon Distribution in the Perturbed FRW Universe

We solve the Boltzmann equation for photons in a perturbed FRW universe by assuming the temperature is in-homogenous and non-isotropic. The local temperature of photons can depend on the position and the direction $\hat{p}$ of the photons.

We can write the temperature as a homogenous background temperature $\bar{T}(\eta)$ and a fluctuation $\Delta T = T(\eta, x^i, \hat{p}^i) - \bar{T}(\eta)$. The photon distribution function in the perturbed FRW universe can be written as

$$f(x^i, p, \hat{p}^i, \eta) = 2\left[\exp\left(\frac{p}{T(x^i, \hat{p}^i, \eta)]}\right) - 1\right]^{-1}$$

$$= 2\left[\exp\left(\frac{p}{\bar{T}(\eta)[1 + \Theta(x^i, \hat{p}^i, \eta)]}\right) - 1\right]^{-1} \quad (4.27)$$

where we have introduced the fractional temperature inhomogeneity also called the brightness function defines as,

$$\Theta(x^i, \hat{p}^i, \eta) \equiv \frac{\Delta T}{T} = \frac{T(\eta, x^i, \hat{p}^i) - \bar{T}(\eta)}{\bar{T}(\eta)}. \quad (4.28)$$

We can Taylor expand (4.27) upto linear order in Θ as

$$f(x^i, p, \hat{p}^i, \eta) = f^0(\bar{T}) + \frac{\partial f}{\partial T}\bigg|_{T=\bar{T}} \bar{T}\Theta \quad (4.29)$$

Using the property

$$\bar{T}\frac{\partial f^0}{\partial \bar{T}} = -p\frac{\partial f^0}{\partial p} \quad (4.30)$$

we can write (4.29) as

$$f(x^i, p, \hat{p}^i, \eta) = f^0(\bar{T}) - \frac{\partial f^0}{\partial p}\Theta \quad (4.31)$$

therefore the first order perturbation in the photon distribution can be written in terms of Θ as

$$f^1(x^i, p, \hat{p}^i, \eta) = -\frac{\partial f^0}{\partial p}\Theta. \quad (4.32)$$

Now we can write the Boltzmann equation for photon distribution in the inhomogeneous universe neglecting collisions,

$$\frac{df(\eta, x^i, p, \hat{p}^i)}{d\eta} = \frac{\partial f}{\partial \eta} + \frac{dx^i}{d\eta}\frac{\partial f}{\partial x^i} + \frac{dp}{d\eta}\frac{\partial f}{\partial p} + \frac{d\hat{p}^i}{d\eta}\frac{\partial f}{\partial \hat{p}_i} = 0 \quad (4.33)$$

In the homogenous universe $\frac{d\hat{p}_i}{d\eta} = 0$ and $\frac{\partial f}{\partial \hat{p}_i} = 0$ so the last term in (4.33) (which represents gravitational lensing of photons) is at least second order in gravitational perturbations and can be dropped to the leading order.

To the leading order $\frac{dx^i}{d\eta} = \frac{p^i}{p^0} \simeq \hat{p}^i$. Using this and the geodesic equation (4.24) for p we can write (4.33) as

$$\frac{df(\eta, x^i, p, \hat{p}_i)}{d\eta} = \frac{\partial f}{\partial \eta} + \hat{p}^i \frac{\partial f}{\partial x^i} + p \frac{\partial f}{\partial p}\left[-\mathcal{H} - \frac{d\Psi}{d\eta} + \frac{\partial}{\partial \eta}(\Psi + \Phi)\right] = 0.$$
(4.34)

We now solve the Boltzmann equation for photons order by order in the perturbation variables (Θ, Ψ, Φ).

To the zeroth order in perturbations (Θ, Ψ, Φ) the Boltzmann equation (4.34) reduces to

$$\left.\frac{df}{d\eta}\right|^{(0)} = \frac{\partial f^0}{\partial \eta} - \mathcal{H}p\frac{\partial f^0}{\partial p} = 0$$
(4.35)

The first term can be written as

$$\frac{\partial f^0}{\partial \eta} = \frac{\partial f^0}{\partial \bar{T}}\frac{d\bar{T}}{d\eta} = -\frac{p}{\bar{T}}\frac{\partial f^0}{\partial p}\frac{d\bar{T}}{d\eta}$$
(4.36)

Substituting in (4.35) we obtain

$$\left(-\frac{1}{\bar{T}}\frac{dT}{d\eta} - \frac{1}{a}\frac{da}{d\eta}\right)\frac{\partial f^0}{\partial p} = 0$$
(4.37)

Equating the terms in the square brackets to zero we obtain the relation

$$\bar{T} \propto 1/a.$$
(4.38)

So in the expanding universe the zeroth order distribution function $f^0 = (\exp(p/\bar{T}) - 1)^{-1}$ remains time invariant as both p and $\bar{T}$ red-shift $\propto 1/a$.

The Boltzmann equation for the terms linear order in (Θ, Φ, Ψ) can be obtained by substituting (4.31) in (4.34) and collecting terms which are first order in the perturbations

$$\left.\frac{df}{d\eta}\right|^{(1)} = -p\frac{\partial}{\partial \eta}\left[\frac{\partial f^0}{\partial p}\Theta\right] - p\hat{p}^i\frac{\partial \Theta}{\partial x^i}\frac{\partial f^0}{\partial p} - p\mathcal{H}\Theta\frac{\partial}{\partial p}\left[p\frac{\partial f^0}{\partial p}\right] - p\frac{\partial f^0}{\partial p}\left[\frac{d\Psi}{d\eta} + \frac{\partial}{\partial \eta}(\Psi + \Phi)\right] = 0$$
(4.39)

We can simplify this equation by making use of (4.30) and (4.36) and we obtain

$$\left(\frac{\partial\Theta}{\partial\eta} + \hat{p}^i\frac{\partial\Theta}{\partial x^i} - \frac{d\Psi}{d\eta} + \frac{\partial}{\partial\eta}(\Psi + \Phi)\right)\left(-p\frac{\partial f^0}{\partial p}\right) = 0. \tag{4.40}$$

which implies that

$$\left(\frac{\partial\Theta}{\partial\eta} + \hat{p}^i\frac{\partial\Theta}{\partial x^i} - \frac{d\Psi}{d\eta} + \frac{\partial}{\partial\eta}(\Psi + \Phi)\right) = 0 \tag{4.41}$$

Now

$$\frac{\partial\Theta}{\partial\eta} + \hat{p}^i\frac{\partial\Theta}{\partial x^i} = \frac{\partial\Theta}{\partial\eta} + \frac{dx^i}{d\eta}\frac{\partial\Theta}{\partial x^i} = \frac{d\Theta}{d\eta} \tag{4.42}$$

therefore from (4.41) we obtain the equation for time evolution of Θ given by

$$\frac{d\Theta}{d\eta} = -\frac{d\Psi}{d\eta} + \frac{\partial}{\partial\eta}(\Psi + \Phi). \tag{4.43}$$

Alternately, opening the total derivative of Θ and Ψ in terms of partial derivatives (4.43) can also be written in the form

$$\frac{\partial\Theta}{\partial\eta} + \hat{p}^i\frac{\partial\Theta}{\partial x^i} + \hat{p}^i\frac{\partial\Psi}{\partial x^i} - \frac{\partial}{\partial\eta}\Phi = 0 \tag{4.44}$$

For cosmic anisotropies at large angles, the perturbations which are measured are the ones which have just entered the horizon at the time of decoupling $z_{dec} \simeq 1100$. Perturbations with shorter wavelength entered the horizon earlier in matter and modes with $k > k_{eq}$ entered in the radiation era.

4.3   Large Angle CMB Anisotropy: Sachs Wolfe Effect

Photons from the last scattering will show a temperature anisotropy which will depend on the Θ, Ψ, Φ at the last scattering surface which we will denote by the conformal time η_* (see Fig. 4.1). Starting from Eq. (4.43) and integrating w.r.t the conformal time η from η_* to the present η_0 we have

$$\int_{\eta_*}^{\eta_0}\frac{d\Theta}{d\eta}d\eta = \int_{\eta_*}^{\eta_0}\left(-\frac{d\Psi}{d\eta} + \frac{\partial}{\partial\eta}(\Psi + \Phi)\right)d\eta \tag{4.45}$$

Fig. 4.1 Angular anisotropy of the CMB due from the inhomogeneity in the matter and radiation density and the gravitational potentials on the last scattering surface (LSS). Over-dense regions appear as hot-spots while under-dense regions appear as hot-spots

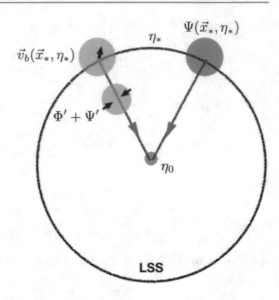

Integral over total derivatives depend only on the end-points and we obtain

$$\Theta(\eta_0) = \Theta(\eta_*) + \Psi(\eta_*) - \Psi(\eta_0) + \int_{\eta_*}^{\eta_0} \frac{\partial}{\partial \eta} (\Psi + \Phi) \, d\eta \qquad (4.46)$$

Here $\Theta(\eta_0) = \Delta T / \bar{T}$ measured in the present epoch. The third term on the r.h.s is gravitational potential in the present epoch. This term is identical to temperature measured in different directions and thus only contributes a constant monopole term and can be dropped. A dipole temperature anisotropy arises due to the local velocity of the matter at the last scattering surface and the motion of the earth w.r.t the CMB. We will add this contribution next. The last term in (4.46) denotes the change in photon energy as it traverses through the changing local potential from the LSS to the observer. The temperature fluctuation at the LSS $\Theta(\eta_*)$ is related to the radiation density perturbation

$$\Theta(\eta_*) = \frac{1}{4}\delta_r(\eta_*). \qquad (4.47)$$

Further if the perturbations are adiabatic the radiation perturbation is related to the (CDM + baryons) density perturbation

$$\delta_r(\eta_*) = \frac{4}{3}\delta_m(\eta_*) \qquad (4.48)$$

For the super-horizon perturbations the matter density perturbations are related to the gravitational potential (3.125) and the primordial comoving curvature perturbation as

$$\delta_m(\eta_*) = -2\Phi(\eta_*). \tag{4.49}$$

Therefore temperature perturbation at the LSS is related to the gravitational potential as

$$\Theta(\eta_*) = \frac{1}{4}\delta_r(\eta_*) = \frac{1}{3}\delta_m(\eta_*) = -\frac{2}{3}\Phi(\eta_*) \tag{4.50}$$

The temperature anisotropy is therefore from (4.46)

$$\Theta(\eta_0) = \frac{\Delta T(\eta_0, \hat{n})}{\bar{T}} = -\frac{2}{3}\Phi(\eta_*) + \Psi(\eta_*) + \int_{\eta_*}^{\eta_0} \frac{\partial}{\partial \eta}(\Psi + \Phi)\,d\eta \tag{4.51}$$

Here $\mathbf{n} = -\hat{\mathbf{p}}$ is the direction of from which photons are observed, the contribution of the dominant first term is called the Sachs-Wolfe (SW) and the second term the integrated Sachs-Wolfe (ISW) effect [23].

SW effect shows up as a plateau in the multipole expansion of $\Delta T/T$ at low multipoles. This then provides a measurement of the super-horizon gravitational potential which in turn is related to the primordial comoving perturbation as $\Phi_{MD} \simeq \Psi_{MD} \simeq (3/5)\mathcal{R}$ therefore the large angle temperature anisotropy can be related to the primordial curvature perturbation generated by inflation as $\Delta T/\bar{T} \simeq (1/5)\mathcal{R}$.

Doppler Contribution

There is an additional contribution to the temperature anisotropy due to the Doppler shift in the frequency of the photons due to both the velocity of the observer as well the emitter (the electrons in the last scattering surface) w.r.t the CMB photons.

The sun has a relative velocity w.r.t the CMB which imparts a dipole anisotropy to the temperature anisotropy. The photon observed in the direction $\mathbf{n}$ has momentum direction $\hat{\mathbf{p}} = -\mathbf{n}$. If the photon momentum has magnitude p at the last scattering surface, its observed magnitude will be

$$p_{obs} = \frac{p}{\gamma(1 - \mathbf{n} \cdot \mathbf{v}_{obs})} \tag{4.52}$$

where $\mathbf{v}_{obs}$ is the velocity of the observer w.r.t the CMB and $\gamma = (1 - v_{obs}^2)^{-1/2} \simeq 1$ is the Lorentz factor. In a coordinate transformation p/T remains constant (4.182), therefore the change in observed photon momentum will result in a temperature change of magnitude

$$\frac{\Delta T(\mathbf{n})}{\bar{T}} = \frac{T(\mathbf{n}) - \bar{T}}{\bar{T}} = \frac{p_{obs} - p}{p} \simeq \mathbf{n} \cdot \mathbf{v}_{obs} \, . \tag{4.53}$$

The Planck collaboration [12] measured the dipole anisotropy $\Delta T = 3362.08 \pm 0.89\mu K$. This along with the monopole temperature $\bar{T} = 2.7255 \pm 0.0006 K$ gives the velocity of the Solar System w.r.t the CMB frame as $v_{obs} = (1.233575 \pm 0.0036) \times 10^{-3} = 369.82 \pm 0.11 \, \text{km s}^{-1}$. The motion of the Solar System gives a dipole contribution to the CMB and does not carry information of the perturbations at the last scattering surface.

There is another Doppler shift of temperature due to the velocity of the electron which last scattered the photon we observe. This term will arise when we add the collision term in the Boltzmann equation for photons. This velocity field is inhomogenous and the resulting temperature anisotropy

$$\left. \frac{\Delta T(\mathbf{n})}{\bar{T}} \right|_{Doppler} = -\mathbf{n} \cdot \mathbf{v}_b(\mathbf{x}, \eta_*) \tag{4.54}$$

therefore contributes also to the higher (than just dipole) moments of the CMB anisotropy.

Putting together all the contributions to the total temperature anisotropy at large angular scales we have

$$\frac{\Delta T(\mathbf{n})}{\bar{T}} = -\frac{2}{3}\Phi(\eta_*) + \Psi(\eta_*) - \int_{\eta_*}^{\eta_0} \frac{\partial}{\partial \eta}(\Psi + \Phi) \, d\eta - \mathbf{n} \cdot \mathbf{v}_b(\mathbf{x}, \eta_*) \, . \tag{4.55}$$

In the standard cosmology–inflation generated superhorizon perturbations followed by reheating and a ΛCDM universe, there is no anisotropic stress tensor at super-horizon scales and $\Psi(\eta_*) = \Phi(\eta_*)$. In this case the observed temperature anisotropy is

$$\frac{\Delta T(\mathbf{n})}{\bar{T}} = \frac{1}{3}\Psi(\eta_*) + \int_{\eta_*}^{\eta_0} \frac{\partial}{\partial \eta}(\Psi + \Phi) \, d\eta - \mathbf{n} \cdot \mathbf{v}_b(\mathbf{x}, \eta_*) \, . \tag{4.56}$$

The dominant contribution comes from the first term. The hot-spots in the CMB map with $\Delta T/T > 0$ have gravitational potential $\Psi = \Phi > 0$ which have $\delta_r, \delta_m < 0$. Therefore hot spots in the CMB map correspond to under-dense regions in the surface of last scattering.

4.3.1 From Inhomogeneity to Anisotropy

The CMB map is a scalar function $\delta T(\hat{n}) = \Delta T(\hat{n})/\bar{T}$ which is defined on a 2-dimensional surface. Observations from COBE, WMAP and Planck measured the temperature anisotropy in different directions and then give a two point correlation

$$\langle \delta T(\mathbf{n}) \, \delta T(\mathbf{n}') \rangle \tag{4.57}$$

where the conical brackets stand for averaging over the all sky for each value of the angle defined by $\mathbf{n} \cdot \mathbf{n}' = \cos\theta$. It is assumed that there is no preferred direction in the sky at large scales other that the dipole arising from the motion of the solar system through the CMB. The two point correlation can then be expressed as a multipole expansion in Legendre polynomials

$$\langle \delta T(\mathbf{n}) \, \delta T(\mathbf{n}') \rangle = \sum_l \frac{(2l+1)}{4\pi} C_l \, P_l(\cos\theta) \tag{4.58}$$

and the coefficients C_l are called the angular power of the TT correlation. The legendre polynomials $P_l(\cos\theta)$ have maxima at $\theta = \pi/l$. This gives a relation between the observation angle in the sky of some characteristic feature in the anisotropy and the corresponding multipole where it will be seen. For example the baryon acoustic oscillations (BAO) have an angular separation between maxima and minima of $\theta_s \sim 1°$ which corresponds to the peak seen in C_l at $l \simeq 200$.

The angular power spectrum can be expressed in terms of the power spectrum of density, potential and velocity perturbations. To do this we first express $\delta T(\mathbf{n})$ in the Fourier space,

$$\delta T(\mathbf{n}, \mathbf{x}_*, \eta_*) = \int \frac{d^3k}{(2\pi)^3} e^{i(kr_*)\hat{\mathbf{k}} \cdot \mathbf{n}} \delta T(\mathbf{k}) \tag{4.59}$$

where $r_* = \eta_0 - \eta_*$ the comoving distance to the surface of last scattering. Using the relation

$$e^{i(kr_*)\hat{\mathbf{k}} \cdot \mathbf{n}} = \sum_l (-i)^l (2l+1) j_l(kr_*) P_l(\hat{\mathbf{k}} \cdot \mathbf{n}) \tag{4.60}$$

we can write the two point temperature correlation as

$$\langle \delta T(\mathbf{n}) \delta T(\mathbf{n}') \rangle = \int \frac{d^3k}{(2\pi)^3} \frac{d^3k'}{(2\pi)^3} \sum_{l,l'} (-i)^{l+l'} (2l+1)(2l'+1) j_l(kr_*) j_{l'}(k'r_*)$$

$$\times P_l(\hat{\mathbf{k}} \cdot \mathbf{n}) P_l(\hat{\mathbf{k}}' \cdot \mathbf{n}') \langle \delta T(\mathbf{k}) \delta T(\mathbf{k}') \rangle \tag{4.61}$$

Now using (4.56) we can write the $T(k)$ correlation in terms of the power spectrum of curvature perturbations. Keeping the leading order Sachs-Wolfe term we have

$$\langle \delta T(\mathbf{k}) \delta T(\mathbf{k'}) \rangle = \frac{1}{9} \langle \Psi(\mathbf{k}) \Psi(\mathbf{k'}) \rangle = \frac{1}{9} T(k) T(k') \langle \mathcal{R}(\mathbf{k}) \mathcal{R}(\mathbf{k'}) \rangle$$

$$= \frac{1}{9} \frac{2\pi^2}{k^3} T(k)^2 \mathcal{P}_{\mathcal{R}} \delta(\mathbf{k} - \mathbf{k'}) \tag{4.62}$$

where $T(k)$ is the transfer function which takes into account the amplification of the modes after horizon entry.

Substituting (4.62) in (4.61), the delta function takes care of integration over $\mathbf{k'}$. The integration over the angular part of $\mathbf{k}$ can be performed using the relation

$$\int d^2\hat{\mathbf{k}} \, P_l(\hat{\mathbf{k}} \cdot \mathbf{n}) P_{l'}(\hat{\mathbf{k}} \cdot \mathbf{n'}) = \frac{4\pi}{2l+1} P_l(\mathbf{n} \cdot \mathbf{n'}) \delta_{ll'} \tag{4.63}$$

and we finally obtain

$$\langle \delta T(\mathbf{n}) \delta T(\mathbf{n'}) \rangle = \sum_l \frac{(2l+1)}{4\pi} P_l(\cos\theta) \left(4\pi \int \frac{dk}{k} j_l^2(kr_*) \frac{1}{9} T(k)^2 \mathcal{P}_{\mathcal{R}} \right). \tag{4.64}$$

From this we see that the term in the brackets is the angular power C_l,

$$C_l^{TT} = 4\pi \int \frac{dk}{k} j_l^2(kr_*) \frac{1}{9} T(k)^2 \mathcal{P}_{\mathcal{R}}. \tag{4.65}$$

The function $j_l(kr_*)$ has maxima when $l \simeq kr_* = k(\eta - \eta_*) \simeq k\eta_0$. The co-moving distance to the LSS which approximately the size of the co-moving horizon in the present epoch is

$$\eta_0 = \int_0^{t_0} \frac{dt}{a(t)} = \int_0^{a_0} \frac{da}{a^2 H(a)} = \int_0^\infty \frac{dz}{H_0 \left[(\Omega_m^2 (z+1)^3 + \Omega_\Lambda \right]^{1/2}}$$

$$= 0.98 H_0^{-1} = 14.6 \, \text{Gpc}. \tag{4.66}$$

where we have taken $\Omega_m = 0.3$, $\Omega_\Lambda = 0.7$, $H_0 = 67.4 \, \text{km s}^{-1} \, \text{Mpc}^{-1}$.

On the other hand size of the co-moving horizon at the last scattering surface $(z_* \simeq 1090)$ is

$$\eta_* = \int_{z_*}^\infty \frac{dz}{H_0 \left[(\Omega_m^2 (z+1)^3 + \Omega_r (z+1)^4 \right]^{1/2}}$$

$$= 0.0675 H_0^{-1} = 300.57 \, h^{-1} \text{Mpc}. \tag{4.67}$$

where we have taken $\Omega_r = 3.24h^{-2} \times 10^{-5}$ which is the present radiation density (including three massless neutrinos).

The observed angular size of the horizon sized perturbations on the last scattering is therefore,

$$\theta_* = \frac{2\eta_*}{\eta_0 - \eta_*} = 0.041 \text{ rad} = 2.35°. \tag{4.68}$$

Consider the scales which are of the LSS super-horizon size $k\eta_* < 1$. which correspond to angle $\theta > 2.35°$ and $l < 76$. These scales entered the horizon in the matter era and have not amplified so the $\Psi_{MD} = (3/5)\mathcal{R}$ and therefore $T(k) = 3/5$. If we assume that the primordial power spectrum is almost scale invariant

$$\mathcal{P_R} = A_s \left(\frac{k}{k_0}\right)^{n_s - 1} \tag{4.69}$$

with $n_s \simeq 1$. Then using the relation

$$\int_0^\infty \frac{dk}{k} j_l(kr_*)^2 = \frac{1}{2l(l+1)} \tag{4.70}$$

we obtain from (4.65),

$$\frac{l(l+1)}{2\pi} C_l^{TT} = \frac{1}{25} A_s \tag{4.71}$$

The Sachs-Wolfe term predicts that the quantity $l(l+1)C_l$ is independent of l for $l < 76$ arising from super-horizon gravitational perturbations (which are generated presumably during inflation).

Figure 4.2 is the plot of $l(l+1)C_l^{TT}/(2\pi)$ vs l generated with the CLASS code [24]. We see that for $l < 76$ the curve is flat as predicted to arise from super-horizon inflationary perturbations due to the SW effect. The small rise towards low l is due to the ISW effect.

Planck measurements [12] of CMB anisotropy spectrum gives us the amplitude of curvature perturbations $A_s = (2.101 \pm 0.0034) \times 10^{-9}$ and spectral tilt $n_s = 0.9665 \pm 0.0038$ (68%CL).

4.3.2 Tensor Perturbations

There are tensor perturbations or gravitational waves which are also generated during inflation. These primordial gravitational waves can give rise to a temperature

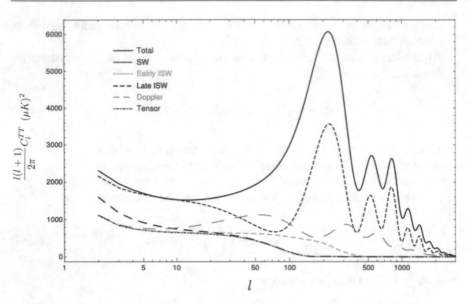

Fig. 4.2 Comparison of contribution to TT angular spectrum from Sachs-Wolfe, Integrated Sachs-Wolfe, Doppler and tensors

anisotropy and also give rise a B-mode polarisation. The tensor perturbations of the FRW metric can be written as

$$ds^2 = a(\eta)^2 \left(-d\eta^2 + \left[\delta_{ij} + h_{ij}(x^i, \eta) \right] dx^i dx^j \right) \tag{4.72}$$

and the transverse, traceless components of h_{ij} are the 2-polarisations of the propagating gravitational waves. Form Einstein's equation $G_{ij} = (8\pi G/3)T_{ij}$ we get

$$h_{ij}'' + 2\mathcal{H}h_{ij}' - \nabla^2 h_{ij} = -8\pi G\pi_{ij} . \tag{4.73}$$

In the absence of anisotropic stress ($\pi_{ij} = 0$), the gravitational waves obey the homogenous equation which in Fourier space is

$$\tilde{h}_{ij}(k, \eta)'' + 2\mathcal{H}\tilde{h}_{ij}(k, \eta)' + k^2\tilde{h}_{ij}(k, \eta) = 0 . \tag{4.74}$$

In the matter era $\mathcal{H} = 2/\eta$ and this equation has the solution

$$\tilde{h}_{ij}(k, \eta) = A_{ij}(k)\frac{3 j_1(k\eta)}{k\eta} + B_{ij}(k)\frac{y_1(k\eta)}{k\eta} \tag{4.75}$$

where j_1 and y_1 are spherical Bessel functions of the first and second kind respectively. In the limit $\eta \to 0$ the solution goes to

$$\tilde{h}_{ij}(k, \eta \to 0) = A_{ij}(k) - \frac{B_{ij}(k)}{(k\eta)^3} . \tag{4.76}$$

Since the second term blows up we choose the initial condition $B_{ij}(k) = 0$. The gravitational wave solution whose primordial amplitude is A_{ij} is

$$\tilde{h}_{ij}(k, \eta) = A_{ij}(k)\frac{3 j_1(k\eta))}{k\eta} = \frac{3A_{ij}}{k\eta}\left[\frac{\sin k\eta}{(k\eta)^2} - \frac{\cos k\eta}{k\eta}\right] . \tag{4.77}$$

At large η the amplitude of the gravitational waves fall off as $h_{ij}(k, \eta) \propto \eta^{-2} \propto a^{-1}$.

The geodesic equation for the photon will get a contribution from the tensor component of the metric. By writing the geodesic equation for photons in the metric given by (4.72) we can derive in a manner similar to (4.24) the equation for photon momentum perturbed by the gravitational waves

$$\frac{1}{p}\frac{dp}{d\eta} = -\mathcal{H} - \frac{1}{2}\hat{p}^i \hat{p}^j \frac{\partial}{\partial\eta}h_{ij} \tag{4.78}$$

Substituting this in the Boltzmann equation (4.33) we obtain at the linear order in perturbations

$$\frac{d\Theta}{d\eta} = -\frac{1}{2}n^i n^j \frac{\partial h_{ij}}{\partial\eta} \tag{4.79}$$

Doing a line of sight integration from the surface of last scattering to the present we obtain the tensor contribution to the temperature anisotropy

$$\Theta(\eta_0) = \Theta(\eta_*) - \frac{1}{2}n^i n^j \int_{\eta_*}^{\eta_0} d\eta \frac{\partial}{\partial\eta}h_{ij}$$

$$= \Theta(\eta_*) - \frac{1}{2}n^i n^j \int_{\eta_*}^{\eta_0} d\eta \frac{\partial}{\partial\eta} \int \frac{d^3k}{(2\pi)^3} e^{ik(\eta_0 - \eta_*)\hat{\mathbf{k}}\cdot\hat{\mathbf{n}}}\tilde{h}_{ij} \tag{4.80}$$

From the solution (4.77) we see that the expression for the time derivative is

$$\frac{\partial\tilde{h}_{ij}(\eta)}{\partial\eta} = -A_{ij}(k)\frac{3k j_2(k\eta))}{k\eta} \tag{4.81}$$

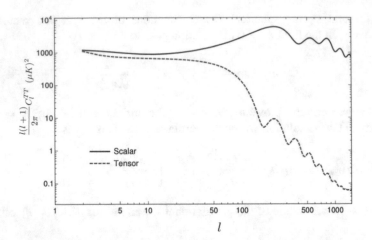

Fig. 4.3 Comparison of TT angular spectrum from scalar and tensor modes. Tensor angular spectrum goes down with increasing l as gravitational waves are damped after horizon entry

And the temperature anisotropy along a direction $\mathbf{n}$ is therefore

$$\delta T(\mathbf{n}, \eta_0) = -\frac{2}{3}\Phi(\mathbf{x}_*, \eta_*) + \frac{1}{2}n^i n^j \int_{\eta_*}^{\eta_0} d\eta \int \frac{d^3 k}{(2\pi)^3} e^{ik(\eta_0 - \eta_*)\hat{\mathbf{k}}\cdot\hat{\mathbf{n}}} A_{ij}(k) \frac{3k j_2(k\eta)}{k\eta} \tag{4.82}$$

The tensor contribution to the TT angular power spectrum C_l^{TT} is proportional to the power spectrum of the tensor modes,

$$C_{l\,(tensor)}^{TT} = \frac{(l+2)!}{(l-2)!} \int \frac{dk}{2\pi} P_t(k) \left[\int_{\eta_*}^{\eta_0} d\eta \frac{3k j_2(k\eta)}{k\eta} \frac{j_l(k(\eta_0 - \eta))}{k^2(\eta_0 - \eta_*)^2} \right]^2 \tag{4.83}$$

In Fig. 4.3 we show the tensor contribution C_l^{TT}. Their contribution to C_l decreases with increasing l since the amplitude of gravitational waves are damped after horizon entry in the matter era.

Here $P_t(k)$ is the initial power spectrum of the two point correlation for each graviton polarization $(+, \times)$,

$$\langle h_+(\mathbf{k}) h_+(\mathbf{k}') \rangle = \langle h_\times(\mathbf{k}) h_\times(\mathbf{k}') \rangle = (2\pi)^3 P_t(k)\delta^3(\mathbf{k} - \mathbf{k}') . \tag{4.84}$$

Inflation predicts a nearly scale invariant tensor power spectrum of the form

$$P_t(k) = \frac{2\pi}{k^3}\mathcal{P}_t \tag{4.85}$$

with

$$\mathcal{P}_t = A_t(k_0) \left(\frac{k}{k_0}\right)^{n_T} \tag{4.86}$$

where $n_T \ll 1$. The ratio of tensor to scalar power is defined as

$$r = \frac{\mathcal{P}_t}{\mathcal{P}_\mathcal{R}} \tag{4.87}$$

and is an important parameter in selecting between theories of inflation. Single field inflation models predict the consistency relation $r = -8n_t$. The tensor contribution to C_l^{TT} has a different l dependence compared to the scalar contribution so the tensor contribution which is larger at smaller l can be compensated for by changing n_s. Since Planck gives a tight constraint on n_s, there is a strong constraint on the $r - n_s$ parameter space as shown in Fig. 4.4. This joint constraint on r and n_s rules out many important models of inflation.

4.3.3 Collisional Boltzmann Equation for Photons

When radiation perturbations are of super-horizon size they are not affected by causal phenomenon like scattering. When they enter the horizon they have causal interactions mainly with electrons which are in turn tightly coupled to the baryons. The evolution of photon distribution function at sub-horizon scales will be governed by the collisional by Boltzmann equation,

$$\frac{df(\eta, x^i, p_i)}{d\lambda} = C[f(p)] \tag{4.88}$$

The dominant photon interaction is the Compton scattering

$$\gamma(\mathbf{p}) + e^-(\mathbf{q}) \leftrightarrow \gamma(\mathbf{p}') + e^-(\mathbf{q}') \tag{4.89}$$

The general principles of deriving the collision term are discussed in Sect. 2.3.2. For the case of Compton scattering (4.88), the net forward reaction will be proportional to $f(\mathbf{p})f_e(\mathbf{q})$ and the reverse reaction will be proportional to $f(\mathbf{p}')f_e(\mathbf{q}')$. The $(1 - f_e)$ Fermi blocking and $(1 + f)$ Bose enhancement factors will be both close to unity as the particle densities are small.

The Boltzmann equation for photons with the collisional term is given by

$$C[f(p)] = \frac{a}{p} \int \frac{d^3\mathbf{q}}{(2\pi)^3 2E_q} \frac{d^3\mathbf{q}'}{(2\pi)^3 2E_{q'}} \frac{d^3\mathbf{q}}{(2\pi)^3 2E_{p'}} \overline{|\mathcal{M}|^2} (2\pi)^4 \delta^3(\mathbf{p} + \mathbf{q} - \mathbf{p} - \mathbf{q})$$

$$\times \delta(E_p + E_q - E_{p'} - E_{q'}) \left\{ f_e(\mathbf{q}')f(\mathbf{p}') - f_e(\mathbf{q})f(\mathbf{p}) \right\} \tag{4.90}$$

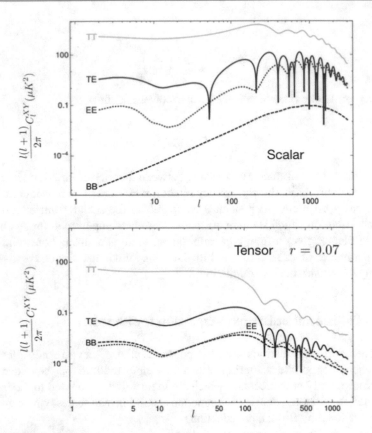

Fig. 4.4 Scalar and tensor perturbation contributions to the CMB polarisation. B-mode polarisation in scalars is generated by lensing of E-modes

where the electron energy $E_q = (m_e^2 + q^2)^{1/2}$ and photon energy $E_p = p$.

In (4.90) the factor of a/p is included as we calculate the collision rate in real time while we have written the homogenous BE in terms of conformal time. In doing this we have converted the Boltzmann equation from the form $df/d\lambda = pdf/d\eta = C[f]$ to $df/dt = \tilde{C}[f]$. Therefore the relation between the two collisional terms is $C[f] = (a/p)\tilde{C}[f]$.

The spin averaged amplitude squared for non-relativistic electrons is given by

$$\overline{|\mathcal{M}|^2} = \frac{1}{4} \sum_{s',s,\lambda,\lambda'} |\mathcal{M}|^2$$

$$= 2(4\pi\alpha)^2 \left(1 + (\hat{p} \cdot \hat{p}')^2\right) + \cdots . \tag{4.91}$$

Where $\alpha = e^2/(4\pi) = 1/137$ is the fine-structure constant and have neglected the terms of order $p/m_e \sim T/m_e \sim 5 \times 10^{-10}(1 + z_*)$ which is a good approximation here. We summed over the final spins and averaged over the initial spins of the particles as we want to follow only the temperature of the photons here. When we calculate the time evolution of average polarization of the photons we will not sum over the photon polarisations ϵ_λ, $\epsilon_{\lambda'}$.

The distribution function for the electrons are taken to be the non-relativistic limit of the Fermi-Dirac distribution $f_e(q) = g_i \left(1 + \exp[(m_e^2 + q^2) - \mu]/T\right)^{-1}$ where the chemical potential μ is related to the number density n_e. We can write the distribution function in terms of number density as

$$f_e(q) = \left(\frac{2\pi}{m_e T}\right)^{3/2} n_e \exp\left(-\frac{|\mathbf{q} - m_e \mathbf{v}_b|^2}{2 m_e T}\right) \tag{4.92}$$

Here we have taken the electron temperature to be the same as the photon temperature $T_e = T$. The electrons follow the baryon fluid due to the Coulomb coupling and they have a bulk velocity which is the same as the baryons but there is on top of this a random velocity $\mathbf{q}/m_e$ of the individual electrons in the distribution. The expression (4.92) is the electron momentum distribution which was isotropic $(f_e \sim e^{-q^2/(2m_e T)})$ in the rest frame of the baryons, transformed to the CMB frame (the frame where the CMB momenta $\mathbf{p}$, $\mathbf{p}'$ and temperature T are defined). Using the electron distribution we can calculate the following multipole moments we will use in calculation of the collision term,

$$\int \frac{d^3q}{(2\pi)^3} f_e(q) = n_e,$$

$$\int \frac{d^3q}{(2\pi)^3} f_e(q) q^i = m_e v_b^i n_e. \tag{4.93}$$

Now Eq. (4.90) can be calculated to the leading order in q/m_e and p/m_e as follows. The d^3q' integral can be done trivially using the delta function for momentum and we are left with the collision term

$$C[f(p)] = \frac{(4\pi\alpha)^2 a}{4 m_e^2} \int \frac{d^3 p'}{(2\pi)^3} \int \frac{d^3 q}{(2\pi)^3} f_e(\mathbf{q}) \frac{1}{p\, p'} \left(1 + \cos^2(\hat{p} \cdot \hat{p}')\right)$$

$$\times (2\pi)\delta\left(p - p' + E(\mathbf{q}) - E(\mathbf{p} + \mathbf{q} - \mathbf{p}')\right)\left[f(\mathbf{p}') - f(\mathbf{p})\right] \tag{4.94}$$

where have taken $f_e(q) \simeq f_e(q')$. We use the approximation

$$E(\mathbf{q}) - E(\mathbf{p} + \mathbf{q} - \mathbf{p}')) \simeq \frac{q^2}{2 m_e} - \frac{(\mathbf{p} + \mathbf{q} - \mathbf{p}')^2}{2 m_e} \simeq \frac{(\mathbf{p}' - \mathbf{p}) \cdot \mathbf{q}}{m_e} \tag{4.95}$$

and expand the delta function in a Taylor series near $(p - p')$,

$$\delta\left(p - p' + \frac{(\mathbf{p}' - \mathbf{p}) \cdot \mathbf{q}}{m_e}\right) = \delta(p - p') + \frac{(\mathbf{p} - \mathbf{p}') \cdot \mathbf{q}}{m_e}\frac{\partial}{\partial p'}\delta(p - p'). \quad (4.96)$$

Substituting this in (4.94) we can perform the integrals over the electron distribution $f_e(q)$ using (4.93). We get n_e from the first term in (4.96) and $n_e m_e v_b^i$ from q^i in the second term and (4.94) reduces to

$$C[f(p)] = \frac{(4\pi\alpha)^2 n_e a}{4m_e^2}\int\frac{d^3 p'}{(2\pi)^3}\frac{1}{p\,p'}\left(1 + (\hat{p}\cdot\hat{p}')^2\right)$$

$$\times (2\pi)\left(\delta(p - p') + (\mathbf{p} - \mathbf{p}')\cdot\mathbf{v}_b\frac{\partial}{\partial p'}\delta(p - p')\right)\left[f(\mathbf{p}') - f(\mathbf{p})\right]$$

$$(4.97)$$

The term in the square bracket can be expanded in terms of the zeroth order and first order terms using (4.31),

$$f(\mathbf{p}') - f(\mathbf{p}) = f^0(p') - f^0(p) - p'\frac{\partial f^0}{\partial p'}\Theta(\hat{p}') + p\frac{\partial f^0}{\partial p}\Theta(\hat{p}) \quad (4.98)$$

Using this in (4.97) we see that the f^0 terms will cancel in the $\mathbf{v}_b$ independent terms due to the delta function. First we calculate the $\mathbf{v_b}$ independent term of (4.97). After performing the dp' integral we obtain,

$$C[f]^{(1)} = \frac{2\pi\alpha^2 n_e a}{m_e^2}\left(p\frac{\partial f^0}{\partial p}\right)\int\frac{d\Omega'}{4\pi}\left(1 + (\hat{p}\cdot\hat{p}')^2\right)\left(\Theta(\hat{p}) - \Theta(\hat{p}')\right)$$

$$(4.99)$$

We can write the coefficient in terms of the total Thomson scattering cross section $\sigma_T = (8\pi/3)(\alpha^2/m_e^2)$ and we get

$$C[f]^{(1)} = \sigma_T n_e a\left(p\frac{\partial f^0}{\partial p}\right)\int\frac{d\Omega'}{4\pi}\frac{3}{4}\left(1 + (\hat{p}\cdot\hat{p}')^2\right)\left(\Theta(\hat{p}) - \Theta(\hat{p}')\right)$$

$$(4.100)$$

The integral of $\Theta(\hat{p})$ is trivial as it is independent of $\hat{p}'$,

$$\int\frac{d\Omega'}{4\pi}\frac{3}{4}\left(1 + (\hat{p}\cdot\hat{p}')^2\right)\Theta(\hat{p}) = \Theta(\hat{p})\int 2\pi\frac{d\cos\theta'}{4\pi}\frac{3}{4}\left(1 + \cos^2\theta'\right) = \Theta(\mathbf{p})$$

$$(4.101)$$

For the other terms its easiest to transform to Fourier space

$$\Theta(\mathbf{x}, \hat{p}, \eta) = \int \frac{d^3k}{(2\pi)^3} \, e^{i\mathbf{k}\cdot\mathbf{x}} \, \Theta(\mathbf{k}, \hat{p}, \eta) \tag{4.102}$$

and do a multipole expansion of the Fourier modes

$$\Theta(\mathbf{k}, \hat{p}, \eta) = \sum_{l=0}^{\infty} (-i)^l \, (2l+1) \, \Theta_l(\mathbf{k}, \eta) P_l(\hat{k} \cdot \hat{p}) \, . \tag{4.103}$$

The multipole moments can be extracted using the relation

$$\Theta_l(\mathbf{k}, \eta) = \frac{1}{(-1)^l} \int_{-1}^{1} \frac{d\mu}{2} P_l(\mu)\Theta(\mu) \, , \quad \mu \equiv \hat{k} \cdot \hat{p} \, . \tag{4.104}$$

We use (4.103) to write $\Theta(\hat{p}')$ in (4.99) as a multipole expansion and do the angular integral over $d\Omega' = 2\pi d\mu'$. Then using (4.104) we see that only the monopole and quadrupole terms survive and the result of the angular integration is

$$\int \frac{d\Omega'}{4\pi} \frac{3}{4} \left(1 + (\hat{p} \cdot \hat{p}')^2\right) \left(\Theta(\mathbf{k}, \hat{p}) - \Theta(\mathbf{k}, \hat{p}')\right)$$

$$= \left(\Theta(\mathbf{k}, \hat{p}) - \Theta_0(\mathbf{k}) + \frac{1}{2} P_2(\hat{k} \cdot \hat{p})\Theta_2(\mathbf{k})\right) \, . \tag{4.105}$$

Therefore the velocity independent collision term is of the form

$$C[f]^{(1)} = \sigma_T n_e a \left(p\frac{\partial f^0}{\partial p}\right) \left(\Theta(\mathbf{k}, \hat{p}) - \Theta_0(\mathbf{k}) + \frac{1}{2} P_2(\mu) \, \Theta_2(\mathbf{k})\right), \quad \mu \equiv \hat{k} \cdot \hat{p} \, . \tag{4.106}$$

Now moving to the velocity dependent term in (4.99), we can transfer the derivative from the delta function to $f(p')$ by doing integration by parts. The dp' can then be trivially done using the delta function and we will be left with

$$C[f]^{(2)} = -\sigma_T n_e a \left(p\frac{\partial f}{\partial p}\right) \int \frac{d\Omega'}{4\pi} \frac{3}{4} \left(1 + (\hat{p} \cdot \hat{p}')^2\right) (\hat{p} - \hat{p}') \cdot \mathbf{v}_b(\mathbf{k})$$

$$= -\sigma_T n_e a \left(p\frac{\partial f^0}{\partial p}\right) \hat{p} \cdot \mathbf{v}_b(\mathbf{k}) \tag{4.107}$$

where we have kept the zeroth order in $f(p)$ as $\mathbf{v}_b$ is already a first order in perturbation.

Therefore the total collision term the sum of (4.106) and (4.107) is,

$$C[f] = \sigma_T n_e a \left(\Theta(\mathbf{k}, \hat{p}) - \Theta_0(\mathbf{k}) + \frac{1}{2} P_2(\mu) \Theta_2(\mathbf{k}) - \hat{p} \cdot \mathbf{v}_b(\mathbf{k}) \right) \left(p \frac{\partial f^0}{\partial p} \right),$$

$$(4.108)$$

The homogenous Boltzmann equation (4.40) as well as the collision term (4.108) have the common factor $(p \partial f^0 / \partial p)$ which cancel leaving a factor (-1) and we obtain the Boltzmann equation (in Fourier space) of the form

$$\frac{\partial \Theta(\mathbf{k}, \hat{p}, \eta)}{\partial \eta} + ik\mu \, \Theta(\mathbf{k}, \hat{p}, \eta) + ik\mu \, \Psi(\mathbf{k}, \eta) - \frac{\partial \Phi(\mathbf{k}, \eta)}{\partial \eta} = \left(\frac{df}{d\eta} \right)_C \quad (4.109)$$

with the collisional term due to Compton scattering given by

$$\left(\frac{df}{d\eta} \right)_C = -\sigma_T n_e a \left(\Theta(\mathbf{k}, \hat{p}) - \Theta_0(\mathbf{k}) + \frac{1}{2} P_2(\mu) \Theta_2(\mathbf{k}) - i\mu v_b(\mathbf{k}) \right). \quad (4.110)$$

where we have assumed that $\nabla \times \mathbf{v}_b = 0$ which implies that $\hat{v}_b = \hat{k}$ and therefore $\hat{p} \cdot \hat{v}_b = i\mu v_b$.

The optical depth of the photon propagating through a region with free electron density n_e is defined as

$$\kappa(\eta) = \int_\eta^{\eta_0} d\eta' a(\eta') n_e(\eta') \sigma_T = -\int_\eta^{\eta_0} d\eta' \dot{\kappa}(\eta'). \quad (4.111)$$

The probability of photon to survive without being scattered. from some initial time η to the present η_0 is $e^{-\kappa}$. From this point on due to a plethora of primes we will switch to the notation of denoting $\partial / \partial \eta$ by an overdot.

The coefficient of the collision term is therefore the differential optical depth $-\dot{\kappa} = a n_e \sigma_T$ and the Boltzmann equation can be written in the form

$$\dot{\Theta}(\mathbf{k}, \hat{p}, \eta) + ik\mu \, \Theta(\mathbf{k}, \hat{p}, \eta) + ik\mu \, \Psi(\mathbf{k}, \eta) - \dot{\Phi}(\mathbf{k}, \eta)$$

$$= (-\dot{\kappa}) \left(-\Theta(\mathbf{k}, \hat{p}, \eta) + \Theta_0(\mathbf{k}, \eta) + \frac{1}{2} P_2(\mu) \Theta_2(\mathbf{k}, \eta) + i\mu v_b(\mathbf{k}, \eta) \right).$$

$$(4.112)$$

This equation can be solved in several approximation schemes. The line of light integration method [25] is used in the numerical codes as it is more computationally efficient as we shall see below. The Boltzmann hierarchy [26] is useful for solving for the first few monopoles analytically.

4.3.4 Line of Sight Integration

Following the line integration method [25], we collect $\Theta(\mathbf{k}, \hat{p}, \eta)$ terms of Eq. (4.112) on the left,

$$\dot{\Theta}(\mathbf{k}, \hat{p}, \eta) + \left(ik\mu - \dot{\kappa}\right)\Theta(\mathbf{k}, \hat{p}, \eta)$$
$$= \dot{\Phi}(\mathbf{k}, \eta) - ik\mu\,\Psi(\mathbf{k}, \eta) + (-\dot{\kappa})\left(\Theta_0(\mathbf{k}, \eta) + \frac{1}{2}P_2(\mu)\,\Pi + i\mu v_b(\mathbf{k}, \eta)\right).$$

(4.113)

where $\Pi = \Theta_2 + \Theta_{P2} + \Theta_{P0}$ and we have included the last two terms which are contributions from the polarisation for completeness. Equation (4.112) can be written as

$$\frac{\partial}{\partial\eta}\left(\Theta\, e^{ik\mu\eta - \kappa}\right) = e^{ik\mu\eta - \kappa}\left[\dot{\Phi} - ik\mu\,\Psi + (-\dot{\kappa})\left(\Theta_0 + \frac{1}{2}P_2(\mu)\,\Pi + i\mu v_b\right)\right]$$

(4.114)

We can formally integrate this equation from some initial time $\eta_i \sim 0$ deep in the last scattering surface ($\eta_i \ll \eta_*$) to the present time η_0 and obtain

$$\Theta(\eta_0)e^{ik\mu\eta_0}\,{}^{\kappa(\eta_0)} - \Theta(\eta_i)e^{ik\mu\eta_i - \kappa(\eta_i)}$$
$$= \int_{\eta_i}^{\eta_0} d\eta\, e^{ik\mu\eta - \kappa}\left[\dot{\Phi} - ik\mu\,\Psi + (-\dot{\kappa})\left(\Theta_0 + \frac{1}{2}P_2(\mu)\,\Pi + i\mu v_b\right)\right]$$

(4.115)

Since η_i was much before the time η_* when the photon started free streaming we can take $e^{-\kappa(\eta_i)} = 0$. We can also take $e^{-\kappa(\eta_0)} \simeq 1$ and we have the solution for the temperature anisotropy expressed as

$$\Theta(\eta_0, \mathbf{k}, \hat{p}) = \int_{\eta_i}^{\eta_0} d\eta\, e^{ik\mu(\eta - \eta_0)}\, e^{-\kappa}\left[\dot{\Phi} - ik\mu\,\Psi + (-\dot{\kappa})\left(\Theta_0 + \frac{1}{2}P_2(\mu)\,\Pi + i\mu v_b\right)\right]$$

(4.116)

We can remove the μ terms in the square brackets integration by parts which replaces μ with a time derivative

$$\mu \rightarrow \frac{1}{ik}\frac{\partial}{\partial\eta}$$

(4.117)

The boundary terms can be dropped as at $\eta_i \simeq 0$, $e^{-\kappa(\eta_i)} = 0$ and at η_0 it adds only a monopole term which can be dropped. It is also useful to define a visibility function $g = -\dot{\kappa}e^{-\kappa}$. This function peaks at the last scattering surface and in approximation schemes one can take $g(\eta) = \delta(\eta - \eta_*)$. The visibility function has a width in η of

80 Mpc around the last scattering surface. It has the property

$$\int_0^\infty g(\eta)d\eta = 1 \tag{4.118}$$

from which we can give $g(\eta, \eta_0)$ the interpretation that it is the probability of a photon scattered at time η to reach us at time η_0.

The solution $\Theta(\eta_0)$ can be expressed as an integral over a source function

$$\Theta(\eta_0, \mathbf{k}, \hat{p}) = \int_0^{\eta_0} d\eta \, e^{ik\mu(\eta-\eta_0)} S_T(\mathbf{k}, \eta) \tag{4.119}$$

where the source function is

$$S_T(\mathbf{k}, \eta) = g\left(\Theta_0 + \Psi - \frac{\dot{v}_b}{k} - \frac{\Pi}{4} - \frac{3}{4}\frac{\ddot{\Pi}}{k^2}\right) + e^{-\kappa}\left(\dot{\Phi} + \dot{\Psi}\right) - \dot{g}\left(\frac{v_b}{k} + \frac{3\dot{\Pi}}{4k^2}\right) - \frac{3\ddot{g}\Pi}{4k^2} \tag{4.120}$$

We see the some terms which we came across when we solved for Θ in the large angle approximation. $g(\Theta_0 + \Psi)$ is the Sachs-Wolfe term, $e^{-\kappa}\left(\dot{\Phi} + \dot{\Psi}\right)$ is the integrated Sachs-Wolfe the terms proportional to v_b and $\dot{v}_b$ are the Doppler terms.

Now we can do a multipole expansion of the $e^{ik\mu(\eta-\eta_0)}$ as in (4.60) and a multipole expansion of $\Theta(\eta_0, \mathbf{k}, \hat{p})$),

$$\Theta(\eta_0, \mathbf{k}, \hat{p}) = \sum_{l=0}^{\infty} (-i)^l (2l+1) P_l(\mu) \, \Theta_l(\eta_0, \mathbf{k}). \tag{4.121}$$

We multiply both sides of (4.119) with $P_l(\mu)$ and integrate over μ by using the relation

$$\int_{-1}^{1} \frac{d\mu}{2} P_l(\mu) e^{ix\mu} = (-i)^{-l} j_l(x) \tag{4.122}$$

and we arrive at the equation for each multipole as an integral over the source function

$$\Theta_l(\mathbf{k}, \eta_0) = \int_0^{\eta_0} d\eta \, S_T(\mathbf{k}, \eta) \, j_l(k(\eta_0 - \eta)). \tag{4.123}$$

Here the angular part and the radial part have factorised. The source function $S_T(\mathbf{k}, \eta)$ is the same for all multipole and has to be computed only once. The angular part $j_l(k(\eta_0 - \eta))$ is a mathematical function independent of the source so it also needs to be computed only once for each multipole. This is computationally simpler than solving the Boltzmann hierarchy where each multipole solution depends upon the nearest neighbour multipoles. Codes like CAMB [27] and CLASS [24] are available for solving the Boltzmann equations relevant for CMB and LSS.

4.3.5 Boltzmann Hierarchy for Photons

We can make use of the orthogonality of $P_l(\mu)$ to convert Eq. (4.113) into separate coupled equations for the multipoles. The properties of $P_l(\mu)$ which we will use are,

$$\int_{-1}^{1} \frac{d\mu}{2} P_l(\mu) P_{l'}(\mu) = \frac{1}{2l+1} \delta_{ll'} , \tag{4.124}$$

$$\mu P_l(\mu) = \frac{1}{2l+1} \left((l+1) P_{l+1}(\mu) + l P_l(\mu) \right) , \tag{4.125}$$

Start with the equation

$$\dot{\Theta}(\mathbf{k}, \mu, \eta) + ik\mu\, \Theta(\mathbf{k}, \mu, \eta) + ik\mu\, \Psi(\mathbf{k}, \eta) - \dot{\Phi}(\mathbf{k}, \eta)$$

$$= (-\dot{\kappa}) \left(-\Theta(\mathbf{k}, \mu, \eta) + \Theta_0(\mathbf{k}, \eta) + i\mu v_b(\mathbf{k}, \eta) \right). \tag{4.126}$$

where we have dropped the source term $P_2(\mu)\Pi$ as it is subdominant for the lower multipoles for which we will solve these equations. We take the first moment by multiplying with $P_0(\mu)$ and integrating w.r.t μ to obtain

$$\dot{\Theta}_0 + k\Theta_1 = \dot{\Phi} \tag{4.127}$$

Now we take the second moment by multiplying by $P_1(\mu)$ and integrating. We obtain

$$\dot{\Theta}_1 - \frac{2k}{3} \Theta_2 - \frac{k}{3} \Theta_0 = \frac{k}{3} \dot{\Psi} + \dot{\kappa} \left(\Theta_1 - \frac{1}{3} v_b \right) \tag{4.128}$$

where we made use of the relation $\mu P_1 = \frac{2}{3} P_2 + \frac{1}{3} P_0$. Continuing this procedure, for all $l \geq 2$ we get

$$\dot{\Theta}_l + k \frac{l+1}{2l+1} \Theta_{l+1} - k \frac{l}{2l+1} \Theta_{l-1} = \dot{\kappa}\Theta_l \tag{4.129}$$

In numerical codes these hierarchy of BE is truncated at some $l \sim 1500$–200.

We can find some analytical solutions for different epochs in time. Consider the time $\eta \leq \eta_*$ just prior to decoupling. In the free electron density is high and the opacity is large $\kappa \gg 1$. Taking time derivatives in Eq. (4.129) as $1/\eta$ and $l \gg 1$, in the tight-coupling limit we have

$$\Theta_l \simeq \frac{k\eta}{2\kappa}\Theta_{l-1} \qquad (4.130)$$

Therefore only the monopole and dipole survive and higher moments are damped as κ becomes large. So in the $\eta \le \eta_*$ epoch we have two coupled equations (4.127) and (4.128) supplemented by the equation for v_b which comes from the Boltzmann equation for baryons. Taking the moments of the Boltzman equation for the baryons gives us the continuity and Euler equations we derived using $\nabla_\mu T^{\mu\nu} = 0$, (Eqs. (3.84) and (3.85)) with an extra baryon-photon coupling term,

$$\dot{\delta}_b + k v_b = 3\dot{\Phi} \qquad (4.131)$$

$$\dot{v}_b + \mathcal{H} v_b = k\Psi + \frac{\dot{\kappa}}{R}(v_b - 3\Theta_1) \qquad (4.132)$$

where the ratio

$$R = \frac{3}{4}\frac{\bar{\rho}_b}{\bar{\rho}_\gamma} = \frac{3}{4}\frac{\Omega_b}{\Omega_\gamma}a \simeq 600a \qquad (4.133)$$

determines the strength of the photon-baryon coupling as a function of a. At the time of decoupling $z_{dec} = 1090$, $R_* = 0.6$.

The baryon photon coupling term on the r.h.s of the baryon Euler equation (4.132) can be derived from the conservation of stress tensor for the baryon-photon fluid. The energy and momentum density of the baryons plus photons is

$$\delta T^0{}_0 = \delta\rho = (\delta\rho_\gamma + \delta\rho_b) = \bar{\rho}_r\left(\delta_\gamma + \frac{4}{3}R\,\delta_b\right),$$

$$\delta T^i{}_0 = \left(\bar{\rho}_\gamma + \bar{P}_\gamma\right)v^i_\gamma + \left(\bar{\rho}_b + \bar{P}_b\right)v^i_b = \frac{4}{3}\left(v^i_\gamma + R\,v^i_b\right). \qquad (4.134)$$

From $\nabla_\mu(\bar{T}^{\mu\nu} + \delta T^{\mu\nu}) = 0$ we can derive the continuity and Euler equation for the baryon-photon fluid with stress tensor components (4.134). The photon density and velocity perturbations are the moments of Θ, with $\delta_\gamma = 4\Theta_0$ and $v^i_\gamma = 3\Theta_1$. We can then use photon Boltzmann equations (4.127) and (4.128) to eliminate Θ_0 and Θ_1 and we will obtain the continuity (4.131) and Euler (4.132) equations with the coupling term with the photons.

4.3.6 Baryon Acoustic Oscillations

The baryon-photon fluid which is tightly coupled at $\eta \sim \eta_*$ falls into the gravitational potential Φ (generated presumably by dark matter) and then bounces back due to photon pressure. This generates oscillations in the baryon-photon fluid which is called Baryon Acoustic Oscillations (BAO) . In the CMB this is seen as the successive peaks in the C_l^{TT} spectrum while in the large scale structures these oscillations are seen in the matter power spectrum at $k^{-1} \simeq 115\,\text{Mpc}$ [28].

In this section we will derive the analytic solutions which describes this phenomenon. We first collect the relevant equations from the last section,

$$\dot{\Theta}_0 + k\Theta_1 = \dot{\Phi} , \tag{4.135}$$

$$\dot{\Theta}_1 - \frac{k}{3}\Theta_0 = \frac{k}{3}\Psi + \frac{\dot{\kappa}}{3}(3\Theta_1 - v_b) , \tag{4.136}$$

$$\dot{v}_b + \mathcal{H}v_b = k\Psi + \frac{\dot{\kappa}}{R}(v_b - 3\Theta_1) . \tag{4.137}$$

with $\dot{\kappa} = -n_e\sigma_T a$. We have truncated the BE series at $l = 1$ and therefore dropped the Θ_2 term from the second equation. The value of $R = 0.6$ so one can solve these equation keeping only linear order term in R, We see that to the zeroth order in R we get from the velocity equation

$$v_b = 3\Theta_1 \tag{4.138}$$

With this approximations we can solve the first two equation in (4.137). Eliminating Θ_1 we obtain the equation for Θ_0 given by

$$\ddot{\Theta}_0 + \frac{k^2}{3}\Theta_0 = -\frac{k^2}{3}\Psi - \ddot{\Phi} \tag{4.139}$$

In the matter era the potentials Ψ and Φ are constants. Therefore $\ddot{\Phi} = 0$ With this (4.139) can be written as

$$\ddot{\Theta}_0 + \frac{k^2}{3}\Theta_0 = -\frac{k^2}{3}\Psi \tag{4.140}$$

This is the equation of a harmonic oscillator with a constant restoring force. The baryon-photon fluid oscillates with the sound speed $c_s \sim 1/\sqrt{3}$ which is the speed of sound of photons. Next we solve the equation to first order in R. The equation for v_b in (4.137) can be written in the form

$$v_b = 3\Theta_1 + \frac{R}{\dot{\kappa}}(\dot{v}_b + \mathcal{H}v_b - k\Psi) \tag{4.141}$$

To solve the equations to the first in R we substitute the zeroth order solution $v_b = 3\Theta_1$ on the r.h.s to obtain

$$v_b = 3\Theta_1 + \frac{R}{\dot{\kappa}}(3\dot{\Theta}_1 + 3\mathcal{H}\Theta_1 - k\Psi) \tag{4.142}$$

We substitute this expression for v_b in the second equation in (4.137) and combine with the first to eliminate Θ_1. We obtain the second order equation for Θ_0,

$$\ddot{\Theta}_0 + \mathcal{H}\frac{R}{1+R}\dot{\Theta}_0 + \frac{k^2}{3(1+R)}\Theta_0 = -\frac{k^2}{3}\Psi + \ddot{\Phi} + \mathcal{H}\frac{R}{1+R}\dot{\Phi} \quad (4.143)$$

Now we see that we have oscillations and from the third term we see that the speed of sound (the coefficient of k^2) of the baryon-photon fluid is now reduced due to the load of the baryons

$$c_s^2 = \frac{1}{3(1+R)} \quad (4.144)$$

The peaks of the CMB will correspond to the wave-numbers k_* where

$$\frac{k_* \eta_*}{\sqrt{3(1+R)}} = n\pi \quad n = 1, 2 \ldots \quad (4.145)$$

where $n = 1, 2 \ldots$ denote the first peak, second peak etc. The two point correlations C_l^{TT} is proportional to the modulus squared of the amplitude and both the peaks and valleys of Θ_0 show up as peaks in C_l^{TT}.

The second term of (4.143) tells us that the oscillations are damped by the baryons. Due to this baryon damping each successive peak in the C_l^{TT} gets smaller.

The acoustic oscillations produce a compression and rarefaction of the baryon density. From (4.132) we can derive the equation for baryon density by using Eq. (4.141) for v_b,

$$\delta_b = 3\Theta_0 + \frac{R}{\dot{\kappa}(1+R)}\left[-k^2\Theta_0 + 3\mathcal{H}(\dot{\Theta}_0 - \dot{\Phi})\right] \quad (4.146)$$

Oscillations in Θ_0 will give rise to oscillations in $\delta_b \sim \cos(kr_s(\eta_*))$ where r_s is the comoving sound horizon defined as

$$r_s = \int_0^\eta c_s(\eta')d\eta' \quad (4.147)$$

The wavenumber of acoustic oscillations peak is given by $k_s(\eta^*)^{-1} = r_s(\eta_*)/\pi = 150h^{-1}$ Mpc and the excess density of baryon shows as oscillatory feature in the galaxy correlations [28] at $k^{-1} = 150\,h^{-1}$Mpc.

4.3.7   Diffusion Damping

We neglected the contribution of the quadrupole Θ_2 in the last section. If we include the quadrupole term then we will get a damping of the monopole term called Silk damping [29]. The BE of photons including quadrupole are

$$\dot{\Theta}_0 + k\Theta_1 = 0 \tag{4.148}$$

$$\dot{\Theta}_1 + \frac{2k}{3}\Theta_2 - \frac{k}{3}\Theta_0 = 0 \tag{4.149}$$

$$\dot{\Theta}_2 - \frac{2}{5}k\Theta_1 = \frac{9}{10}\dot{\kappa}\Theta_2 \tag{4.150}$$

where we have taken $v_b = 3\Theta_1$ In the last equation we see that when $-\dot{\kappa} = n_e a\sigma_T \ll \mathcal{H}$ we can drop the time derivative term in that last equation compared to the scattering term. Then the equation reduces to

$$\Theta_2 = -\frac{4k}{9\dot{\kappa}}\Theta_1 \tag{4.151}$$

Since $\Theta_1 = (1/3)v_b$ we see that in the tight coupling regime the baryon velocity generates a temperature dipole which further generates a quadrupole by scattering. Now substitute Θ_1 in the second equation and combine with the first equation to eliminate Θ_1. We get an equation for the monopole given by

$$\ddot{\Theta}_0 + \left(\frac{-8k^2}{27\dot{\kappa}}\right)\dot{\Theta}_0 + \frac{k^2}{3}\Theta_0 = 0 \tag{4.152}$$

We see that the second term is a damping term on the oscillation of Θ_0. The time evolution of Θ_0 can be written as

$$\Theta_0 \propto e^{ik\eta/\sqrt{3}} e^{-k^2/k_d^2} \tag{4.153}$$

where k_D represents the Silk damping,

$$k_d^2 = \frac{-27\dot{\kappa}}{4\eta} \tag{4.154}$$

The damping occurs due to the diffusion of the photons out of the over-dense regions after multiple scatterings. The mean free path of photons is $\Lambda_{mfp} = 1/n_e\sigma_T a$. In time t the photons undergo $N = t/\Lambda_{mfp}$ scatterings. Therefore they diffuse out over a co-moving distance $\lambda_D = (N)^{1/2}\lambda_{mfp} = (tn_e\sigma_T)^{1/2}$. Diffusion damping suppresses over-densities at length scales lower than ~ 10 Mpc.

4.4 Cosmic Microwave Background Polarisation

The light from the last scattering surface is scattered by electrons near the LSS and a temperature anisotropy results in a polarisation of the CMB in the E mode polarisation . Photons propagating in a background of gravitational waves can also have a B mode polarisation. This polarisation mode also arises from the lensing of E-modes but from the difference in C_l^{BB} expected from GR waves and that from lensing one can extract the B mode due to primordial gravitational waves. This primordial B-mode polarisation is still not observed and remains the holy grail for the ongoing ground based CMB experiments and the planned satellite based experiments in the future. In this section we define the polarisation variables which are observed in the CMB. In the next section we discuss the Boltzmann equations for polarisation.

4.4.1 E and B Mode Polarisation

Consider electromagnetic wave propagation in a direction $\hat{z}$. Its electric field at a given point of observation z_0 can be described by

$$\mathbf{E} = \left(E_x e^{i\delta_x} \hat{\epsilon}_x + E_y e^{i\delta_y} \hat{\epsilon}_y \right) e^{i(kz_0 - \omega t)} \tag{4.155}$$

here $\hat{\epsilon}_x = (1, 0, 0)$ and $\hat{\epsilon}_y = (0, 1, 0)$ are the two polarisation vectors which are orthogonal to each other and to the direction of propagation. The electric field has an amplitude and a phase associated with each polarisation.

The intensity of light, the amount linear and circular polarisation can be quantified through the parameters I, Q, U, V which are defined in terms of $\mathbf{E}$ as

$$I = \langle |\hat{\epsilon}_x \cdot \mathbf{E}|^2 \rangle + \langle |\hat{\epsilon}_y \cdot \mathbf{E}|^2 \rangle = E_x^2 + E_y^2$$

$$Q = \langle |\hat{\epsilon}_x \cdot \mathbf{E}|^2 \rangle - \langle |\hat{\epsilon}_y \cdot \mathbf{E}|^2 \rangle = E_x^2 - E_y^2$$

$$U = 2\, Re\, \langle (\hat{\epsilon}_x \cdot \mathbf{E})^* (\hat{\epsilon}_y \cdot \mathbf{E}) \rangle = 2 E_x E_y \cos(\delta_x - \delta_y)$$

$$V = 2\, Im\, \langle (\hat{\epsilon}_x \cdot \mathbf{E})^* (\hat{\epsilon}_y \cdot \mathbf{E}) \rangle = 2 E_x E_y \sin(\delta_x - \delta_y) \tag{4.156}$$

The angular brackets denote time averaging over times much longer than the time period of the photons. Here I is the intensity which is related to the temperature. Intensity fluctuation from the mean $\Delta I / I = 4\Delta T / T$. If Q or U is non-zero then light is linearly polarised and if $V \neq 0$ then light is circularly polarised. The four parameters (I, Q, U, V) are not all independent and they obey the condition $I^2 = Q^2 + U^2 + V^2$.

In the CMB polarisation is caused by Thomson scattering from the electrons in the last scattering surface. Polarisation is also caused by the tensor modes. These two mechanisms give rise to linear polarisation and CMB is not expected to have any circular polarisation. So we take $V = 0$ and study the evolution of the Q and U polarisations.

The values of Q and U characterise the polarisation of radiation. However these quantities are dependent on the orientation of the coordinate system around the propagation axis. Consider a rotation of the coordinate system by an angle φ around the propagation direction $\hat{k} = \mathbf{z}$. The polarisation vectors $\hat{\epsilon}_x$ and $\hat{\epsilon}_y$ transform like vectors and in the new coordinate system the polarisations $\hat{\epsilon}'_x$ and $\hat{\epsilon}'_y$ will be

$$
\begin{aligned}
\hat{\epsilon}'_x &= \cos\varphi\,\hat{\epsilon}_x + \sin\varphi\,\hat{\epsilon}_y \\
\hat{\epsilon}'_y &= -\sin\varphi\,\hat{\epsilon}_x + \cos\varphi\,\hat{\epsilon}_y .
\end{aligned}
\tag{4.157}
$$

Now we use the rotated $(\epsilon'_x, \epsilon'_y)$ in (4.156) to calculate the (I', Q', U', V') in the new coordinates we find $I' = I$ and $V' = V$ while Q and U transform as

$$
\begin{aligned}
Q' &= \cos 2\varphi\,Q + \sin 2\varphi\,U \\
U' &= -\sin 2\varphi\,Q + \cos 2\varphi\,U
\end{aligned}
\tag{4.158}
$$

The quantity $Q \pm iU$ transforms as

$$
(Q \pm iU)' = e^{\mp 2i\varphi}\,(Q \pm iU)
\tag{4.159}
$$

A general field $X_{\mathbf{k}}$ has a definite helicity λ if on rotating the coordinates by angle φ around $\mathbf{k}$, $X_{\mathbf{k}} \to e^{-i\lambda\varphi} X_{\mathbf{k}}$. Therefore $(Q \pm iU)$ has helicity ± 2. Hence we can expand the polarisation anisotropy $(Q \pm iU)(\hat{n})$ in terms of the spin-2 spherical harmonics $_{\pm 2}Y_{lm}$. The spin-s spherical harmonics $_{s}Y_{lm}$ [30] obey the orthogonality relations and completeness relations similar to the spin-0 Y_{lm}. The orthogonality and completeness relations of $_{s}Y_{lm}$ are

$$
\int d\Omega \;_{s}Y_{lm}(\theta, \phi) \;_{s}Y^{*}_{l'm'}(\theta, \phi) = \delta_{ll'}\delta_{mm'} ,
$$

$$
\sum_{lm} {}_{s}Y_{lm}(\theta, \phi) \;_{s}Y^{*}_{lm}(\theta'\phi') = \frac{1}{\sin\theta}\delta(\theta - \theta')\delta(\phi - \phi') .
\tag{4.160}
$$

Under complex conjugation and parity the $_{s}Y_{lm}(\hat{n})$ transform as

$$
{}_{s}Y^{*}_{lm}(\hat{n}) = (-1)^{m+s} \;_{-s}Y_{l,m}(\hat{n}) , \qquad {}_{s}Y_{lm}(-\hat{n}) = (-1)^{l} \;_{-s}Y_{l,-m}(\hat{n}).
\tag{4.161}
$$

We expand $(Q \pm iU)(\hat{n})$ in spin-2 spherical harmonics as

$$
(Q \pm iU)(\hat{n}) = \sum_{lm} {}_{\pm 2}a_{lm} \;_{\pm 2}Y_{lm}(\hat{n})
\tag{4.162}
$$

where

$$
{}_{\pm 2}a_{lm} = \int d\Omega \;_{\pm 2}Y^{*}_{lm}(\hat{n})\,(Q \pm iU)(\hat{n}) .
\tag{4.163}
$$

This is different compared to the temperature anisotropy, which is a spin-0 field and is expanded in terms of the spin-0 spherical harmonics

$$\Delta T(\hat{n}) = \sum_{lm} a_{lm}^T \; Y_{lm}(\hat{n}) . \tag{4.164}$$

and then the angular two point correlation of temperature anisotropy is given by

$$C_l^{TT} = \frac{1}{2l+1} \sum_m a_{lm}^T \, a_{lm}^{T*} \tag{4.165}$$

We can construct spin-0 variables from $Q \pm iU$ by using spin raising and lowering operators $\eth$ (eth) and $\bar{\eth}$ (ethbar) respectively which have the properties[1]

$$\eth(_s Y_{lm}) = \sqrt{(l-s)(l+s+1)} \; _{s+1}Y_{lm} ,$$
$$\bar{\eth}(_s Y_{lm}) = \sqrt{(l+s)(l-s+1)} \; _{s-1}Y_{lm} . \tag{4.166}$$

One can therefore construct spin-0 variables which can be expanded in $Y_{lm} = \, _0 Y_{lm}$ from the helicity ± 2 polarisations $Q \pm iU$ by operating with $\eth^2$ or $\bar{\eth}^2$. Following this procedure we have

$$\bar{\eth}^2 (Q + iU) (\hat{n}) = \sum_{lm} \, _2 a_{lm} \, \bar{\eth}^2 \, _2 Y_{lm} = \sum_{lm} \mathcal{N}_l \, _2 a_{lm} \; Y_{lm}(\hat{n}) ,$$

$$\eth^2 (Q - iU) (\hat{n}) = \sum_{lm} \, _{-2} a_{lm} \, \eth^2 \, _{-2} Y_{lm} = \sum_{lm} \mathcal{N}_l \, _{-2} a_{lm} \; Y_{lm}(\hat{n}) . \tag{4.167}$$

where

$$\mathcal{N}_l = \left(\frac{(l+2)!}{(l-2)!} \right)^{1/2} . \tag{4.168}$$

Now we can express the expansion coefficients $\pm 2 a_{lm}$ in terms of Y_{lm}^* as follows

$$_2 a_{lm} = \int d\Omega \; _2 Y_{lm}^*(\hat{n}) \; \left(Q + iU \right)(\hat{n})$$

---

[1] The explicit forms of $\eth$ and $\bar{\eth}$ operators are given in [30]

$$\eth(_s Y_{lm}) = -(\sin\theta)^s \left[\frac{\partial}{\partial\theta} + \frac{i}{\sin\theta} \frac{\partial}{\partial\phi} \right] ((\sin\theta)^{-s} \, _s Y_{lm}) ,$$

$$\bar{\eth}(_s Y_{lm}) = -(\sin\theta)^{-s} \left[\frac{\partial}{\partial\theta} - \frac{i}{\sin\theta} \frac{\partial}{\partial\phi} \right] ((\sin\theta)^{s} \, _s Y_{lm}) .$$

$$= N_l^{-1} \int d\Omega \ Y_{lm}^*(\hat{n}) \ \bar{\eth}^2 \Big(Q + iU\Big)(\hat{n}), \tag{4.169}$$

and

$$_{-2}a_{lm} = \int d\Omega \ _{-2}Y_{lm}^*(\hat{n}) \ \Big(Q - iU\Big)(\hat{n})$$

$$= N_l^{-1} \int d\Omega \ Y_{lm}^*(\hat{n}) \ \eth^2\Big(Q - iU\Big)(\hat{n}) \tag{4.170}$$

This procedure is well defined for multipoles $l > 2$. Its useful to organise these coefficients in parity-even and parity-odd combinations [31, 32],

$$a_{lm}^E \equiv -\frac{1}{2} \left(_2a_{lm} + \ _{-2}a_{lm}\right),$$

$$a_{lm}^B \equiv \frac{i}{2} \left(_2a_{lm} - \ _{-2}a_{lm}\right). \tag{4.171}$$

Now we can define spin-0 polarisation variables in real space as

$$E(\hat{n}) = \sum_{lm} a_{lm}^E \ Y_{lm}(\hat{n})$$

$$B(\hat{n}) = \sum_{lm} a_{lm}^B \ Y_{lm}(\hat{n}) \tag{4.172}$$

$E(\hat{n})$ and $B(\hat{n})$ have different properties under parity transformation $\hat{n} \rightarrow -\hat{n}$. Under parity we have $_sY_{lm}(-\hat{n}) = (-1)^l \ _{-s}Y_{lm}(\hat{n})$. Using this can see that under parity transformation

$$a_{lm}^T \rightarrow (-1)^l \ a_{lm}^T,$$

$$a_{lm}^E \rightarrow (-1)^l \ a_{lm}^E,$$

$$a_{lm}^B \rightarrow -(-1)^l a_{lm}^B. \tag{4.173}$$

Therefore E is a scalar as $E(-\hat{n}) = E(\hat{n})$ while B is a pseudo-scalar, $B(-\hat{n}) = -B(\hat{n})$.

The two point correlations which can be calculated to compare with experiments will depend upon $\hat{n} \cdot \hat{n}'$ due to the statistical isotropy of space—there is no preferred direction at large scales. This implies that after averaging over a all sky the two point correlations must be of the form

$$\langle a_{lm}^X a_{l'm'}^{*Y}\rangle = \delta_{ll'}\delta_{mm'} \ a_{lm}^X a_{l'm'}^{*Y}, \qquad X, Y = T, E, B. \tag{4.174}$$

The angular spectrum of the two point correlation are obtained from the a_{lm}'s by summing over m

$$\frac{1}{2l+1} \sum_{m=-l}^{l} a_{lm}^X a_{lm}^{*Y} = C_l^{XY}, \qquad X, Y = T, E, B \,. \tag{4.175}$$

In the CMB it is assumed that there is no parity violation in the propagation or in scattering of photons so the only non-zero correlations expected are $C_l^{TT}, C_l^{ET}, C_l^{EE}$ and C_l^{BB}. Parity conservation predicts that C_l^{TB} and C_l^{EB} are both zero. So far in experiments only $C_l^{TT}, C_l^{EE}, C_l^{ET}$ (of primordial origin) have been observed. Thompson scattering gives rise to only E mode polarisation while primordial gravitational waves generate both E and B mode polarisation. The first observation of E mode polarisation was by DASI [13] while B modes which are generated by lensing of E modes has been observed by SPTPol [14]. The observation of primordial B mode polarisation produced by tensor perturbations generated during inflation is one of the targets of the next generation of proposed CMB experiments [15].

4.4.2 Polarisation from Thomson Scattering

Let us calculate the Thomson scattering of photons by electrons to see how polarisation is generated by quadrupole anisotropy in the incident light intensity [33, 34]. The differential cross section of Thomson scattering is

$$\frac{d\sigma}{d\Omega} = \frac{3\sigma_T}{8\pi} |\epsilon \cdot \epsilon'|^2 \tag{4.176}$$

Here ϵ is the polarisation of the incident photon and ϵ' the polarisation of the outgoing photon. Let us set the z-axis along the direction $\mathbf{k}$ of the incoming photon. Then $\hat{\mathbf{k}} = (0, 0, 1)$ and we can choose the two possible polarisation directions of the in-coming photon as $\epsilon_x = (1, 0, 0)$ and $\epsilon_y = (0, 1, 0)$. Let the out photon be along any general direction $\hat{\mathbf{k}}' = (\sin\theta \sin\phi, \cos\theta \cos\phi, \cos\theta)$ and we choose the polarisation directions to be $\epsilon_1' = (-\sin\phi, -\cos\phi, 0)$ and $\epsilon_2' = (\cos\theta \cos\phi, \cos\theta \sin\phi, -\sin\theta)$. We will start with unpolarised radiation so that $I_1 = I_2 = I/2$. Now we can use (4.176) to calculate the intensities of the scattered light in the two polarisations I_x and I_y. The intensities of the scattered light in the two polarisation directions are

$$I_x' = \frac{3\sigma_T}{8\pi} \left(|\epsilon_1 \cdot \epsilon_x'|^2 I_1 + |\epsilon_2 \cdot \epsilon_x'|^2 I_2 \right) = \frac{3\sigma_T}{8\pi} \frac{I}{2},$$

$$I_y' = \frac{3\sigma_T}{8\pi} \left(|\epsilon_1 \cdot \epsilon_y'|^2 I_1 + |\epsilon_2 \cdot \epsilon_y'|^2 I_2 \right) = \frac{3\sigma_T}{8\pi} \frac{I}{2} \cos^2\theta \,. \tag{4.177}$$

where $\cos\theta = \hat{k}\cdot\hat{k}'$. Here I is the incident beam while I' is intensity of the scattered light (at a unit distance from the scatterer).

Therefore the intensity and polarisations of the scattered light are

$$I' = I'_x + I'_y = \frac{3\sigma_T}{8\pi}\frac{1}{2}\left(1 + \cos^2\theta\right) I,$$

$$Q' = I'_x - I'_y = \frac{3\sigma_T}{8\pi}\frac{1}{2}\sin^2\theta\, I,$$

$$U' = 0 \tag{4.178}$$

Therefore an initial unpolarised light becomes polarised due to Thomson scattering if the integrand on the r.h.s of the second equation is non-zero. For this to be non-zero the incident intensity must have a non-zero quadrupole moment as we shall see next.

We can rotate the coordinates around z-axis by some angle ϕ then Q' and U' will change. The combination $Q \pm iU \to e^{\mp\phi}(Q \pm iU)$.

Now we do a multipole expansion of incident intensity $I(\theta, \phi)$,

$$I(\theta, \phi) = \sum_{lm} a_{lm} Y_{lm}(\theta, \phi) \tag{4.179}$$

We substitute (4.179) in (4.178) and do the angular integration

$$I'(\hat{n}) = \frac{3\sigma_T}{16\pi}\int d\Omega \left(1 + \cos^2\theta\right) I,$$

$$(Q' - iU')(\hat{n}) = \frac{3\sigma_T}{16\pi}\int d\Omega\, e^{2\phi}\,\sin^2\theta\, I, \tag{4.180}$$

and we obtain

$$I'(\hat{n}) = \frac{3\sigma_T}{16\pi}\left[\frac{8}{3}\sqrt{\pi}\, a_{00} + \frac{4}{3}\sqrt{\frac{\pi}{5}}\, a_{10}\right], \tag{4.181}$$

$$(Q' - iU')(\hat{n}) = \frac{3\sigma_T}{4\pi}\sqrt{\frac{2\pi}{15}}\, a_{22} \tag{4.182}$$

To find the change in polarisation one must rotate I to the propagation axis $\hat{n}$ of the outgoing photons to compute $\Delta_Q(\hat{n})$ and $\Delta_U(\hat{n})$. Let β be the angle between the in-photon and the out photon directions. The coefficients of I in the rotated coordinate system $\tilde{a}_{lm}$ are related to the coefficients a_{lm} in Eq. (4.179) by the Wigner $D^l_{m'm}$ matrix elements,

$$\tilde{a}_{lm} = \int d\Omega\, Y^*_{lm}\,(R\Omega)\, I(\Omega)$$

$$= \sum_{m'=-m}^{m} D^{l*}_{m'm}(R) \int d\Omega\, Y^*_{lm'}(\Omega)\, I(\Omega) \qquad (4.183)$$

where R is the rotation matrix $R(\beta)\hat{z} = \hat{n}$. From (4.180) we know that the multipole coefficients of I that contribute to polarisation have $l = 2$. Also if the incoming $I(\theta, \omega)$ has a ϕ symmetry around the direction of propagation which will be the case if the incident beam is unpolarised, then (4.183) reduces to

$$\tilde{a}_{22} = a_{20}d^{2*}_{02}(\beta) = \frac{\sqrt{6}}{4}\, a_{20} \sin^2 \beta \qquad (4.184)$$

where $d^l_{mm'}$ is the reduced D-matrix element. The outgoing polarisation which is also the change in polarisation can now be obtained by replacing a_{22} in (4.182) by (4.184) and we have

$$\Delta Q(\hat{n}) - i\,\Delta U(\hat{n}) = \frac{3\sigma_T}{4\pi}\sqrt{\frac{2\pi}{15}}\, \tilde{a}_{22} = \frac{3\sigma_T}{8\pi}\sqrt{\frac{\pi}{5}}\, a_{20} \sin^2 \beta \qquad (4.185)$$

On averaging over all incoming directions β the factor $\sin^2 \beta$ will be replaced by $1/2$.

We see that the quadrupole moment a_{20} of the incident intensity I contributes to a change in polarisation. Due to azimuthal symmetry of I around the propagation direction, $\tilde{a}_{22}$ is real, which implies that $\Delta U = 0$. This means that the scattered photons have only E mode polarisation.

When there are tensor perturbations the incident beam not have azimuthal symmetry which will give imaginary contributions to $\tilde{a}_{22}$. Therefore we have B-mode polarisation generated by Thomson scattering of photons propagating in a metric with tensor perturbations.

The Boltzmann equations for the scalar contribution to temperature perturbation Θ^S and polarisation $\Theta^S_P = \Delta Q \pm i\,\Delta U$ are [25],

$$\dot{\Theta}^S + (ik\mu - \dot{\kappa})\,\Theta^S - \dot{\Phi} + ik\mu\Psi = -\dot{\kappa}\left[\Theta^S_0 - i\mu v_b - \frac{1}{2}P_2(\mu)\Pi\right],$$

$$\dot{\Theta}^S_P + (ik\mu - \dot{\kappa})\,\Theta^S_P = -\dot{\kappa}\left[\frac{1}{2}\,(1 - P_2(\mu))\,\Pi\right],$$

$$\Pi = \Theta^S_2 + \Theta^S_{P,2} + \Theta^S_{P,0}\ . \qquad (4.186)$$

The quadrupole anisotropy of temperature Θ_2^S is the source for generating polarisation Θ_P^S. next we will study the contribution of gravitational waves to polarisation.

4.4.3 Polarisation by Gravitational Waves

Photons as they propagate through tensor background have an anisotropy pattern in temperature by the tensors. These modes when they Thomson scatter get polarised in the way we saw in the last section. Gravitational waves contribute to the C_l^{TT} but also to the C_l^{EE} and C_l^{BB}. The Boltzmann equation for temperature in the tensor background was worked out earlier Eq. (4.79) and where we now add the collision term on the r.h.s, to obtain

$$\frac{\partial \Theta^T}{\partial \eta} + ik\mu\Theta^T + \frac{1}{2}\frac{\partial h_{ij}}{\partial \eta}\hat{p}^i \hat{p}^j = \dot{\kappa}\Theta^T \tag{4.187}$$

Consider the two gravitational wave modes $h_+ = h_{11} = -h_{22}$ and $h_\times = h_{12} = h_{21}$. Now take the components of $\hat{\mathbf{p}} = (\cos\theta\cos\phi, \sin\theta\sin\phi, \cos\theta) \equiv (p_1, p_2, p_3)$. Then we can calculate

$$\frac{1}{2}\dot{h}_{ij}\hat{p}_i\hat{p}_j = \dot{h}_+(p_1^2 - p_2^2) + \dot{h}_\times p_1 p_2$$

$$= \dot{h}_+(\sin^2\theta\cos 2\phi) + \dot{h}_\times \sin^2\theta \sin 2\phi$$

$$= \sqrt{\frac{2\pi}{15}}\left[Y_2^2(\dot{h}_+ - ih_\times) + Y_2^{-2}(\dot{h}_+ + ih_\times)\right]$$

$$= \left(\frac{\pi}{15}\right)^{1/4}\left[Y_2^2 \dot{h}_{\lambda=2} + Y_2^{-2}\dot{h}_{\lambda=-2}\right] \tag{4.188}$$

We can separate the temperature and polarisation generated by each gravitational helicity mode $\lambda = \pm 2$ separately by expanding [35],

$$\Theta^T = \left(\frac{\pi}{15}\right)^{1/4}\sum_{\lambda=\pm 2}\Theta_\lambda^T Y_2^\lambda \tag{4.189}$$

and similarly for the polarisation $\Theta_P = \Delta_Q \pm i\Delta_U$, Then the coupled equations for Θ_λ and $\Theta_{P\lambda}$ which are the temperature and polarisation generated by each helicity of the gravitational waves are of the form

$$\dot{\Theta}_\lambda^T + (ik\mu - \dot{\kappa})\Theta_\lambda^T + \frac{\dot{h}_\lambda}{2} = \frac{-\dot{\kappa}}{10}\int d\Omega'\, Y_2^{*\lambda}\, \Theta_\lambda^T\, Y_2^\lambda \tag{4.190}$$

and

$$\dot{\Theta}_{P\lambda}^T + (ik\mu - \dot{\kappa})\Theta_{P\lambda}^T = \frac{\dot{\kappa}}{10} \int d\Omega' \; Y_2^{*\lambda} \; \Theta_\lambda^T \; Y_2^\lambda \tag{4.191}$$

where we have kept the leading order contributions to each equation on the r.h.s. The temperature anisotropy generated by the gravitational waves source the polarisation. The complete source term for tensor contribution to temperature perturbation Θ_λ^T and polarisation $\Theta_{P\lambda}^T$ for each gravitational polarisation mode ($\lambda = +, \times$) is given in [36],

$$\dot{\Theta}_\lambda^T + (ik\mu - \dot{\kappa}) \, \Theta_\lambda^T + \frac{1}{2}\dot{h}_\lambda = -\dot{\kappa}\,\Psi_T \, ,$$

$$\dot{\Theta}_{P\lambda}^T + (ik\mu - \dot{\kappa}) \, \Theta_{P\lambda}^T = \dot{\kappa}\,\Psi_T \, ,$$

$$\Psi_T = \frac{1}{10}\Theta_{\lambda,0}^T + \frac{1}{7}\Theta_{\lambda,2}^T + \frac{3}{70}\Theta_{\lambda,4}^T - \frac{3}{5}\Theta_{P\lambda,0}^T + \frac{6}{7}\Theta_{P\lambda,2}^T - \frac{3}{70}\Theta_{P\lambda,4}^T \, ,$$

$$\tag{4.192}$$

Where the source term Ψ_T is expressed in multipoles moments of Θ_λ^T and $\Theta_{P\lambda}^T$.

These equations can be solved numerically to obtain the two point correlations $C_l^{TT}, C_l^{TE}, C_l^{EE}$ and C_l^{BB}. The different two point correlation spectrum obtained by using the CLASS code [24] are shown in Fig. 4.4.

4.5 Neutrinos in Cosmology

Neutrinos decouple from the other particles when the weak interaction rate $\Gamma = G_F^2 T^5$ falls below the Hubble expansion rate $H + 1.66 M_p \sqrt{g_\rho} T^2$. This occurs at a thentemperature $T_\nu = 1 MeV$ when the neutrinos were relativistic, $T_\nu \gg m_\nu$. So their distribution function at that time with $E \simeq p$ was of the form

$$f_\nu^0(p, \eta) = \frac{1}{e^{p/T_\nu} + 1} \, . \tag{4.193}$$

The temperature of the relativistic particles in the unperturbed FRW universe evolves as $T \propto a^{-1}$ We can write the neutrino temperature as $T_\nu = (T_{\nu 0}/a)$ where $T_{\nu 0} = (4/11)^{1/3} T_0$ is the present neutrino temperature. The photon temperature rises by a factor $(11/4)^{1/3}$ compared to neutrinos after electron-positron annihilation into photons.

Using the unperturbed distribution function (4.193) we get the present energy density of the neutrinos as

$$\bar{\rho}_\nu = \frac{1}{(2\pi)^3} \int d\Omega \, p^2 dp \, \frac{\sqrt{m_\nu^2 + p^2}}{e^{ap/T_{\nu 0}} + 1} \, . \tag{4.194}$$

We calculate the neutrinos background density for relativistic as well as non relativistic neutrinos using (4.194) in different limits. Change the integration variable to $x = ap/T_{\nu 0}$,

$$\bar{\rho}_\nu = \frac{1}{(2\pi)^3} \frac{T_{\nu 0}^3}{a^3} \int d\Omega \, x^2 dx \frac{\sqrt{m_\nu^2 + x^2 T_{\nu 0}^2 / a^2}}{e^x + 1}. \tag{4.195}$$

In the early era neutrinos were relativistic and $m_\nu^2 \ll x^2 T_{\nu 0}^2 / a^2$, for relativistic neutrinos we can drop m_ν and the integral is

$$\bar{\rho}_\nu = \frac{7}{8} \frac{\pi^2}{15} \frac{T_{\nu 0}^4}{a^4}. \tag{4.196}$$

Big bang nucleosynthesis (BBN) which successfully predicts the abundance of light elements abundance is sensitive to the Hubble expansion rate $H = (8\pi G/3)^{1/2}\rho$ at the time of nucleosynthesis and thus gives an upper bound on the number of relativistic particles that can be present at 1 MeV. The BBN bound is presented as effective number of neutrino species that can be present at BBN,

$$N_{\text{eff}} = N_\nu + \left[\left(\frac{T_f}{T_\nu} \right)^4 N_f + \frac{4}{7} \left(\frac{T_s}{T_\nu} \right)^4 N_s \right] \tag{4.197}$$

where N_f is the number of Weyl fermions (Dirac fermions will contribute $2N_f$) and N_s is the number of scalars with temperatures T_f and T_s respectively.

The BBN prediction is based on the baryons to photon ratio η and the baryon density Ω_b at th time of BBN. Taking the Planck numbers $\eta = (6.129 \pm 0.039) \times 10^{-10}$ and $\Omega_b h^2 = 0.02239 \pm 0.00014$, Fields et al. [37] derive $N_{\text{eff}} = 2.86 \pm 0.15$ at BBN. The standard model prediction is $N_{\text{eff}} = 3.046$. The extra 0.046 is due to the fact that neutrinos are not completely decoupled and their temperature rises slightly after $e^+ e^-$ annihilation [38]. Therefore all beyond standard model particles with mass $m_x < 1$ MeV must have temperatures $T_x \ll T_\nu = T_\gamma$ at the time of BBN to evade the BBN bound.

If at some later epoch a neutrino species becomes non-relativistic $m_\nu \gg T_{\nu 0}/a$ then we an evaluate the integral (4.195) by taking $\sqrt{m_\nu^2 + x^2 T_{\nu 0}^2 / a^2} = m_\nu$ and we get

$$\bar{\rho}_\nu = m_\nu \frac{3}{4} \frac{\xi(3)}{\pi^2} \frac{T_{\nu 0}^3}{a^3} = m_\nu n_\nu = m_\nu \left(\frac{113 \, \text{cm}^{-3}}{a^3} \right). \tag{4.198}$$

Here even when the neutrinos are non-relativistic their number density is evaluated as if they are relativistic since their distribution function does not change.

The relation (4.198) gives the Cowsik-MacClelland bound [39,40] on the sum of neutrino masses,

$$\Omega_\nu h^2 = \frac{\sum_i m_{\nu_i}}{93.14\,\text{eV}}\,. \tag{4.199}$$

From neutrino oscillation experiments we know the mass differences of the neutrinos. The solar neutrino observations tell us that $\Delta m^{12} = m_2^2 - m_1^2 = 7.5 \times 10^{-5}\text{eV}^2$ and the atmospheric neutrino observations tell us that $|\Delta m_{23}^2| = 2.4 \times 10^{-3}\text{eV}^2$. Atmospheric neutrino mass difference gives a lower mass on the neutrino energy density $\Omega_\nu h^2 \geq 0.006$.

If each neutrino has mass $\sim 15\,\text{eV}$ then $\Omega_\nu \sim 0.3$ which is the required density of dark matter at cosmological scales. Observations of large scale structures however rule out this possibility of neutrinos being the dark matter.

4.5.1   Neutrino Free-Streaming

Neutrino which are non-relativistic at decoupling will free-stream after they stop scattering particles in the plasma (kinetic decoupling). The neutrino mean velocity after decoupling is

$$v_{th} = \frac{\langle p_\nu \rangle}{m_\nu} = \frac{3T_\nu}{m_\nu} = 151\,(1+z)\left(\frac{1\text{eV}}{m_\nu}\right)\text{km}\,\text{s}^{-1}\,. \tag{4.200}$$

They become non-relativistic at temperature $T_\nu = (1/3)m_\nu$ which occurs at the redshift z_{nr} given by the $T_{\nu 0}(1 + z_{nr}) = (1/3)m_\nu$ which gives

$$1 + z_{nr} = 1988\left(\frac{m_\nu}{\text{eV}}\right)\,. \tag{4.201}$$

After neutrinos become non-relativistic their density perturbations grows. This is governed by the Jeans equation (3.144) where taking the equation of state for non-relativistic neutrinos to be $\omega_\nu \sim 0$ and the speed of sound for neutrinos to be the thermal velocity $c_{s\nu} = v_{th}$ we have

$$\ddot{\delta}_\nu + \mathcal{H}\dot{\delta}_\nu + \left(v_{th}^2 k^2 \delta_\nu - \frac{3}{2}\mathcal{H}^2\delta\right) = 0\,. \tag{4.202}$$

We define the neutrino free-streaming wave-number k_{fs} as

$$k_{\text{fs}} = \sqrt{\frac{3}{2}}\frac{\mathcal{H}}{v_{th}}\,. \tag{4.203}$$

Neutrinos density of wavenumber $k > k_{\text{fs}}$ will oscillate while modes with wavenumber $k < k_{\text{fs}}$ will have a growing mode solution denoting the clustering of neutrino density (in the gravitational potential from all particles with total density perturbation δ). k_{fs} is the co-moving wave-number which implies that the neutrinos will only cluster at physical length scales larger than $\lambda_{\text{fs}} = 2\pi a(t)/k_{\text{fs}}$. We can write (4.203) by taking $\mathcal{H} = aH$ and using (4.200) to obtain

$$k_{\text{fs}}(z) = \sqrt{\frac{3}{2}} \frac{H(z)}{(1+z)v_{th}} = 0.81 \frac{\sqrt{\Omega_\lambda + \Omega_m(1+z)^3}}{(1+z)^2} \left(\frac{m_v}{\text{eV}}\right) h\,\text{Mpc}^{-1}. \qquad (4.204)$$

Therefore as z decreases k_{fs} grows as $(1+z)^{-1/2}$ while the neutrino has the free-streaming velocity (4.200). When neutrinos are non-relativistic then their speed of sound is $c_s \sim 0$ and δ_m has growing mode solution for all k like CDM. The minimum value of k_{fs} is therefore (4.204) evaluated at z_{nr} which gives

$$k_{\text{fs}}(z_{nr}) = 0.018\,\Omega_m^{1/2}\left(\frac{m_v}{eV}\right)^{1/2} h\text{Mpc}^{-1}. \qquad (4.205)$$

A neutrino with mass $m_v = 1\text{eV}$ has $k(z_{nr}) = 0.3\,h\,\text{Mpc}^{-1}$ and it will suppress structures smaller than $\lambda_{\text{fs}} = 20h^{-1}\text{Mpc}$. This is larger than a cluster size so neutrinos as hot dark matter is ruled out. Planck [12] gives an upper bound on the sum of neutrino masses of to be $\sum_i m_{v_i} \leq 0.12\,\text{eV}$ at 95% CL. This bound is obtained from observation of lensing of CMB by the gravitational potential of clusters.

Three neutrinos with mass differences of $\Delta m_{12}^2 7.5 \times 10^{-5}\text{eV}^2$ and $\Delta|m_{23}^2| = 2.4 \times 10^{-3}\text{eV}^2$ can explain the solar and neutrino atmospheric neutrino observations.

MiniBooNE experiment [41] reported another oscillation channel which would require a fourth neutrino.

4.5.2 Boltzmann Equations for Neutrino Perturbations

We now study in detail the perturbations in neutrino energy-momentum to see its effect on CMB and LSS. Boltzmann equation for neutrinos and cold dark matter will be identical to those of photons and baryons respectively with the important difference that there will be no collision term.

In the case of neutrinos we will take the unperturbed distribution function to include the neutrino mass and we will treat perturbations in the distribution function and its moments and not the temperature perturbation like we do for photons. Similarly in the case of CDM we will deal with the moments of the distribution function which give the continuity and Euler equations just like the baryons.

We start by writing the neutrino distribution function as with the unperturbed neutrino distribution function and its first order perturbation

$$f_\nu(p, \hat{p}, \eta) = f^0(p, T_\nu) + f_\nu^1(p, \hat{p}, \eta) \tag{4.206}$$

We construct another variable $F_\nu(\hat{p}, \eta)$ from f_ν^1 by integrating over the magnitude p but not over the directions,

$$F_\nu(\hat{p}, \eta) \equiv \frac{1}{\bar{\rho}_\nu} \int \frac{p^2 dp}{(2\pi^2)} \, p \, f_\nu^1(p, \hat{p}, \eta) \tag{4.207}$$

Now we transform to the Fourier space to obtain $F_\nu(\hat{p}, \mathbf{k}, \eta)$. The quantity $F_\nu(\hat{p}, \mathbf{k}, \eta)$ obeys the same BE as photon temperature perturbation $\Theta(\hat{p}, \mathbf{k}, \eta)$ (without the collision term),

$$\dot{F}_\nu + ik\mu F_\nu + ik\mu \Psi - \dot{\Phi} = 0 \tag{4.208}$$

where $\mu = \hat{k} \cdot \hat{p}$.

Now as we did with the photons, we can expand $F(\mathbf{k}, \mu, \eta)$ in multipole moments and express (4.208) as a hierarchy of BE for the multipoles. We expand $F_\nu(\mathbf{k}, \mu, \eta)$ in Legendre polynomials

$$F_\nu(\mathbf{k}, \mu, \eta) = \sum_{l=0}^{\infty} (-i)^l (2l+1) P_l(\mu) F_{\nu l}(\mathbf{k}, \eta) \tag{4.209}$$

and the multipoles can be constructed from $F_\nu(\mathbf{k}, \mu, \eta)$ by the inverse relation

$$F_{\nu l}(k, \eta) = \frac{1}{(-i)^l} \int_{-1}^{1} \frac{d\mu}{2} P_l(\mu) F_\nu(\mathbf{k}, \mu, \eta) \tag{4.210}$$

Taking the moments of Eq. (4.208) by multiplying with $P_{(\mu)}$ and integrating over μ (and applying the recursion relations (4.125) we obtain the BE equations for the multipoles

$$\dot{F}_{\nu 0} + k F_{\nu 1} = \dot{\Phi}$$

$$\dot{F}_{\nu 1} + \frac{2}{3} F_{\nu 2} - \frac{k}{3} F_{\nu 0} = k\Psi$$

$$\dot{F}_{\nu l} + k \left(\frac{l+1}{2l+1} F_{\nu l+1} - \frac{l}{2l+1} F_{\nu l-1} \right) = 0 \tag{4.211}$$

The various multipoles can be now related to the moments of the stress tensor by using the defining relation (4.207)

$$F_{\nu 0} = \frac{1}{\bar{\rho}_\nu} \int_{-1}^{1} \frac{d\mu}{2} \int \frac{p^2 dp}{(2\pi^2)} \, p \, f_\nu^1 = \frac{\delta\rho_\nu}{\bar{\rho}_\nu} = \delta_\nu \, ,$$

$$F_{\nu 1} = \frac{(\bar{P}_\nu + \bar{\rho}_\nu)}{\bar{\rho}_\nu} i\mathbf{k} \cdot v_\nu \equiv \frac{4}{3}\theta_\nu \, ,$$

$$F_{\nu 2} = \frac{-3}{2\bar{\rho}_\nu} \hat{k}_l \hat{k}_m \pi_\nu^{lm} = \frac{\sigma_\nu}{2} \, . \tag{4.212}$$

We see that the multipole moments $F_{\nu 0}$, F_{nu1} and F_{nu2} are the neutrino density perturbation, velocity gradient and anisotropic stress tensor respectively. We an now write the Eqs. (4.211) in terms of these physical quantities to obtain the set of BE for neutrino perturbations as follows

$$\dot{\delta}_\nu = -\frac{1}{3}\theta_\nu + \dot{\Phi} \, ,$$

$$\dot{\theta}_\nu = \frac{k^2}{4}\delta_\nu - k^2\sigma_\nu + k^2\Psi \, ,$$

$$\dot{\sigma}_\nu = \frac{4}{15}\theta_\nu - \frac{3}{10}kF_{\nu 3} \, . \tag{4.213}$$

Here the first equation is the continuity equation. We have taken the gradient in the second Euler equation to write it in terms of velocity gradient θ_ν. The last equation for the anisotropic stress σ_ν involves a higher moment $F_{\nu 3}$. We can truncate the BE series after $l = 2$ and set $F_{\nu l} = 0$ for all $l \geq 3$ as an initial condition. For a comprehensive treatment of massive neutrinos in cosmology see Ref. [42].

4.6 Boltzmann Equations for CDM

The equations for CDM are easy to write as they are like the baryons minus the scattering. The CDM Boltzmann equations are the continuity and Euler equations like (4.132),

$$\dot{\delta}_c = 3\dot{\Phi} - \theta_c \, ,$$

$$\dot{\theta}_c + \mathcal{H}\theta_c = k^2\Psi \tag{4.214}$$

Here $\theta_c = i\mathbf{k} \cdot \mathbf{v}_c$ is the gradient of the CDM velocity. Although the equations are similar the behaviour of baryons and CDM are totally different. In the matter era CDM perturbations grow as $\delta_c \propto a$ while the barons are tightly coupled to the photons and they oscillate as acoustic waves supported by photon pressure. After decoupling from photons at $z_{dec} \simeq 1090$ baryon density start growing. Large scale structure formation is due to CDM and the baryons fall into the potential wells of the CDM and form galaxies.

We can test new models of particle interactions using CMB and LSS observations. For example if there is an interaction between the two dark sectors neutrinos and CDM there is still an effect on the CDM as all of them are interlinked through gravity. Consider a neutrino-CDM interaction with total spin-independent cross section $\sigma_{\nu\chi}$. The Euler equations of ν and CDM get modified as follows,

$$\dot{\theta}_\nu = \frac{k^2}{4}\delta_\nu - k^2\sigma_\nu + k^2\Psi - \dot{\mu}(\theta_\nu - \theta_c)$$

$$\dot{\theta}_c + \mathcal{H}\theta_c = k^2\Psi + S^{-1}\dot{\mu}(\theta_\nu - \theta_c) \tag{4.215}$$

where

$$\dot{\mu} = an_\nu\sigma_{\nu\chi}, \quad S = \frac{3}{4}\frac{\rho_c}{\rho_\nu} \tag{4.216}$$

This mimics the baryon-photon couplings. The neutrino free-streaming will prevent the infall of dark matter and structure formation at the smaller length scales will be suppressed [43].

4.7   Open Problems

The ΛCDM model has shown a remarkable consistency with CMB temperature and polarisation anisotropy measurements. However there are some issues where there are some discrepancies of the CMB fits assuming ΛCDM with other observations.

Measurements of the expansion rate and distances over the interval $0 < z < 1$, give the value of $H_0 = 73.5 \pm 1.4 \, \text{km s}^{-1} \, \text{Mpc}^{-1}$ [44] while CMB +BAO observations with the ΛCDM assumption gives a value $H_0 = 67.4\pm0.5 \, \text{km s}^{-1} \, \text{Mpc}^{-1}$ [12] which is 4.4 σ discrepancy. New physics like evolving dark energy [45], extra light particles $\Delta N_{eff} = 0.4$ [46], neutrino self interaction [47] or neutrino dark matter interaction [48] can raise the CMB value closer to the local value and may address the Hubble tension [49].

References

1. A.A. Penzias, R.W. Wilson, A measurement of excess antenna temperature at 4080 Mc/s. Astrophys. J. **142**, 419–421 (1965)
2. R. Dicke, P. Peebles, P. Roll, D. Wilkinson, Cosmic black-body radiation. Astrophys. J. **142**, 414–419 (1965)
3. R.A. Alpher, R.C. Herman, Evolution of the universe. Nature **162**(4124), 774–775 (1948)
4. G. Gamow, The evolution of the universe. Nature **162**(4122), 680–682 (1948)

5. A. McKellar, in *Molecular Lines from the Lowest States of Diatomic Molecules Composed of Atoms Probably Present in Interstellar Space*. Publications of the Dominion Astrophysical Observatory, Vancouver, B.C., Canada, vol. 7(6), pp. 251–272 (1941)
6. D.M. Meyer, M. Jura, A precise measurement of the cosmic microwave background temperature from optical observations of interstellar CN. ApJ **297**, 119 (1985)
7. S.T. Staggs, N.T. Jarosik, S.S. Meyer, An absolute measurement of the cosmic microwave background radiation temperature at 10.7 GHz. ApJ **473**, L1 (1996)
8. J.C. Mather, E. Cheng, D. Cottingham, R. Eplee, D. Fixsen, et al., Measurement of the cosmic microwave background spectrum by the COBE FIRAS instrument. Astrophys. J. **420**, 439–444 (1994)
9. D. Fixsen, On the temperature of cosmic microwave background. Astrophys. J. **707**, 916–920 (2009)
10. G.F. Smoot, C. Bennett, A. Kogut, E. Wright, J. Aymon et al., Structure in the COBE differential microwave radiometer first-year maps. Astrophys. J. **396**, L1–L5 (1992)
11. G. Hinshaw, et al., Nine-year Wilkinson Microwave Anisotropy Probe (WMAP) observations: cosmological parameter results (2012). arXiv:1212.5226 [astro-ph.CO]
12. Y. Akrami, et al., [Planck Collaboration], Planck 2018 results. I. Overview and the cosmological legacy of Planck. arXiv:1807.06205 [astro-ph.CO]
13. J.M. Kovac, et al., Detection of polarization in the cosmic microwave background using DASI. Nature **420**(6917), 772–787 (2002)
14. D. Hanson et al., [SPTpol], Detection of B-mode polarization in the cosmic microwave background with data from the south pole telescope. Phys. Rev. Lett. **111**(14), 141301 (2013)
15. K.N. Abazajian et al., [CMB-S4], *CMB-S4 Science Book*, First edition. [arXiv:1610.02743 [astro-ph.CO]]
16. S. Dodelson, *Modern Cosmology* (Academic Press, San Diego, CA, 2003)
17. V. Mukhanov, *Physical Foundations of Cosmology* (Cambridge University Press, Cambridge, 2005)
18. W. Hu, Lecture notes on CMB theory: From nucleosynthesis to recombination. [arXiv:0802.3688 [astro-ph]]
19. R. Durrer, *The Cosmic Microwave Background* (Cambridge University Press, Cambridge, 2008)
20. D.H. Lyth, A.R. Liddle, *The Primordial Density Perturbation: Cosmology, Inflation and the Origin of Structure* (Cambridge University Press, Cambridge, 2009)
21. O. Piattella, *Lecture Notes in Cosmology* (Springer International Publishing, 2018)
22. W. Hu, N. Sugiyama, Anisotropies in the cosmic microwave background: An analytic approach. Astrophys. J. **444**, 489–506 (1995). https://doi.org/10.1086/175624. [arXiv:astro-ph/9407093 [astro-ph]]
23. R. Sachs, A. Wolfe, Perturbations of a cosmological model and angular variations of the microwave background. Astrophys. J. **147**, 73–90 (1967)
24. D. Blas, J. Lesgourgues, T. Tram, The Cosmic Linear Anisotropy Solving System (CLASS) II: Approximation schemes. JCAP **1107**, 034 (2011)
25. U. Seljak, M. Zaldarriaga, A Line of sight integration approach to cosmic microwave background anisotropies. Astrophys. J. **469**, 437–444 (1996)
26. C.P. Ma, E. Bertschinger, Cosmological perturbation theory in the synchronous and conformal Newtonian gauges. Astrophys. J. **455**, 7 (1995). https://doi.org/10.1086/176550. [arXiv:astro-ph/9506072]
27. A. Lewis, A. Challinor, A. Lasenby, Efficient computation of CMB anisotropies in closed FRW models. Astrophys. J. **538**, 473 (2000). [arXiv:astro-ph/9911177].
28. D.J. Eisenstein et al. [SDSS], Detection of the Baryon acoustic peak in the large-scale correlation function of SDSS luminous red galaxies. Astrophys. J. **633**, 560–574 (2005)
29. J. Silk, Cosmic black-body radiation and galaxy formation. Astrophys. J. **151**, 459 (1968)
30. J.N. Goldberg, A.J. Macfarlane, E.T. Newman, F. Rohrlich, E.C.G. Sudarshan, Spin-s spherical harmonics and ð. J. Math. Phys. **8**, 2155 (1966)

31. U. Seljak, M. Zaldarriaga, Signature of gravity waves in polarization of the microwave background. Phys. Rev. Lett. **78**, 2054–2057 (1997)
32. M. Kamionkowski, A. Kosowsky, A. Stebbins, A Probe of primordial gravity waves and vorticity. Phys. Rev. Lett. **78**, 2058–2061 (1997)
33. A. Kosowsky, Introduction to microwave background polarization. New Astron. Rev. **43**, 157 (1999)
34. A. Kosowsky, Cosmic microwave background polarization. Ann. Phys. **246**, 49–85 (1996)
35. A.G. Polnarev, Polarization and anisotropy induced in the microwave background by cosmological gravitational waves. Sov. Astron. **29**, 607–613 (1985)
36. R. Crittenden, J.R. Bond, R.L. Davis, G. Efstathiou, P.J. Steinhardt, The imprint of gravitational waves on the cosmic microwave background. Phys. Rev. Lett. **69**, 1856 (1993)
37. B.D. Fields, K.A. Olive, T.H. Yeh, C. Young, Big-bang nucleosynthesis after Planck. JCAP **03**, 010 (2020)
38. G. Mangano, G. Miele, S. Pastor, T. Pinto, O. Pisanti, P.D. Serpico, Nucl. Phys. B **729**, 221–234 (2005). https://doi.org/10.1016/j.nuclphysb.2005.09.041. [arXiv:hep-ph/0506164 [hep-ph]]
39. S. Gershtein, Y. Zeldovich, Rest mass of muonic neutrino and cosmology. JETP Lett. **4**, 120–122 (1966)
40. R. Cowsik, J. McClelland, An upper limit on the neutrino rest mass. Phys. Rev. Lett. **29**, 669–670 (1972)
41. A. Aguilar-Arevalo et al., [MiniBooNE], Significant excess of electronlike events in the MiniBooNE short-baseline neutrino experiment. Phys. Rev. Lett. **121**(22), 221801 (2018)
42. J. Lesgourgues, S. Pastor, Massive neutrinos and cosmology. Phys. Rept. **429**, 307–379 (2006)
43. R.J. Wilkinson, C. Boehm, J. Lesgourgues, Constraining dark matter-neutrino interactions using the CMB and large-scale structure. JCAP **05**, 011 (2014)
44. A.G. Reiss et al., Large magellanic cloud cepheid standards provide a 1% foundation for the determination of the hubble constant and stronger evidence for physics beyond ΛCDM. ApJ **876**, 1 (2019)
45. E. Di Valentino, A. Melchiorri, O. Mena, S. Vagnozzi, Nonminimal dark sector physics and cosmological tensions. Phys. Rev. D **101**(6), 063502 (2020)
46. F. D'Eramoa et al., Hot axions and the H_0 tension. JCAP **11**, 014 (2018)
47. C.D. Kreisch, F.Y. Cyr-Racine, O. Doré, The neutrino puzzle: anomalies, interactions, and cosmological tensions. Phys. Rev. D **101**(12), 123505 (2020)
48. S. Ghosh, R. Khatri, T.S. Roy, Dark neutrino interactions phase out the Hubble tension. [arXiv:1908.09843 [hep-ph]]
49. L. Knox, M. Millea, Hubble constant hunter's guide. Phys. Rev. D **101**(4), 043533 (2020)

Inflation

5

Abstract

We discuss models of inflation which is consistent with the stringent limits on the tensor to scalar ratio and spectral index placed by PLANCK. There are a few well motivated inflation models which make the cut namely, Natural Inflation, curvature coupled Higgs inflation, R^2 Starobinsky inflation and the No-Scale Supergravity models. We discuss the pros and cons of each of these models and discuss their other phenomenological consequence and future prospects in experiments.

5.1 Introduction

Inflation is the cosmological theory [1–6] where the universe goes through a period of exponential expansion. An e^N fold expansion of the scale factor with $N = 50-70$ can solve the puzzle of the observed homogeneity -why the Cosmic Microwave Background anisotropy $\Delta T/T \sim 10^{-5}$ over regions in the present universe which were never in causal contact in the standard hot big bang era of the universe. The super-horizon anisotropy in the CMB was first observed by COBE [7] and confirmed with increasing precision by WMAP [8] and Planck [9] experiments. Inflation also gives a natural explanation of the observed flatness $\Omega_k = 0.0007 \pm 0.0037 (95\%\text{CL})$ [10] even though in the standard hot-big bang cosmology (Appendix A.1), Ω_k grows with the scale factor as $\Omega_k \propto a^2$ during the radiation era and $\Omega_k \propto a$ during the matter domination.

Inflation also provides an explanation of the observed large scale structure in the universe as it provides the primordial scale invariant perturbations [11] which under gravitational collapse to form galaxies and clusters of galaxies [12–15].

© Springer Nature Switzerland AG 2020
S. Mohanty, *Astroparticle Physics and Cosmology*,
Lecture Notes in Physics 975, https://doi.org/10.1007/978-3-030-56201-4_5

5.2   Inflation

COBE observations proved that the cosmic microwave background (CMB) radiation
(with mean temperature $T_0 = 2.725\,\mathrm{K}$) has an anisotropy of the order $\Delta T/T_0 \sim$
10^{-5} over the entire horizon whose physical radius is $R_h \simeq H_0^{-1}$. The CMB photons
decoupled when the electron and protons combine to form neutral hydrogen at the
time when the temperature was $3000\,\mathrm{K}$ and the age of the universe 4×10^5 years. At
the time of decoupling the there was no causal connection in the region which forms
our horizon today. In the theory of inflation, the size of perturbations grow faster
than the size of the horizon in the inflationary stage of the universe. After inflation
ends there is a phase of reheating and then the standard hot big-bang cosmology
of radiation and matter dominated era follows. In the radiation and matter era the
horizon grows faster than the length scale of perturbations and they *perturbations
re-enter the horizon*. This explains the presense of super-horizon perturbations first
observed by COBE. The amplitude of the (adiabatic) perturbations remain the same
and the large scale perturbations we observe in the CMB are primordial—generated
during inflation and unaffected by any sub-horizon causal physics.

To obtain perturbations of super horizon length scales, the wavelength of
perturbations $\lambda = \lambda_0 a$ during an early era must become larger than the Horizon
$R_h \simeq H^{-1} = a/\dot{a}$. This implies that in some early era we require

$$\frac{d}{dt}\left(\frac{\lambda_0 a}{H^{-1}}\right) > 0 \Rightarrow \ddot{a} > 0 \tag{5.1}$$

i.e there must have been a period of positive acceleration. The Friedman equation (A.12) $\ddot{a} = -(4\pi G/3)(\rho + 3p)$ then implies that in some period in the past the
universe had a dominant component with negative pressure $p < -\frac{1}{3}\rho$. If $p/\rho \simeq -1$
then the universe goes through a period of exponential expansion $a \sim e^{Ht}$. One way
to get a negative pressure is by having a scalar field potential. The energy momentum
tensor of the scalar field is

$$T_{\mu\nu}(\phi) = \partial_\mu\phi\partial_\nu\phi - g_{\mu\nu}\left[\frac{1}{2}\partial^\rho\phi\partial_\rho\phi + V(\phi)\right]. \tag{5.2}$$

For the homogeneous background field, the energy momentum tensor takes the form
of a perfect fluid with energy density and pressure for scalar fields given by

$$\rho_\phi = \frac{\dot{\phi}^2}{2} + V(\phi), \tag{5.3}$$

$$p_\phi = \frac{\dot{\phi}^2}{2} - V(\phi). \tag{5.4}$$

The resulting equation of state is

$$\omega_\phi \equiv \frac{p_\phi}{\rho_\phi} = \frac{\frac{\dot{\phi}^2}{2} - V(\phi)}{\frac{\dot{\phi}^2}{2} + V(\phi)}. \tag{5.5}$$

If the potential energy of the field dominates over its kinetic energy i.e. $\dot{\phi}^2 \ll V(\phi)$, then the above simple relation (5.5) implies that the scalar field can act as a negative pressure source i.e. $\omega_\phi < 0$ and can provide accelerated expansion i.e. $\omega_\phi < -\frac{1}{3}$. The Friedmann equation (A.9) and the equation of motion of the scalar field are respectively

$$H^2 = \frac{8\pi G}{3}\left(\frac{\dot{\phi}^2}{2} + V(\phi)\right), \tag{5.6}$$

$$\ddot{\phi} + 3H\dot{\phi} + V'(\phi) = 0. \tag{5.7}$$

These equations determine the dynamics of the scale-factor and scalar field in a FRW universe.

Slow-roll inflation occurs when $\dot{\phi}^2 \ll V(\phi)$. A second order differentiation of the condition $\dot{\phi}^2 \ll V(\phi)$ implies $\ddot{\phi} \ll V'(\phi)$. In the slow-roll approximation equation (5.6) becomes

$$3H^2 \simeq \frac{8\pi}{M_{pl}^2}V(\phi). \tag{5.8}$$

and (5.7) becomes

$$3H\dot{\phi} \simeq -V'(\phi). \tag{5.9}$$

By differentiating Eq. (5.8) w.r.t. time and combining the result with Eq. (5.9) we obtain

$$\dot{H} \simeq -\frac{4\pi}{M_{pl}^2}\dot{\phi}^2 \tag{5.10}$$

The slow-roll conditions $\dot{\phi}^2 \ll V(\phi)$ and $\ddot{\phi} \ll V'(\phi)$ can be defined in terms of dimensionless parameters as

$$\epsilon = -\frac{\dot{H}}{H^2} \simeq \frac{4\pi}{M_{pl}^2}\left(\frac{\dot{\phi}}{H}\right)^2 \simeq \frac{M_{pl}^2}{16\pi}\left(\frac{V'(\phi)}{V(\phi)}\right)^2, \tag{5.11}$$

$$\eta = -\frac{\ddot{\phi}}{H\dot{\phi}} \simeq \frac{M_{pl}^2}{8\pi}\left[\frac{V''(\phi)}{V(\phi)} - \frac{1}{2}\left(\frac{V'(\phi)}{V(\phi)}\right)^2\right]. \tag{5.12}$$

After the slow roll phase $\epsilon \ll 1$, the period of Inflation ends when $\dot{\phi}^2 \approx V(\phi)$ and $\epsilon(\phi_e) \simeq 1$.

It is convenient to define the slow roll parameters in terms of the potential

$$\epsilon_V \equiv \frac{M_{pl}^2}{16\pi} \left(\frac{V'}{V}\right)^2 = \epsilon,$$

$$\eta_V \equiv \frac{M_{pl}^2}{8\pi} \left(\frac{V''}{V}\right) = \eta + \epsilon. \tag{5.13}$$

During the slow roll phase $V(\phi)$ is nearly constant and solving (5.8), we see that during this period scale factor evolves exponentially $a(t) \sim e^{Ht}$. The number of e-foldings before the inflation ends $\epsilon(\phi_e) = 1$, is given by

$$N(\phi) = \int_{t_i}^{t_e} H \, dt = \int_{\phi_i}^{\phi_e} \frac{H}{\dot{\phi}} d\phi$$

$$\simeq \frac{8\pi}{M_{pl}^2} \int_{\phi_e}^{\phi_i} \frac{V}{V'} d\phi = \frac{2\sqrt{\pi}}{M_{pl}} \int_{\phi_e}^{\phi_i} \frac{d\phi}{\sqrt{\epsilon}} \tag{5.14}$$

where we used the slow-roll equations (5.9) and (5.8). To solve the horizon and flatness problems we required the number of e-foldings during inflation to exceed 60

$$N_{tot} \equiv \ln \frac{a_e}{a_i} \gtrsim 40 - 60. \tag{5.15}$$

the value of N_{tot} depends on the inflation potential. The slow-roll phase, should last a minimum $N = 40 - 60$ e-folds to solve the horizon and flatness problems (there is non phenomenological limit on the maximum duration of inflation). During the slow-roll phase the quantum fluctuations in density and pressure of the scalar field called the inflaton are generated which are then imprinted on the CMB. In the CMB anisotropy measurements we probe the last 8 e-foldings before the end of inflation.

5.3   Perturbations of the FRW Universe

We studied perturbations in the fluid description of $T_{\mu\nu}$ for studying large scale structure in Chap. 3. and we studied the kinetic theory with phase space distributions in Chap. 4. In this section we will study the perturbations of a scalar field and metric and apply it for studying the gauge invariant curvature perturbations generated by the inflaton as it slow-rolls during inflation.

5.3.1 Metric Perturbations

Linear order perturbations in the metric and field around the homogeneous background solutions of the field $\phi(t)$ and the metric $g_{\mu\nu}(t)$ can be written as

$$\delta\phi(t, \mathbf{x}) = \phi(t, \mathbf{x}) - \phi(t), \tag{5.16}$$

$$\delta g_{\mu\nu}(t, \mathbf{x}) = g_{\mu\nu}(t, \mathbf{x}) - g_{\mu\nu}(t). \tag{5.17}$$

The most general linearly perturbed spatially flat FRW metric can be written as

$$ds^2 = -(1 + 2\Psi)dt^2 + 2a(t)B_i dt dx^i + a(t)^2 \left[(1 - 2\Phi)\delta_{ij} + 2E_{ij}\right] dx^i dx^j, \tag{5.18}$$

where Ψ, Φ are the scalar perturbations, B_i are the vector perturbations and E_{ij} are the tensor perturbations. The scalar, vector and tensor perturbations are decoupled during inflation and thereafter evolve independently (*SVT decomposition theorem* [13]).

The vector perturbations are not sourced by inflation and they quickly decay with the expansion of the universe.

One can make use of the general coordinate transformation of GR to further restrict the dynamical d.o.f in the metric perturbations. In Sect. 3.2.2 we discussed the gauge dependence of the metric perturbations and stress tensor perturbations where the stress tensor was in the hydrodynamic form. In this section we discuss gauge transformations of the stress tensor of the scalar field and construct gauge invariant combinations of the metric and scalar field which can be tested in experiments.

5.3.2 Gauge-Invariant Perturbations

An important gauge invariant observable (3.46) is the co-moving curvature perturbation introduced in [16],

$$\mathcal{R} \equiv \Phi - \frac{H}{\rho + p}\delta q, \tag{5.19}$$

The scalar momentum perturbation δq is the $0i$−component of the perturbed energy momentum tensor $\delta T_i^0 = \partial_i \delta q$. During inflation $\delta T_i^0 = -\dot{\phi}\partial_i \delta\phi$, comparing these two relations we see that $\delta q = -\dot{\phi}\delta\phi$. The background density and pressure of the scalar field obey the relation $\rho + p = \dot{\phi}^2$. Therefore the co-moving curvature perturbations (5.19) during inflation becomes

$$\mathcal{R} \simeq \Phi + \frac{H}{\dot{\phi}}\delta\phi. \tag{5.20}$$

Geometrical interpretation of $\mathcal{R}$ is that it measures the spatial curvature of the comoving hypersurface where $\delta\phi = 0$, i.e.

$$\mathcal{R} = \Phi\Big|_{\delta\phi=0}. \tag{5.21}$$

Another important gauge-invariant quantity is curvature perturbations on constant energy density hypersurfaces (3.47) defined as

$$\zeta \equiv \Phi + \frac{H}{\dot{\rho}}\delta\rho, \tag{5.22}$$

Similar to $\mathcal{R}$, the quantity ζ can be constructed by considering the slicing of the spacetime into constant energy density hyperserfaces,

$$\delta\rho \rightarrow \delta\rho - \dot{\rho}\alpha = 0 \quad \Longrightarrow \quad \alpha = \frac{\delta\rho}{\dot{\rho}}, \tag{5.23}$$

substituting this α into the metric transformation relation $\Psi \rightarrow \Psi + H\alpha$ gives the relation (5.22) for ζ. Since during slow-roll (from Eq. (5.3)) one has

$$\delta\rho = \dot{\phi}\dot{\delta\phi} + V'\delta\phi \simeq V'\delta\phi$$

and

$$\dot{\rho} = \dot{\phi}\ddot{\phi} + V'\dot{\phi} \simeq V'\dot{\phi}.$$

These equations imply

$$\frac{\delta\rho}{\dot{\rho}} \simeq \frac{\delta\phi}{\dot{\phi}}.$$

As a consequence, ζ becomes

$$\zeta \simeq \Phi + \frac{H}{\dot{\phi}}\delta\phi. \tag{5.24}$$

The geometrical interpretation of ζ is that it is the spatial curvature of the uniform density hypersurface, i.e.

$$\zeta = \Phi\Big|_{\delta\rho=0}. \tag{5.25}$$

   In a single field slow roll inflation models the perturbations produced are adiabatic. Non-adiabatic perturbations can arise in multi-field inflation models [17] and are defined as the iso-curvature entropy perturbation

$$\Gamma \equiv \frac{\delta p}{\dot{p}} - \frac{\delta \rho}{\dot{\rho}}, \tag{5.26}$$

which is gauge invariant. The non-adiabatic part of the pressure perturbations are defined as

$$\delta p_{nad} \equiv \dot{p}\Gamma \equiv \delta p - \frac{\dot{p}}{\dot{\rho}}\delta \rho, \tag{5.27}$$

In inflationary models with more than one field, the some perturbations can be non-adiabatic [17] and can have observable signatures in the CMB anisotropy. For example, the relative density perturbations (isocurvature or entropy perturbations) between photon and CDM can be defined as

$$S_{m\gamma} \equiv \frac{\delta \rho_{cdm}}{\rho_{cdm}} - \frac{3}{4}\frac{\delta \rho_\gamma}{\rho_\gamma}. \tag{5.28}$$

can have a observable signature in the CMB spectrum. CMB observations suggest that if the iso-curvature perturbations are present, their amplitude is small compared to amplitude of the adiabatic (curvature) perturbations [9].

We have shown in (3.105) that the time evolution of the gauge invariant curvature perturbations is given by

$$\dot{\mathcal{R}} = -\frac{H}{\rho + p}\delta p_{nad} + \left(\frac{k}{aH}\right)^2 \left[\frac{H^2}{3(\rho + p)}\delta q\right], \tag{5.29}$$

therefore if there are no non-adiabatic matter perturbations $\delta p_{nad} = 0$ or no isocurvature perturbations $\Gamma = 0$, the curvature perturbations $\mathcal{R}$ are conserved on superhorizon scales $k \ll aH$. This ensures that when adiabatic perturbations exit the horizon during inflation and re-enter during the radiation or matter era, their amplitude remains the same.

5.3.3   Curvature Perturbation During Inflation

For a metric with small perturbations, the Einstein tensor $G_{\nu\mu} = R_{\mu\nu} - (1/2)g_{\mu\nu}R$ can be written as

$$G_{\nu\mu} = G^{(0)}_{\nu\mu} + \delta G_{\nu\mu} + \dots,$$

where $\delta G_{\nu\mu}$ represents the terms with linear metric perturbations $\delta g_{\mu\nu}$. The stress energy tensor T^ν_μ can be expanded in terms density and pressure perturbations and we can write the first oder perturbations in Einstein field equations as,

$$\delta G^{\nu}_{\mu} = 8\pi G \ \delta T^{\nu}_{\mu}. \tag{5.30}$$

The gauge freedom allows to choose the two functions α and β which provides two conditions on the scalar functions Φ, Ψ, B, E and therefore allows us to remove any two of them. One of the useful gauge choices which makes calculations convenient is the *conformal Newtonian gauge* or *longitudinal gauge* which is defined by the conditions

$$B = 0, \quad E = 0.$$

In this gauge the perturbed FRW line element is

$$ds^2 = -(1 + 2\Psi)dt^2 + a(t)^2(1 - 2\Phi)\delta_{ij}dx^i dx^j. \tag{5.31}$$

and in this metric the components of the perturbed Einstein tensor are,

$$\delta G^0_0 = -2\nabla^2\Phi + 6H^2\Psi + 6H\dot{\Phi}, \tag{5.32}$$

$$\delta G^0_i = -2\partial_i(H\Psi + \dot{\Phi}), \tag{5.33}$$

$$\delta G^i_j = \partial^i\partial_j(\Phi - \Psi) + [\nabla^2(\Psi - \Phi) + 2\ddot{\Phi} + (4\dot{H} + 6H^2)\Psi + H(2\dot{\Psi} + 2\dot{\Phi})]\delta^i_j. \tag{5.34}$$

The stress-energy-momentum tensor for the scalar field ϕ as defined in Eq. (5.2), the components of the perturbed $T_{\mu\nu}$ are given by

$$\delta T^0_0 = -\delta\rho = \dot{\phi}^2\Psi - \dot{\phi}\delta\dot{\phi} - V'\delta\phi, \tag{5.35}$$

$$\delta T^0_i = \delta q = -\dot{\phi}\partial_i\delta\phi, \tag{5.36}$$

$$\delta T^i_j = \delta p = [-\dot{\phi}^2\Psi + \dot{\phi}\delta\dot{\phi} - V'\delta\phi]\delta^i_j. \tag{5.37}$$

where we have used the relation

$$\delta T^{\nu}_{\mu} = \delta(g^{\nu\tau}T_{\mu\tau}) = \delta g^{\nu\tau}T_{\mu\tau} + g^{\nu\tau}\delta T_{\mu\tau}.$$

To compute the curvature perturbation $\mathcal{R}$, one first considers the ij-component of the perturbed Einstein field equation (5.30). First taking the off-diagonal components, $i \neq j$, of the Eqs. (5.34) and (5.37), we have

$$\partial^i\partial_j(\Phi - \Psi) = 0 \quad \Longrightarrow \quad \Phi = \Psi, \tag{5.38}$$

Now consider the diagonal components, $i = j$, of Eqs. (5.34) and (5.37),

$$\ddot{\Phi} + 4H\dot{\Phi} + (2\dot{H} + 3H^2)\Phi = -\dot{\phi}^2\Psi + \dot{\phi}\dot{\delta\phi} - V'\delta\phi. \tag{5.39}$$

Since ϕ is background quantity which is only time dependent, Eqs. (5.33) and (5.36) for $0i$-components give

$$\dot{\Phi} + H\Phi = 4\pi G \dot{\phi}\delta\phi = \epsilon H^2 \frac{\delta\phi}{\dot{\phi}}, \tag{5.40}$$

where $\epsilon = 4\pi G \frac{\dot{\phi}^2}{H^2}$ (slow-roll parameter). Similarly the Eqs. (5.32) and (5.35) for 00-component gives

$$\nabla^2\Phi - 3H\dot{\Phi} - 3H^2\Phi = 4\pi G(\dot{\phi}\dot{\delta\phi} - \dot{\phi}^2\Phi + V'\delta\phi). \tag{5.41}$$

To describe the spatial distribution of the perturbations it is convenient to work in terms of the Fourier decomposition of the metric and the field perturbations, which relates perturbation corresponding to a given comoving wavenumber k in Fourier space with with corresponding comoving wavelength $\lambda = \frac{2\pi}{k}$. The physical wavelength is $\lambda_{phy} = a(t)\lambda$ and the physical wavenumber of the the perturbations is $k_{phy} = k/a(t)$.

We can express the perturbations Ψ as a sum of Fourier modes as comoving wavenumber $\mathbf{k}$:

$$\Phi(t, \mathbf{x}) = \int \frac{d^3\mathbf{k}}{(2\pi)^{3/2}} \Phi_{\mathbf{k}}(t) e^{i\mathbf{k}\cdot\mathbf{x}}. \tag{5.42}$$

With a similar expression for the Fourier transformation of $\delta\phi$.

We now add the Fourier transforms of Eqs. (5.39) and (5.41) to arrive at the equation of motion of gravitational potential Φ_k as

$$\ddot{\Phi}_k + \left(H - 2\frac{\ddot{\phi}}{\dot{\phi}}\right)\dot{\Phi}_k + 2\left(\dot{H} - H\frac{\ddot{\phi}}{\dot{\phi}}\right)\Phi_k + \frac{k^2}{a}\Phi_k = 0, \tag{5.43}$$

where we have used the background equation for scalar field $V' \simeq -3H\dot{\phi}$ and the relation $\dot{H} \simeq -4\pi G \dot{\phi}^2$. Using the slow-roll parameter relation

$$\delta = \eta - \epsilon = \frac{-\ddot{\phi}}{H\dot{\phi}},$$

the Eq. (5.43) can also be written as

$$\ddot{\Phi}_k + H(1 - 2\epsilon + 2\eta)\dot{\Phi}_k + 2H^2(\eta - 2\epsilon)\Phi_k + \frac{k^2}{a}\Phi_k = 0, \tag{5.44}$$

Since the slow-roll parameters satisfy $\epsilon \ll 1$ and $\eta \ll 1$, we see from Eq. (5.44) that on superhorizon scales $k \ll (aH)$,

$$\dot{\Phi}_k \simeq 2(2\epsilon - \eta)H\Psi_k \quad \Rightarrow \quad \dot{\Psi}_k \ll H\Psi_k \tag{5.45}$$

which implies that on super-horizon scales the time variations of the perturbations Ψ_k can be neglected compared to $H\Psi_k$. This condition also holds true for field perturbations $\delta\dot{\phi}_k \ll H\delta\phi_k$. Therefore on super-horizon scales, from Eq. (5.40), we can relate the gravitational potential and field perturbations as

$$\Phi_k \simeq \epsilon H \frac{\delta\phi}{\dot{\phi}}, \tag{5.46}$$

This can be used to compute the comoving curvature perturbation $\mathcal{R}_k$ on superhorizon scale (5.20) as

$$\mathcal{R}_k \simeq \Phi_k + \frac{H}{\dot{\phi}}\delta\phi_k \tag{5.47}$$

$$\simeq (1+\epsilon)\frac{H}{\dot{\phi}}\delta\phi_k$$

$$\approx \frac{H}{\dot{\phi}}\delta\phi_k . \tag{5.48}$$

The comoving curvature perturbations are related to the scalar field perturbations whose origin is quantum fluctuations as we will discuss in the following sections.

Power Spectrum of Curvature Perturbation

The power spectrum of comoving curvature perturbation $\mathcal{R}$ is defined as

$$\mathcal{P}_\mathcal{R}(k) = \frac{k^3}{2\pi^2}\langle|\mathcal{R}_k|^2\rangle . \tag{5.49}$$

where $\langle|\mathcal{R}_k|^2\rangle$ denotes the statistical average of the quantity $|\mathcal{R}_k|^2$.

Using (5.47), the power spectrum of comoving curvature perturbation on superhorizon scale $k \ll (aH)$ becomes

$$\mathcal{P}_\mathcal{R}(k) \simeq \frac{k^3}{2\pi^2}\frac{H^2}{\dot{\phi}^2}\langle|\delta\phi_k|^2\rangle \tag{5.50}$$

$$\simeq \frac{k^3}{4\pi^2\epsilon M_{pl}^2}\langle|\delta\phi_k|^2\rangle .$$

where $\langle |\delta\phi_k|^2 \rangle$ stands for the expectation value of the quantum fluctuations of the inflaton $|\delta\phi_k|^2$ in the vacuum quantum state $|0\rangle$.

5.3.4 Quantum Fluctuations of the Inflaton

We now calculate the the scalar field perturbation mode amplitudes $\delta\phi_k$. Consider the perturbation of (5.7) for scalar field ϕ,

$$\delta\ddot{\phi}_k + 3H\delta\dot{\phi}_k + \frac{k^2}{a^2}\delta\phi_k + V''\delta\phi_k = -2V'\Psi_k + 4\dot{\phi}\dot{\Psi}_k \qquad (5.51)$$

where we have used the background equation (5.9). Since on superhorizon scales $|2V'\Psi_k| \gg |4\dot{\phi}\dot{\Psi}_k|$ (which follows from the condition $\dot{\Psi}_k \ll H\Psi_k$ upon using the relation $V' \simeq -3H\dot{\phi}$). From Eqs. (5.46) and (5.9), the e.o.m or the scalar field is

$$\delta\ddot{\phi}_k + 3H\delta\dot{\phi}_k + (V'' + 6\epsilon H^2)\delta\phi_k + \frac{k^2}{a^2}\delta\phi_k = 0. \qquad (5.52)$$

Substituting $\delta\phi_k = \delta\sigma_k/a$ and introducing the conformal time[1] $d\tau = dt/a$, we get

$$\delta\sigma_k'' - \frac{1}{\tau^2}\left(\nu^2 - \frac{1}{4} \right)\delta\sigma_k + k^2\delta\sigma_k = 0, \qquad (5.53)$$

where prime denotes the derivatives w.r.t. conformal time τ and

$$\nu^2 = \frac{9}{4} + 9\epsilon_V - 3\eta_V. \qquad (5.54)$$

and we have used the relation

$$\frac{a''}{a} = \frac{1}{\tau^2}\left(\nu^2 - \frac{1}{4} \right) \simeq \frac{1}{\tau^2}(2 + 3\epsilon).$$

We see that the solution of scalar field perturbation equation (5.53) can be written as Hankel functions

$$\delta\sigma_k = \sqrt{-\tau}[c_1(k)H_\nu^{(1)}(-k\tau) + c_2(k)H_\nu^{(2)}(-k\tau)], \qquad (5.55)$$

where $H_\nu^{(1)}$ and $H_\nu^{(2)}$ are the Hankel's functions of the first and second kind, respectively. We follow the *Bunch–Davies initial condition* [18, 19] that in the ultraviolet regime, $k \gg aH$ ($-k\tau \gg 1$) the solutions goes to the positive energy

---

[1] In quasi de-Sitter expansion the Hubble rate is not a constant and $\dot{H} = -\epsilon H^2$, the scale factor goes as $a(\tau) = -\frac{1}{H\tau}\frac{1}{1-\epsilon}$.

plane wave solution $e^{-ik\tau}/\sqrt{2k}$. In the limit $-k\tau \gg 1$ Hankel's functions are given by

$$H_\nu^{(1)}(-k\tau \gg 1) \sim \sqrt{\frac{2}{-k\tau\pi}}e^{i\left(-k\tau-\frac{\pi}{2}\nu-\frac{\pi}{4}\right)}, \tag{5.56}$$

$$H_\nu^{(2)}(-k\tau \gg 1) \sim \sqrt{\frac{2}{-k\tau\pi}}e^{i\left(-k\tau-\frac{\pi}{2}\nu-\frac{\pi}{4}\right)}. \tag{5.57}$$

Imposing the Bunch–Davies initial condition we have

$$c_1(k) = \frac{\sqrt{\pi}}{2}e^{i\left(\nu+\frac{1}{2}\right)\frac{\pi}{2}}, \qquad c_2(k) = 0,$$

from Eq. (5.55) we get the exact solution for $\delta\sigma_k$

$$\delta\sigma_k = \frac{\sqrt{\pi}}{2}e^{i\left(\nu+\frac{1}{2}\right)\frac{\pi}{2}}\sqrt{-\tau}H_\nu^{(1)}(-k\tau). \tag{5.58}$$

As we are interested in the modes which have become superhorizon $k \ll aH$ ($-k\tau \ll 1$) during inflation, knowing that in the limit $-k\tau \ll 1$ Hankel's function have solution

$$H_\nu^{(1)}(-k\tau \ll 1) \sim \sqrt{\frac{2}{\pi}}\frac{\Gamma(\nu)}{\Gamma(3/2)}2^{\nu-\frac{3}{2}}e^{-i\frac{\pi}{2}}(-k\tau)^{-\nu}, \tag{5.59}$$

the solution (5.58) on superhorizon scales becomes

$$\delta\sigma_k \simeq \frac{\Gamma(\nu)}{\Gamma(3/2)}2^{\nu-\frac{3}{2}}e^{i\left(\nu+\frac{1}{2}\right)\frac{\pi}{2}}\frac{1}{\sqrt{2k}}(-k\tau)^{\frac{1}{2}-\nu}. \tag{5.60}$$

Since $\epsilon \ll 1$ and $\eta \ll 1$, we can set $\nu \sim \frac{3}{2}$ in the factors but will not do the same in the exponent because exponent term $(-k\tau)^{\frac{1}{2}-\nu}$ gives the small scale dependence of the power spectrum of perturbations. Going back to original variable $\delta\phi_k$, we find the the fluctuations on superhorizon scales in cosmic time

$$|\delta\phi_k(t)| \simeq \frac{H}{\sqrt{2k^3}}\left(\frac{k}{aH}\right)^{\frac{3}{2}-\nu}. \tag{5.61}$$

5.3.5 Scalar Power Spectrum

The power spectrum of scalar fluctuations (5.50) can therefore be written as,

$$\mathcal{P}_\mathcal{R}(k) \simeq \frac{1}{8\pi^2\epsilon} \frac{H^2}{M_p^2} \left(\frac{k}{aH}\right)^{n_s-1} \tag{5.62}$$

$$\equiv \Delta_\mathcal{R}^2 \left(\frac{k}{aH}\right)^{n_s-1},$$

where we have defined the spectral index n_s of the comoving curvature perturbations, which determines the tilt of the power spectrum as

$$n_s - 1 \equiv \frac{d\ln\mathcal{P}_\mathcal{R}}{d\ln k} \tag{5.63}$$

$$= 3 - 2\nu = -6\epsilon_V + 2\eta_V.$$

where we have used ν from Eq. (5.54). Since the slow-roll parameters ϵ_V and η_V are much smaller than unity, therefore $n_s - 1 \simeq 0$ which implies that in slow roll inflation curvature perturbations are generated with with an almost scale invariant spectrum. For comparison with the observations, the power spectrum (5.62) can be given as

$$\mathcal{P}_\mathcal{R}(k) = \Delta_\mathcal{R}^2(k_0) \left(\frac{k}{k_0}\right)^{n_s-1}, \tag{5.64}$$

where $k_0 = a_0 H_0$ is the pivot scale. The pivot scale corresponds to a wavelength $\lambda_0 = 2\pi/k_0^{-1}$ at which the CMB anisotropy measurement has the maximum sensitivity. $\Delta_\mathcal{R}^2(k_0)$ is the amplitude of the power spectrum at the pivot scale k_0.

5.3.6 Tensor Power Spectrum

Along with density fluctuations (or scalar perturbations), inflation also predicts the existence of gravitational waves which are identified with the tensor perturbations in the metric [20]. The equation of motion for h_{ij} can be obtained from second order expansion of the Einstein–Hilbert action [12, 15]

$$S^{(2)} = \frac{M_p^2}{2} \int dx^4 \sqrt{-g} \frac{1}{2} \partial_\rho h_{ij} \partial^\rho h_{ij} \tag{5.65}$$

$$= \frac{M_p^2}{4} \int d\eta dx^3 \frac{a^2}{2} [(h'_{ij})^2 - (\partial_l h_{ij})^2].$$

This is the same action as for the conformally-coupled massless scalar field in FRW universe. We define the Fourier expansion

$$h_{ij} = \int \frac{d^3k}{(2\pi)^3} \sum_{s=+,\times} h_{\mathbf{k}}^s(\tau) e_{ij}^s(k) e^{i\mathbf{k}.\mathbf{x}}, \qquad (5.66)$$

where e_{ij}^s are the polarization tensors which satisfy the properties

$$e_{ij}(k) = e_{ji}(k), \quad e_{ii}(k) = 0, \quad k^i e_{ij}(k) = 0, \quad e_{ij}^s(k) e_{ij}^{s'}(k) = 2\delta_{ss'}. \qquad (5.67)$$

Using (5.66) and (5.67), the action (5.65) leads to the e.o.m. for the quantity $h_{\mathbf{k}}$

$$h_{\mathbf{k}}^{s''} + 2\frac{a'}{a} h_{\mathbf{k}}^{s'} + k^2 h_{\mathbf{k}}^s = 0. \qquad (5.68)$$

Defining the canonically normalized field $v_{\mathbf{k}}^s \equiv \frac{1}{2} a h_{\mathbf{k}}^s M_p$, the e.o.m. (5.68) becomes

$$v_{\mathbf{k}}^{s''} + \left(k^2 - \frac{a''}{a} \right) v_{\mathbf{k}}^s = 0, \qquad (5.69)$$

where

$$\frac{a''}{a} = \frac{2}{\tau^2}(2 + 3\epsilon)$$

during quasi de-Sitter epoch when $\dot{H} = -\epsilon H$. The solution of (5.68) can be written as (as in the scalar case (5.61)) as,

$$|v_{\mathbf{k}}^s| = \frac{1}{M_p} \frac{aH}{\sqrt{2k^3}} \left(\frac{k}{aH} \right)^{\frac{3}{2} - \nu_T}. \qquad (5.70)$$

Here the quantity ν_T, given by

$$\nu_T \simeq \frac{3}{2} + \epsilon,$$

has been obtained using the relation

$$\frac{a''}{a} = \frac{1}{\tau^2}\left(\nu_T^2 - \frac{1}{4} \right) \simeq \frac{1}{\tau^2}(2 + 3\epsilon).$$

The power spectrum of tensor perturbations is

$$\mathcal{P}_T \equiv \frac{k^3}{2\pi^2} \sum_{s=+,\times} |h_{\mathbf{k}}^s|^2 = 2 \times \frac{k^3}{2\pi^3} \frac{4|v_{\mathbf{k}}^s|^2}{a^2}, \tag{5.71}$$

where the factor of 2 is due to the sum over the two polarization states of the gravitational wave. Substituting the solution (5.70), we get the amplitude of the tensor power spectrum

$$\mathcal{P}_T = \frac{2}{\pi^2} \frac{H^2}{M_p^2} \left(\frac{k}{aH} \right)^{n_T} \tag{5.72}$$

$$\equiv \Delta_T^2 \left(\frac{k}{aH} \right)^{n_T} .$$

We define the *tensor spectral index* n_T as

$$n_T = \frac{d \ln \mathcal{P}_T}{d \ln k} \tag{5.73}$$

$$= 3 - 2v_T = -2\epsilon .$$

As the action for tensor perturbations (5.65) is that of a massless scalar field, there will be no appearance of slow-roll parameter η in v_T.

Tensor-to-Scalar Ratio and Energy Scale of Inflation The *tensor-to-scalar ratio* r is defined as the ratio of the two amplitudes

$$r \equiv \frac{\Delta_T^2}{\Delta_{\mathcal{R}}^2} = 16\epsilon , \tag{5.74}$$

which determines the relative contribution of the tensor modes to mean squared low multipole CMB anisotropy. where in the last equality in (5.74), we have used the amplitude relations (5.62) and (5.72) for scalar and tensor perturbations. The scalar amplitude is fixed from the observations $\Delta_{\mathcal{R}}^2 \simeq 1.95 \times 10^{-9}$ and, from (5.72), amplitude of the tensor perturbations $\Delta_T^2 \propto H^2 \approx V(\phi)$, therefore the value of tensor-to-scalar ratio is a direct measure of the energy scale of inflation potential

$$V(\phi)^{1/4} \sim \left(\frac{r}{0.01} \right)^{1/4} 10^{16} \, \text{GeV}. \tag{5.75}$$

Therefore measurement of the tensor amplitudes will give us a direct handle on the energy scale of inflation potential.

5.3.7   Inflationary Observables in CMB

The physical quantities/observables are: the amplitude of the power spectrum of curvature perturbations $\Delta_{\mathcal{R}}^2$, spectral index n_s, running of spectral index α_s and tensor-to-scalar ratio r. These are obtained by measuring the temperature and E-mode polarization anisotropy spectrum. The tensor to scalar ratio r is determined from the B-mode polarization (after removing the B-mode polarization from dust in the foreground and the B-model due to lensing of E-mode polarization). The inflationary observables as given by PLANCK-2018 [9], from the the combination of *PLANCK TT, TE, EE, low E +lensing* are, $\ln(10^{10}\Delta_{\mathcal{R}}^2) = 3.044 \pm 0.014$, 68% CL and $n_s = 0.9670 \pm 0.0037$, 68% CL

The value of amplitude and spectral index are given at $68\%CL$ at the pivot scale $k = 0.05\,\mathrm{Mpc}^{-1}$. Whereas the upper bound on tensor-to-scalar ratio is determined at $95\%CL$ at $k = 0.002\,\mathrm{Mpc}^{-1}$ [9].

The bound on the tensor to scalar ratio r (at $95\%CL$ at the pivot scale $k = 0.002\,\mathrm{Mpc}^{-1}$) becomes more stringent by combining the PANCK data with that of BICEP and Keck (BKP) [9, 21] $r_{0.002} < 0.064$ 95% CL.

The running of the spectral index is consistent with zero [9],

$$\frac{dn_s}{d\ln k} = -0.0045 \pm 0.0067, \quad 68\% \text{ CL.} \tag{5.76}$$

The stringent bound on the n_s and r severely rules out many models of inflation.

5.4   Inflation Models

In this section, we discuss inflation models in different extended theories of gravity and supergravity theory [22, 23] which are still consistent with the stringent bounds from CMB observations.

5.4.1 $f(R)$ Models of Inflation

One modifications to GR which plays a role in both inflation and dark energy models is $f(R)$−gravity in which the Lagrangian density can be a general function of the Ricci scalar R [24–26],

$$S = \frac{1}{8\pi G} \int d^4x \sqrt{-g}\, f(R) + S_{(m)}, \tag{5.1}$$

In $f(R)$ theories, one can make a conformal transformation of the metric such that the $f(R)$ theory transforms actions into an Einstein gravity action with an extra scalar field [25, 27]. Some useful identities related to conformal transformations are,

$$\tilde{g}_{\mu\nu}(x) = \Omega^2(x) g_{\mu\nu}(x),$$ (5.2)

$$\tilde{g}^{\mu\nu} = \Omega^{-2} g^{\mu\nu},$$

$$ds^2 = \Omega^{-2} d\tilde{s}^2,$$

$$\sqrt{-g} = \Omega^{-4} \sqrt{-\tilde{g}},$$

$$R = \Omega^2 \left[\tilde{R} + 6\frac{\Box\Omega}{\Omega} - 12\frac{\tilde{g}^{\mu\nu}\partial_\mu\Omega\partial_\nu\Omega}{\Omega^2} \right].$$

As the line element ds changes in conformal transformation, it is not a general coordinate transformation and going from the $f(R)$ (Jordan frame) to $\tilde{R}$ (Einstein frame) is not a gauge transformation.

By making a conformal transformation from the Jordan to the Einstein frame we obtain an explicit potential term for the scalar degree component of the metric. Choose the conformal factor

$$\Omega^2 = f' \equiv \frac{\partial f}{\partial R}$$ (5.3)

Using the identities (5.2) transform the Lagrangian to the Einstein frame

$$\sqrt{-g} f(R) = \sqrt{-\tilde{g}} \left(-\frac{1}{2}\tilde{R} + \frac{1}{2}\tilde{g}^{\mu\nu}\partial_\mu\varphi\partial_\mu\varphi - V \right).$$ (5.4)

where

$$\varphi \equiv \sqrt{\frac{3}{2}} M_P \ln|f'|$$ (5.5)

and the potential is

$$V = M_P^2 \frac{(Rf' - f)}{2f'^2}$$ (5.6)

This is the inflation potential in $f(R)$ theory models of inflation. Equation (5.5) can be inverted to write $R = R(\varphi)$ and the potential (5.6) as a function of φ..

5.4.2 Starobinsky Model of Inflation

The simplest model of inflation in $f(R)$ gravity is the Starobinsky model of inflation
[28]. The Starobinsky action is

$$S = \frac{M_P^2}{2} \int d^4x \sqrt{-g} \left(R + \frac{1}{6M^2} R^2 \right) \tag{5.7}$$

To go to the Einstein frame make the conformal transformations (in units $M_P = 1$)

$$g_{\mu\nu} \rightarrow \tilde{g}_{\mu\nu} = f' g_{\mu\nu} = \frac{1}{2} \left(1 + \frac{R}{3M^2} \right) g_{\mu\nu} \tag{5.8}$$

and define the field φ

$$\varphi = \sqrt{\frac{3}{2}} \ln |f'| = \sqrt{\frac{3}{2}} \ln \left(\frac{1}{2} + \frac{R}{6M^2} \right),$$

$$\Rightarrow R = 6M^2 \left(e^{\sqrt{2/3}\varphi} - \frac{1}{2} \right) \quad \text{and} \quad f' = e^{\sqrt{2/3}\varphi} \tag{5.9}$$

then (5.7) assumes the form

$$S = \frac{1}{2} \int d^4x \sqrt{-\tilde{g}} \left[\tilde{R} + (\partial_\mu \varphi)^2 - V(\varphi) \right], \tag{5.10}$$

where the potential $V(\varphi)$ is given by (putting back the M_P)

$$V(\varphi) = M_P^2 \frac{(Rf' - f)}{2f'^2}$$

$$= \frac{3}{4} M^2 M_P^2 \left(1 - e^{-\sqrt{2/3}\varphi/M_P} \right)^2. \tag{5.11}$$

In Fig. 5.1 we show the plot the potential $V(\varphi)$ as a function of the field φ.
Inflation end at $\epsilon_V = 1$ at $\varphi_{end} \simeq 0.1 M_{pl}$ and value of φ 60-efoldings before
the end of inflation is $\varphi_{60} \simeq M_{pl}$.
The slow roll parameters are

$$\epsilon_V = \frac{4}{3} \frac{e^{-\sqrt{\frac{2}{3}} \frac{\varphi}{M_{pl}}}}{\left(1 - e^{-\sqrt{\frac{2}{3}} \frac{\varphi}{M_{pl}}} \right)^2}, \qquad \eta_V = \frac{4}{3} \frac{2e^{-2\sqrt{\frac{2}{3}} \frac{\varphi}{M_{pl}}} - e^{-\sqrt{\frac{2}{3}} \frac{\varphi}{M_{pl}}}}{\left(1 - e^{-\sqrt{\frac{2}{3}} \frac{\varphi}{M_{pl}}} \right)^2} \tag{5.12}$$

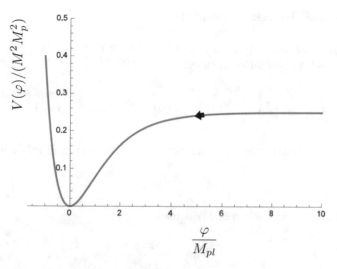

Fig. 5.1 The Starobinsky potential scaled by $M^2 M_{pl}^2$ as a function of field φ/M_{pl}

and the number of e-foldings from any point φ till the end of inflation at φ_f is

$$N = \int H \, dt = \int \frac{H}{\dot{\varphi}} d\varphi = \frac{8\pi}{M_{pl}^2} \int_{\varphi}^{\varphi_f} \frac{V}{V_\varphi} d\varphi$$

$$\simeq \frac{3}{4} e^{-\sqrt{\frac{2}{3}} \frac{\varphi}{M_{pl}}} \tag{5.13}$$

The number of e-foldings is insensitive to the the field value φ_f at the end of inflation. Using these relations we can write the slow roll parameters in terms of N as

$$\epsilon_V = \frac{12}{(4N^2 - 3)^2} \simeq \frac{3}{4N^2}, \qquad \eta_V = \frac{9 - 4N}{(4N - 3)^2} \simeq \frac{1}{N} \tag{5.14}$$

Therefore the Starobinsky model predicts

$$n_s - 1 = 1 - \frac{2}{N}, \qquad r = \frac{12}{N^2}, \tag{5.15}$$

where N is the number of e-foldings before the end of inflation. For $N = 60$ the Starobinsky model predicts $n_s = 0.967$ and $r = 0.0033$ thus satisfying the observations from CMB.

5.4.3 $f(\phi)R$ Models of Inflation

Another class of models with a coupling between field and curvature scalar are the
non-minimally coupled inflation models with action [29],

$$S_J = \int d^4x \sqrt{-g} \left[\frac{M_p^2}{2} f(\phi)R - k(\phi)\frac{1}{2}g^{\mu\nu}\partial_\mu\phi\partial_\nu\phi - V(\phi) \right]. \tag{5.16}$$

The action (5.16) can be cast in the Einstein frame with the canonical transformation

$$\tilde{g}_{\mu\nu} = f(\phi)g_{\mu\nu} \tag{5.17}$$

The potential in theEinstein frame becomes

$$\tilde{V}(\phi) = \frac{V(\phi)}{f(\phi)^2} \tag{5.18}$$

After these transformations the kinetic term can be made canonical by transforming
to a new field σ which is related to ϕ by

$$\left(\frac{d\tilde{\phi}}{d\phi}\right) = \left(\frac{3M_p^2}{2f^2}\left(\frac{\partial f}{\partial\phi}\right)^2 + \frac{k(\phi)}{f(\phi)}\right)^{1/2}. \tag{5.19}$$

we get the Einstein Frame action

$$S_E = \int d^4x \sqrt{-\tilde{g}} \left[\frac{M_p^2}{2}\tilde{R} - \frac{1}{2}\tilde{g}^{\mu\nu}\partial_\mu\tilde{\phi}\partial_\nu\tilde{\phi} - \tilde{V}(\tilde{\phi}) \right], \tag{5.20}$$

5.4.4 Higgs Inflation

The standard model Higgs can play the roll of inflation in the early universe if there
is a large curvature coupling [30]. The action for the Higgs inflation model is

$$S_J = \int d^4x \sqrt{-g} \left[\frac{M_p^2 + \xi\phi^2}{2}R - \frac{1}{2}g^{\mu\nu}\partial_\mu\phi\partial_\nu\phi - \frac{\lambda}{4}(\phi^2 - v^2)^2) \right]. \tag{5.21}$$

The field excursions during inflation or at the Planck scale $\phi \sim M_P \gg v = 246\,\text{GeV}$ so the potential during the inflationary phase is $V = \frac{\lambda}{4}\phi^4$.
This action (5.21) can be transformed to the Einstein frame with a conformal

transformation

$$\tilde{g}_{\mu\nu} = f(\phi)g_{\mu\nu} \quad \text{with} \quad f(\phi) = \frac{1 + \xi\phi^2}{M_P^2} \tag{5.22}$$

The potential in the Einstein frame is

$$V_E(\phi) = \frac{\lambda}{4} \frac{\phi^4}{f(\phi)^2} \tag{5.23}$$

The kinetic is canonical in the new field χ which obeys

$$\left(\frac{d\chi}{d\phi}\right) = \left(\frac{3M_P^2}{2f^2}\left(\frac{\partial f}{\partial \phi}\right)^2 + \frac{1}{f(\phi)}\right)^{1/2}. \tag{5.24}$$

When the field values are large $\phi \gg \sqrt{2/3}M_P/\xi$ the first term in the bracket in (5.24) dominates, and we can solve for the canonical field in terms of ϕ as

$$\chi = \sqrt{\frac{3}{2}}M_P \ln\left(\frac{1 + \xi\phi^2}{M_P^2}\right)^{1/2} \tag{5.25}$$

The Einstein frame potential (5.23) in terms of the canonical χ field becomes

$$V_E(\chi) = \frac{\lambda M_P^4}{4\xi^2}\left(1 - e^{\sqrt{2/3}\chi/M_P}\right)^2 \tag{5.26}$$

This is the same potential as in the Starobinsky model. To give the correct amplitude of the temperature anisotropy we require $\xi = 49000\sqrt{\lambda}$.

5.5 Future Prospects

The 2018 data from Planck observation [9] gives the amplitude, spectral index and tensor-to-scalar ratio as $10^{10}\ln(\Delta_{\mathcal{R}}^2) = 3.044 \pm 0.014$, $n_s = 0.9670 \pm 0.0037$ at (68% CL) and $r_{0.002} < 0.11$ at (95%CL), respectively [9]. Also the latest results by BKP collaboration put r at $r_{0.05} < 0.064$ at (95%CL) [21]. The tight constraint on n_s and r have ruled out many of the standard inflation potentials like the quartic and quadratic potential as shown in Fig. 5.2.

It must be kept in mind that the Planck results are derived with a background cosmological model namely the ΛCDM model. If the cosmological model is different then the fits to n_s etc can be different. For example if one introduces sterile neutrinos with the mass and mixings required from MiniBooNE [32] then we have to introduce self interaction of the sterile neutrinos in order to avoid the constraint

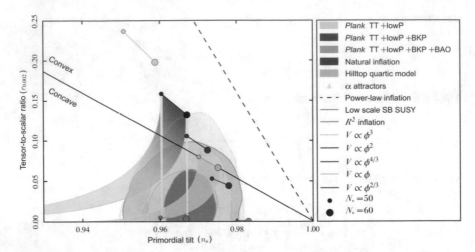

Fig. 5.2 Joint constraint on $r - n_s$ from Planck [9]. This assumes the relation $n_t = -r/8$ and no-running of n_s. Many inflation potentials like the quartic and quadratic potentials are ruled out. Plot taken from PDG-2018 [31]

on N_{eff} from BBN. In this cosmological model we fit the CMB parameters we have a shift in n_s. This allows many new inflation models and actually disallows the Starobinsky model [33].

There are several experiments, like the ground based Keck/BICEP3, SPT-3G, AdvACT, CLASS, Simons Array, balloon based Spider,EBEX and satellites based CMBPol, LiteBIRD and COrE experiments in operation or being planned [34]. They aim to probe the tensor-to-scale ratio at the level of $r \sim 10^{-3}$. High precision measurements of small-scale temperature anisotropies along with the observations of $B-$mode will not only validate the inflationary paradigm but may also help in pinning down the correct inflation model.

References

1. A.H. Guth, Inflationary universe: a possible solution to the horizon and flatness problems. Phys. Rev. D **23**, 347 (1981)
2. A.A. Starobinsky, Spectrum of relict gravitational radiation and the early state of the universe. JETP Lett. **30**, 682 (1979)
3. D. Kazanas, Dynamics of the universe and spontaneous symmetry breaking. Astrophys. J. **241**, L59 (1980)
4. A.D. Linde, A new inflationary universe scenario: a possible solution of the horizon, flatness, homogeneity, isotropy and primordial monopole problems. Phys. Lett. **B108**, 389 (1982)
5. K. Sato, First order phase transition of a vacuum and expansion of the universe. Mon. Not. Roy. Astron. Soc. **195**, 467 (1981)
6. A. Albrecht, P.J. Steinhardt, Cosmology for grand unified theories with radiatively induced symmetry breaking. Phys. Rev. Lett. **48**, 1220 (1982)

7. G.F. Smoot, Summary of results from COBE. AIP Conf. Proc. **476**, 1–10 (1999). [arXiv:astro-ph/9902027]
8. C.L. Bennet et al., WMAP coll., Nine-year Wilkinson microwave anisotropy probe (WMAP) observations: final maps and results. Astrophys. J. Suppl. **208**, 20 (2013)
9. Y. Akrami, et al., [Planck Collaboration], Planck 2018 results. X. Constraints on inflation (2018). arXiv:1807.06211 [astro-ph.CO]
10. N. Aghanim, et al. [Planck Collaboration], Planck 2018 results. VI. Cosmological parameters (2018). arXiv:1807.06209
11. V.F. Mukhanov, H.A. Feldman, R.H. Brandenberger, Theory of cosmological perturbations. Part 1. Classical perturbations. Part 2. Quantum theory of perturbations. Part 3. Extensions. Phys. Rept. **215**, 203 (1992)
12. A. Riotto, Inflation and the theory of cosmological perturbations. ICTP Lect. Notes Ser. **14**, 317 (2003) [hep-ph/0210162]
13. S. Dodelson, Modern Cosmology. Academic Press, Amsterdam (2003) 440pp.
14. V.F. Mukhanov, Physical Foundations of Cosmology. Cambridge University Press, Cambridge (2005)
15. D. Baumann, TASI lectures on inflation (2009). arXiv:0907.5424
16. J.M. Bardeen, P.J. Steinhardt, M.S. Turner, Spontaneous creation of almost scale-free density perturbations in an inflationary universe. Phys. Rev. D **28**, 679 (1983)
17. C. Gordon, et al., Adiabatic and entropy perturbations from inflation. Phys. Rev. D **63**, 023506 (2001)
18. T.S. Bunch, P.C.W. Davies, Quantum field theory in de sitter space: renormalization by point splitting. Proc. Roy. Soc. Lond. A 360, 117 (1978)
19. N.A. Chernikov, E.A. Tagirov, Quantum theory of scalar fields in de Sitter spacetime. Annales Poincare Phys. Theor. A **9**, 109 (1968)
20. L.F. Abbott, M. Wise, Constraints on generalized inflationary cosmologies. Nucl. Phys. B **244**, 541 (1984)
21. P.A.R. Ade, et al., [BICEP2 and Planck Collaborations], joint analysis of BICEP2/$KeckArray$ and $Planck$ data. Phys. Rev. Lett. **114**, 101301 (2015)
22. J. Martin, C. Ringeval, R. Trotta, V. Vennin, The best inflationary models after planck. JCAP **1403**, 039 (2014)
23. S. Tsujikawa, Distinguishing between inflationary models from cosmic microwave background. Prog. Theor. Exp. Phys. **2014**(6), (2014)
24. T.P. Sotiriou, V. Faraoni, $f(R)$ theories of gravity. Rev. Mod. Phys. **82**, 451 (2010)
25. A. De Felice, S.Tsujikawa, $f(R)$ theories. Living Rev. Rel. **13**, 3 (2010)
26. T. Clifton, P.G. Ferreira, A. Padilla, C. Skordis, Modified gravity and cosmology. Phys. Rep. **513**, 1 (2012)
27. T. Chiba, M. Yamaguchi, Conformal-Frame (In)dependence of cosmological observations in scalar-tensor theory. JCAP **1310**, 040 (2013)
28. A.A. Starobinsky, A new type of isotropic cosmological models without singularity. Phys. Lett. B **91**, 99 (1980)
29. G.K. Chakravarty, S. Mohanty, G. Lambiase, Testing theories of gravity and supergravity with inflation and observations of the cosmic microwave background. Int. J. Mod. Phys. D **26**(13), 1730023 (2017)
30. F.L. Bezrukov, M. Shaposhnikov, The standard model higgs boson as the inflaton. Phys. Lett. B **659**, 703 (2008)
31. M. Tanabashi, et al., Reviews of particle physics. Phys. Rev. D **98**, 030001 (2018)
32. A. Aguilar-Arevalo, et al. [MiniBooNE], Significant excess of ElectronLike events in the MiniBooNE short-baseline neutrino experiment. Phys. Rev. Lett. **121**(22), 221801 (2018)
33. A. Mazumdar, S. Mohanty, P. Parashari, Inflation models in the light of self-interacting sterile neutrinos. Phys. Rev. D **101**(8), 083521 (2020)
34. K.N. Abazajian, et al. [CMB-S4], CMB-S4 science book, first edition (2016). [arXiv:1610.02743 [astro-ph.CO]]

Effective Potential and Phase Transitions

6

Abstract

We study the one loop effective potentials in field theory at zero and finite temperatures. The evolution of the Higgs potential at finite temperature shows that the electroweak symmetry breaking was not a first order phase transition which implies that any existing baryon asymmetry which was present in the universe was washed out by sphalerons in the electroweak era. We explore possible extensions of the standard model which can solve this problem. We also explore other consequences of early universe phase transitions like generation of gravitational waves and production of large mass primordial black holes.

6.1 Introduction

The discovery of the Higgs boson in 2012 experimentally supports the idea of an universal Higgs field which provides masses to the elementary particles. This measurement of the Higgs mass which is equivalent to measurement of the quartic coupling λ has in turn raised many intriguing questions about the Higgs potential. In the standard model the RG running of the Higgs coupling shows that λ becomes negative at energy scales $\mu = 10^8 - 10^{11}$ GeV which makes the Higgs potential negative at large field values. This issue of vacuum stability of the Higgs potential raises many problems. For example during inflation the Higgs field like all other fields are expected to be of the order of the Hubble scale. If the Hubble scale during infration is larger than 10^8 GeV then the Higgs field will be in the unstable regime and this will disrupt the inflation. Another issue with the Higgs potential is that the phase transition that occurred in the early universe from the symmetric to the electroweak symmetry breaking phase was not a first order phase transition. This creates a problem that electroweak sphalerons can wash out any existing baryon asymmetry in the universe.

© Springer Nature Switzerland AG 2020

S. Mohanty, *Astroparticle Physics and Cosmology*,
Lecture Notes in Physics 975, https://doi.org/10.1007/978-3-030-56201-4_6

These issues with the Higgs quartic coupling and Higgs potential at finite temperature opens the opportunity to extend the standard model to test theories of grand unification, supersymmetry or neutrino masses which could solve these problems of the standard model Higgs potential. First order phase transitions in the early universe also generate gravitational waves. The frequency of the peak energy of these stochastic gravitational waves depends on the temperature of phase transitions. This then opens the possibility of testing early universe phase transitions by measuring the gravitational waves in future space based experiments like LISA, BBO and DECIGO.

6.2   Coleman-Weinberg One Loop Effective Potential

Given a theory of scalars, fermions and vector particles we can examine the effective potential of the classical scalar field when the other fields are treated as quantum fluctuations. For this we need to compute the diagrams scalar external legs the other particles running in the loops. The one loop diagrams with scalar legs at all orders is called the Coleman-Weinberg potential [1]. Consider the path integral representation of the generating function of the scalar field action

$$Z = \int D\phi e^{\frac{i}{\hbar} S[\phi]} \tag{6.1}$$

We can evaluate the path integral for a general action

$$S[\phi] = \int d^4 x [-\frac{1}{2}\phi(\partial^2)\phi - V(\phi)] \tag{6.2}$$

by the method of steepest descent. The integrand makes the largest contribution where $S[\phi]$ is a minimum and that occurs at classical trajectories of the scalar field $\phi = \phi_c$ where ϕ_c obeys the equation of motion

$$\partial^2 \phi_c + V'(\phi_c) = 0 \tag{6.3}$$

where prime denotes derivative w.r.t. ϕ_c. We expand the field around its classical solution $\phi = \phi_c + \varphi$ by writing the action as

$$S[\phi] = S[\phi_c] + \frac{\delta S[\phi]}{\delta \phi}\bigg|_{\phi=\phi_c} \varphi + \frac{1}{2}\frac{\delta^2 S[\phi]}{\delta \phi^2}\bigg|_{\phi=\phi_c} \varphi^2 + \cdots$$

$$= S[\phi_c] - \frac{1}{2}\varphi\left(\partial^2 - V''(\phi_c)\right)\varphi \tag{6.4}$$

The effective action $\Gamma[\phi_c]$ which is obtained by integrating out the quantum fluctuation φ can be written as

$$e^{\frac{i}{\hbar}\Gamma[\phi_c]} \equiv e^{\frac{i}{\hbar}S[\phi_c]} \int D\varphi e^{-\frac{i}{\hbar}\frac{1}{2}\varphi(\partial^2 - V''(\phi_c))\varphi} \tag{6.5}$$

The effective potential can be expressed in terms of the effective action as

$$\Gamma[\phi_c] \equiv \int d^4x \left(-V_{eff}(\phi_c) + \frac{1}{2}Z(\phi_c)(\partial\phi_c)^2 + \cdots \right) \tag{6.6}$$

where the dots denote possible higher order corrections of $(\partial\phi_c)^2$.

The evaluation of the path integral (6.5) integration of the Gaussian function of φ can be done exactly with the result

$$\int D\varphi e^{-\frac{i}{\hbar}\frac{1}{2}\varphi(\partial^2 - V''(\phi_c))\varphi} = C \left(\frac{1}{\det(\partial^2 + V'')} \right)^{1/2} = Ce^{-\frac{1}{2}\text{Tr}\log(\partial^2 + V'')} \tag{6.7}$$

where[1] C is of order $\hbar$ and will contribute only in a logarithm in the effective action. From (6.5) and (6.7) we obtain the relation

$$e^{\frac{i}{\hbar}\Gamma[\phi_c]} = C e^{\frac{i}{\hbar}S[\phi_c] - \frac{1}{2}\text{Tr}\log(\partial^2 + V'')} \tag{6.8}$$

From (6.6) and (6.8) we get the expression for the order $\hbar$ correction to the effective potential

$$\int d^4x V_{eff}(\phi_c) = \int d^4x \left[V(\phi_c) - \frac{i\hbar}{2}\text{Tr}\log\left(\partial^2 + V''\right) \right] \tag{6.9}$$

The order $\hbar$ correction can be evaluated as follows

$$\text{Tr}\log\left(\partial^2 + V''\right) = \int d^4x \langle x| \log\left(\partial^2 + V''\right)|x\rangle$$

$$= \int d^4x \int \frac{d^4k}{(2\pi)^4} \int \frac{d^4k'}{(2\pi)^4} \langle x|k\rangle\langle k| \log\left(\partial^2 + V''\right)|k'\rangle\langle k'|x\rangle$$

$$= \int d^4x \int \frac{d^4k}{(2\pi)^4} \log\left(-k^2 + V''\right) \tag{6.10}$$

---

[1] $\det(M) = e^{\text{Tr}\log M}$.

From (6.9) and (6.10) we see that the effective potential to one loop (order $\hbar$) is of the form

$$V_{eff}(\phi_c) = V(\phi_c) - \frac{i\hbar}{2} \int \frac{d^4k}{(2\pi)^4} \log\left(-k^2 + V''\right) \tag{6.11}$$

This integral is divergent. One can regulate the integral by isolating the singularities and removing them by adding suitable counter terms in the Lagrangian.

6.2.1   Dimensional Regularization of Scalar Loop

Here we will use the dimensional regularization [2] to regulate the integrals in the one loop potential (6.11). To isolate the singularities in the integral, which can be removed by suitable counter-terms in this we first change the momentum $k = (k_0, \mathbf{k})$ to Euclidean space $k_E = (ik_0, \mathbf{k})$,

$$V_1(\phi_c) = \frac{1}{2} \int \frac{d^4k_E}{(2\pi)^4} \log\left(k_E^2 + V''\right). \tag{6.12}$$

In the dimensional regularization method the number of dimension of the Euclidean space is analytically continued from 4 to a complex number n dimensions. The integral has singularities which arise as $1/(2 - \frac{n}{2})$ poles. The renormalization prescription is to absorb these poles in the counter-terms.

$$V_1(\phi) = \frac{1}{2}(\mu^2)^{2-\frac{n}{2}} \int \frac{d^n k_E}{(2\pi)^n} \log\left(k_E^2 + m^2(\phi_c)\right) \tag{6.13}$$

where the mass scale μ has to be introduced to restore the mass dimension of V_1 to 4 and we have defined the field dependent mass $V'' = m^2(\phi_c)$ defined from the tree-level potential. In order to apply the standard dimensional regularization relations for the integrals we first evaluate

$$\frac{\partial V_1(\phi_c)}{\partial m^2(\phi_c)} = \frac{1}{2}(\mu^2)^{2-\frac{n}{2}} \int \frac{d^n k_E}{(2\pi)^n} \frac{1}{k_E^2 + m^2(\phi_c)} \tag{6.14}$$

We use the general formula for the n dimensional integral

$$\int d^n p_E \frac{(p_E^2)^\alpha}{(p_E^2 + M^2)^\beta} = \pi^{\frac{n}{2}} (M^2)^{\frac{n}{2}+\alpha-\beta} \times \frac{\Gamma\left(\alpha + \frac{n}{2}\right)\Gamma\left(\beta - \alpha - \frac{n}{2}\right)}{\Gamma\left(\frac{n}{2}\right)\Gamma(\beta)} \tag{6.15}$$

and evaluate

$$\frac{\partial V_1(\phi_c)}{\partial m^2(\phi_c)} = \frac{1}{2}(\mu^2)^{2-\frac{n}{2}} \frac{\pi^{\frac{n}{2}}}{(2\pi)^n} (m^2)^{\frac{n}{2}-1} \Gamma\left(1 - \frac{n}{2}\right) \tag{6.16}$$

to get the one-loop correction to the effective potential in the dimensional regularization procedure

$$V_1(\phi_c) = \frac{1}{2}(\mu^2)^{2-\frac{n}{2}}\frac{1}{(4\pi)^{\frac{n}{2}}}\frac{1}{(\frac{n}{2})}(m^2(\phi_c))^{\frac{n}{2}}\Gamma\left(1-\frac{n}{2}\right)$$

$$= \frac{m^4(\phi_c)}{32\pi^2}\left(\frac{m^2(\phi_c)}{4\pi\mu^2}\right)^{\frac{n}{2}-2}\frac{1}{(\frac{n}{2})(1-\frac{n}{2})}\Gamma\left(2-\frac{n}{2}\right) \qquad (6.17)$$

where we have used the identity $\Gamma(z+1) = z\Gamma(z)$. We now take the limit $\epsilon = 2 - \frac{n}{2} \Longrightarrow 0$ and use the series expansions

$$\Gamma(\epsilon)\left(\frac{m^2(\phi_c)}{4\pi\mu^2}\right)^{-\epsilon}\frac{1}{(\frac{n}{2})(1-\frac{n}{2})}$$

$$= \left(\frac{1}{\epsilon}-\gamma_E\right)\left(1-\epsilon\log\left(\frac{m^2(\phi_c)}{4\pi\mu^2}\right)\right)\frac{1}{(-2+\epsilon)(\epsilon-1)},$$

$$= \frac{1}{2}\left(\log\left(\frac{m^2(\phi_c)}{\mu^2}\right)-\frac{3}{2}-\left[\frac{1}{\epsilon}-\gamma_E+\log(4\pi)\right]\right) \qquad (6.18)$$

(where $\gamma_E = 0.5772..$ is the Euler-Mascheroni constant) to obtain from (6.17) in the dimensional regularization method

$$V_1(\phi_c) = \frac{m^4(\phi_c)}{64\pi^2}\left(\log\left(\frac{m^2(\phi_c)}{\mu^2}\right)-\frac{3}{2}-\left[\frac{1}{\epsilon}-\gamma_E+\log(4\pi)\right]\right)$$

$$(6.19)$$

In the $\overline{\text{MS}}$ prescription the terms in the square brackets

$$C_{UV} = -\frac{m^4(\phi_c)}{64\pi^2}\left[\frac{1}{\epsilon}-\gamma_E+\log(4\pi)\right] \qquad (6.20)$$

is absorbed in the counter-terms and the one-loop potential is given by

$$V_\phi(\phi_c) = \frac{m^4(\phi_c)}{64\pi^2}\left(\log\frac{m^2(\phi_c)}{\mu^2}-\frac{3}{2}\right) \qquad (6.21)$$

On the other hand in the MS renormalization scheme only the $1/\epsilon$ term is absorbed in the counter-terms and the $(\gamma_E - \log(4\pi))$ terms remains in the effective potential. Since the same prescription (MS or $\overline{\text{MS}}$) is used for defining the renormalized couplings (λ, m etc.) and the effective potential, the effective potential is independent of the renormalization scheme used.

6.2.2   Fermion Loop

If the scalar field has Yukawa couplings with fermions ψ

$$\mathcal{L} = i\bar{\psi}\gamma^{\mu}\partial_{\mu}\psi - y\phi\,\bar{\psi}\psi \tag{6.22}$$

then the classical field generates a fermion mass

$$M_f(\phi_c) = y\phi_c \tag{6.23}$$

The one-loop effective potential obtained by integrating out the fermion can be written as

$$e^{-iV_f(\phi_c)} = \int D[\bar{\psi}]D[\psi]\exp\left(-i\int d^4x\,\bar{\psi}(i\slashed{\partial} - M_f(\phi_c))\psi\right) \tag{6.24}$$

Going to momentum space

$$
\begin{aligned}
e^{-iV_f(\phi_c)} &= \int D[i\bar{\psi}(p)]D[\psi(p)]e^{\left(-i\int\frac{d^4p}{(2\pi)^4}\bar{\psi}(p)(\slashed{p}-M_f)\psi(p)\right)} \\
&= \prod_p \int d[i\bar{\psi}(p)]d[\psi(p)]e^{(-i\bar{\psi}(p)(\slashed{p}-M_f)\psi(p))} \\
&= \prod_p det\left[\slashed{p} - M_f\right] \\
&= \exp\left[\int\frac{d^4p}{(2\pi)^4}tr\ln(\slashed{p} - M_f)\right] \tag{6.25}
\end{aligned}
$$

The trace over the sponsorial indices can be taken as follows,

$$
\begin{aligned}
tr\ln(\slashed{p} - M_f) &= tr\frac{1}{2}\ln(\slashed{p} - M_f)^2 \\
&= \frac{1}{2}4\ln(-p^2 + M_f^2) \tag{6.26}
\end{aligned}
$$

Going to Euclidean momenta $p = (p_0, \mathbf{p}) \to (ip_0, \mathbf{p})$, we find that to the effective scalar potential due to a fermion loop is

$$V_f(\phi_c) = -4\frac{1}{2}\int\frac{d^4p}{(2\pi)^4}Tr\left(\ln\left(p^2 + M_f^2(\phi_c)\right)\right) \tag{6.27}$$

which has relative negative sign compared to the case of bosons and an extra factor of 4 coming from two helicities for particles and anti-particle. For Weyl fermions (like neutrinos where the particles and anti-particles have 1 chirality each, the factor of 4 is replaces by a factor of 2

We evaluate the integral (6.27) by dimensional regularisation method and subtracting counter-terms in the $\overline{MS}$ scheme in the same way as for the scalars and

$$V_f(\phi_c) = -2g_f \frac{M_f^4(\phi_c)}{64\pi^2} \left(\log \frac{M_f^2(\phi_c)}{\mu^2} - \frac{3}{2} \right) \tag{6.28}$$

where $g_f = 2$ for Dirac fermions and $g_f = 1$ for Weyl fermions.

6.2.3   Gauge Bosons Loop

If the scalar field has gauge couplings and its classical part contributes to the mass of the gauge bosons then the gauge boson loops will contribute to the scalar potential. The vector boson propagator is of the form (in the Landau gauge)

$$\Pi^\mu{}_\nu = \frac{-i}{p^2 + i\epsilon} \Delta^\mu{}_\nu \tag{6.29}$$

where $\Delta^\mu{}_\nu = g^\mu{}_\nu - p^\mu p_\nu / p^2$ and has the property necessary gauge invariance $p_\mu \Delta^\mu{}_\nu - 0$ and since it is a projection operator $((\Delta^\mu{}_\nu))^n = \Delta^\mu{}_\nu$ and $tr(\Delta^\mu{}_\nu) = \Delta^\mu{}_\mu = 3$ which reflects that the massive gauge boson has three scalar degrees of freedom. The gauge boson vertices are from the mass term

$$\mathcal{L} = \frac{1}{2} M^2_{V\,ab}(\phi_c) V^a{}_\mu V^{b\mu} \tag{6.30}$$

where

$$M^2_{V\,ab}(\phi_c) = g_a g_b Tr\left[\left(T^i_{a\,k} \phi_i \right) \left(T^k_{b\,j} \phi^j \right) \right] \tag{6.31}$$

where a, b denote the gauge groups with gauge couplings g_a, g_b and generators T_a, T_b. The one-loop potential from gauge boson loops is given by

$$V_V(\phi_c) = Tr\left(\Delta^\mu{}_\mu \right) \frac{1}{2} \int \frac{d^4 p}{(2\pi)^4} Tr\left(\ln\left(p^2 + M^2_V(\phi_c) \right) \right) \tag{6.32}$$

The integrating and subtracting according to the $(\overline{MS})$ scheme as before the contribution of massive gauge bosons to the scalar potential at one loop is

$$V_V(\phi_c) = 3 \frac{M^2_{V\,ab}(\phi_c)}{64\pi^2} \left(\log \frac{M^2_{V\,ab}(\phi_c)}{\mu^2} - \frac{5}{6} \right) \tag{6.33}$$

The constant term in the parenthesis is $5/6$ for the gauge bosons as in n space-time dimensions, $tr(\Delta^{\mu}{}_{\mu}) = n - 1$ which then gives a different constant compared to the scalar and fermion cases in the $(\overline{MS})$ scheme.

6.3   Effective Potential at Finite Temperature

In the early universe the scalar potential of a field in thermal equilibrium with the heat bath is described by the free energy which is related to the partition function

$$F = -\frac{T}{V} \ln Z$$

The partition function of free scalar, gauge boson and fermion fields is evaluated in Appendix C. The free theory of scalars, gauge bosons and fermions at finite temperature contributes to the scalar potential through the scalar field dependence of the masses of scalars, gauge bosons and fermions. The leading order thermal corrections to the Coleman-Weinberg one-loop potential can be calculated from the partition function of the free theory. The interactions contribute through hard thermal loops through the so called ring or daisy diagrams and we shall add the ring contribution to the thermal potential in the following section. Other aspects of quantum field theory at finite temperature can be found in literature [3–6].

6.3.1   Scalar Loop at Finite Temperature

The potential at finite temperature is the thermodynamic potential (C.23) derived in Appendix C.2.1,

$$V_{\phi}^{\beta}(\phi_c) = F = \int \frac{d^3 p}{(2\pi)^3} \left[\frac{1}{2} E_p + \frac{1}{\beta} \ln \left(1 - e^{-\beta E_p} \right) \right] \tag{6.34}$$

The first term in the integral is the zero-temperature one loop potential (6.12),

$$\frac{1}{2} \int \frac{d^3 p}{(2\pi)^3} E_p = \frac{1}{2} \int \frac{d^4 p}{(2\pi)^4} \log \left(p^2 + m^2(\phi_c) \right) \tag{6.35}$$

To see this start with the integral which can be evaluated by the method of residues,

$$E_p \int \frac{dp_0}{2\pi i} \frac{1}{-p_0^2 + E_p^2 - i\epsilon} = \frac{1}{2} . \tag{6.36}$$

Integrate both sides w.r.t. E_p

$$-\frac{i}{2} \int \frac{dp_0}{2\pi} \frac{1}{\ln} \left[-p_0^2 + E_p^2 - i\epsilon \right] = \frac{E_p}{2} \qquad (6.37)$$

Wick rotate to Euclidean space $p_0 \to i p_0$ to get

$$\frac{1}{2} \int \frac{dp_0}{2\pi} \frac{1}{\ln} \left[p_0^2 + E_p^2 - i\epsilon \right] = \frac{E_p}{2} \qquad (6.38)$$

From which we get (6.35).

The zero-temperature one-loop integral can be evaluated by dimensional regularization to give the scalar-loop contribution to the scalar potential (6.21)

$$
\begin{aligned}
V_\phi^{T=0} &= \frac{1}{2} \int \frac{d^3 p}{(2\pi)^3} E_p = \frac{1}{2} \int \frac{d^4 p}{(2\pi)^4} \ln \left(p^2 + m^2(\phi_c) \right) \\
&= \frac{m^4(\phi_c)}{64\pi^2} \left(\ln \frac{m^2(\phi_c)}{\mu^2} - \frac{3}{2} \right).
\end{aligned}
\qquad (6.39)
$$

The temperature dependent contribution of the scalars to the effective potential is the second term in (6.34). This integral is finite and by changing variables to $x = \beta |\mathbf{p}|$ can be written as

$$
\begin{aligned}
\Delta V_\phi^\beta (\phi_c) &= \int \frac{d^3 p}{(2\pi)^3} \frac{1}{\beta} \ln \left(1 - e^{-\beta \sqrt{(\mathbf{p}^2 + m^2(\phi_c))}} \right) \\
&= \frac{1}{2\pi^2 \beta^4} J_B (\beta^2 m^2(\phi_c))
\end{aligned}
\qquad (6.40)
$$

where

$$J_B(\beta^2 m^2) \equiv \int_0^\infty dx \, x^2 \ln \left[1 - e^{-\sqrt{x^2 + \beta^2 m^2}} \right]. \qquad (6.41)$$

The high temperature ($y^2 = \beta^2 m^2 \ll 1$) series expansion of J_B is

$$J_B(y^2) = -\frac{\pi^4}{45} + \frac{\pi^2}{12} y^2 - \frac{\pi}{6} y^3 - \frac{1}{32} y^4 \ln \left(\frac{y^2}{a_b} \right) + \frac{\xi(3)}{3(4\pi)^4} y^4 + O(y^6) \qquad (6.42)$$

where $a_b = 16\pi^2 \exp(\frac{3}{2} - 2\gamma_E)$ (and $\ln a_b = 5.4076$).

The low temperature ($y^2 \gg 1$) series expansion of bosonic thermal integral (6.41) is

$$
J_B(y^2) = -\sum_{n=1}^{\infty} \frac{1}{n^2} y^2 K_2(ny) .
\tag{6.43}
$$

where $K_2(ny)$ is the modified Bessel's function of the second kind.

The low temperature approximation of ΔV_ϕ^β is

$$
\Delta V_\phi^\beta = -T^4 \left(\frac{m}{2\pi T} \right)^{3/2} e^{-\frac{m}{T}}
\tag{6.44}
$$

and for high temperatures ($\beta m \ll 1$) we have

$$
\Delta V_\phi^\beta = -\frac{\pi^2}{90} T^4 + \frac{1}{24} m^2 T^2 - \frac{1}{12\pi} m^3 T
$$
$$
- \frac{m^4}{64\pi^2} \ln \left(\frac{m^2}{a_b T^2} \right) + \frac{1}{6\pi^2} \frac{m^6 \xi(3)}{(4\pi)^4 T^2} + O\left(\frac{m^8}{T^4} \right)
$$
$$
\tag{6.45}
$$

6.3.2 Fermion Loop at Finite Temperature

Fermions with Yukawa couplings to scalars will contribute to the scalar potential via the scalar field dependence of the fermion mass (6.23). The free energy of fermions is from (C.35)

$$
V_F^\beta(\phi_c) = 2g_f \int \frac{d^3 p}{(2\pi)^3} \left[-\frac{1}{2} E_p - \frac{1}{\beta} \ln \left(1 + e^{-\beta E_p} \right) \right]
\tag{6.46}
$$

where $E_f = (M_f^2(\phi_c) + \mathbf{p}^2)^{1/2}$ and the helicity factor $g_f = 2$ for Dirac fermions and $g_f = 1$ for Weyl fermions.

The first term is the zero-temperature fermion-loop contribution (6.28)

$$
V_F^{T=0} = 2 g_f \int \frac{d^3 p}{(2\pi)^3} \left[-\frac{1}{2} E_p \right]
$$
$$
= -2 g_f \frac{M_f^4(\phi_c)}{64\pi^2} \left(\log \frac{M_f^2(\phi_c)}{\mu^2} - \frac{3}{2} \right) .
\tag{6.47}
$$

The second term of (6.28) is the finite temperature contribution from fermions

$$\Delta V_F^\beta(\phi_c) = -2\,g_f \int \frac{d^3 p}{(2\pi)^3} \frac{1}{\beta} \ln\left(1 + e^{-\beta\sqrt{(\mathbf{p}^2+M_f^2(\phi_c))}}\right)$$

$$= -2\,g_f \frac{1}{2\pi^2\beta^4} J_F(\beta M_f(\phi_c)) \tag{6.48}$$

where

$$J_F(\beta M_f) \equiv \int_0^\infty dx\, x^2 \ln\left[1 + e^{-\sqrt{x^2+\beta^2 M_f^2}}\right]. \tag{6.49}$$

The high temperature ($y^2 = \beta^2 m^2 \ll 1$) series expansion of J_F is

$$J_F(y^2) = \frac{7}{8}\frac{\pi^4}{45} - \frac{\pi^2}{24} y^2 - \frac{1}{32} y^4 \ln\left(\frac{y^2}{a_f}\right) + \frac{\xi(3)}{3(4\pi)^4} y^4 + O(y^6) \tag{6.50}$$

where $a_f = \pi^2 \exp(\frac{3}{2} - 2\gamma_E)$ (and $\ln a_f = 2.6351$).

The low temperature ($y^2 \gg 1$) series expansion of J_F is

$$J_F(y^2) = -\sum_{n=1}^\infty \frac{(-1)^n}{n^2} y^2 K_2(ny). \tag{6.51}$$

where $K_2(ny)$ is the modified Bessel's function of the second kind.

The low temperature ($\beta M_f \gg 1$) approximation of ΔV_F^β is

$$\Delta V_F^\beta = T^4 \left(\frac{M_f}{2\pi T}\right)^{3/2} e^{-\frac{M_f}{T}} \tag{6.52}$$

and for the high temperatures ($\beta M_f \ll 1$) we have

$$\Delta V_F^\beta = \frac{7}{8}\frac{\pi^2}{90} T^4 - \frac{1}{48} M_f^2 T^2 - \frac{M_f^4}{64\pi^2} \ln\left(\frac{M_f^2}{a_f^2 T^2}\right) + \frac{7}{6\pi^2}\frac{M_f^6\,\xi(3)}{(4\pi)^4 T^2} + O\left(\frac{m^8}{T^4}\right). \tag{6.53}$$

There is no term proportional to M_f^3 in the high temperature expansion of J_F.

6.3.3 Gauge Boson Loop at Finite Temperature

Gauge bosons contribution is of the same form as the scalar contribution with an extra factor of 3 due to the three degrees of freedom of massive vectors.

$$V_V^\beta(\phi_c) = g_V \int \frac{d^3p}{(2\pi)^3} \left[\frac{1}{2} E_p + \frac{1}{\beta} \ln\left(1 - e^{-\beta E_p} \right) \right] \tag{6.54}$$

where $E_f = (M_V^2(\phi_c) + \mathbf{p}^2)^{1/2}$.

The zero-temperature one-loop gauge boson contribution to the scalar potential is the first term in (6.54)

$$V_V^{T=0} = 3 \frac{M_V^4(\phi_c)}{64\pi^2} \left(\ln \frac{M_V^2(\phi_c)}{\mu^2} - \frac{5}{6} \right). \tag{6.55}$$

The temperature dependent contribution of gauge bosons to the effective potential is the second term in (6.54)

$$\Delta V_V^\beta(\phi_c) = 3 \int \frac{d^3p}{(2\pi)^3} \frac{1}{\beta} \ln \left(1 - e^{-\beta\sqrt{(\mathbf{p}^2 + M_V^2(\phi_c))}} \right)$$

$$= 3 J_B(\beta M_V(\phi_c)) \tag{6.56}$$

6.4   Standard Model Higgs Potential at Finite Temperature

In the Standard Model the Higgs field is a SU(2) doublet with four scalar degrees of freedom,

$$\Phi = \begin{pmatrix} \chi_1 + i\chi_2 \\ \frac{1}{\sqrt{2}}(\phi_c + h + i\chi_3) \end{pmatrix}$$

where ϕ_c is the classical Higgs background field whose effective potential we wish to compute, h is the quantum excitations over the Higgs background which is the Higgs boson whose mass in the present universe is $m_h = 125.26 \pm 0.21\,\text{GeV}$ and χ_a $(a = 1, 2, 3)$ are the three Goldstone bosons. The SU(2) invariant Higgs potential at the tree level is

$$V_0(\Phi) = -\frac{\mu_h^2}{2} \Phi^\dagger \Phi + \frac{\lambda}{4} \left(\Phi^\dagger \Phi \right)^2 \tag{6.57}$$

From the tree level potential we can determine the potential for the Higgs field

$$V_0(\phi_c) = -\frac{\mu_h^2}{2}\phi_c^2 + \frac{\lambda}{4}\phi_c^4. \tag{6.58}$$

From the Higgs potential (6.57) we can determine the field dependent masses of the Higgs boson h and the Goldstone bosons χ_a as

$$m_h^2(\phi_c) = \frac{\partial^2 V_0(\Phi)}{(\partial h)^2}\bigg|_{\Phi=\phi_c} = 3\lambda\phi_c^2 - \mu_h^2$$

$$m_{\chi_a}^2(\phi_c) = \frac{\partial^2 V_0(\Phi)}{(\partial \chi_a)^2}\bigg|_{\Phi=\phi_c} = \lambda\phi_c^2 - \mu_h^2 \tag{6.59}$$

In the present universe when the Higgs field is at the minima of its potential (6.58), $\phi_c = v = \sqrt{\frac{\mu_h^2}{\lambda}} = 246.22\,\text{GeV}$ we have the Higgs mass given by $m_h^2(v) = 2\lambda v^2 = 2\mu_h^2$ and Goldstone bosons are massless, $m_{\chi_a}^2(v) = 0$. For the general values of the Higgs field ϕ_c the Goldstone bosons masses are non-zero and they contribute to the one-loop potential.

The main contribution to the one-loop potential of the standard model Higgs comes from $(h, \chi_a, W^\pm, Z, t)$ and we neglect the contribution from the lighter fermions. The field dependent gauge boson and top quark masses are

$$m_W^2(\phi_c) = \frac{g^2}{4}\phi_c^2, \quad m_Z^2(\phi_c) = \frac{g^2 + g'^2}{2}\phi_c^2 \quad \text{and} \quad m_t^2 = \frac{h_t^2}{2}\phi_c^2. \tag{6.60}$$

The one-loop Coleman-Weinberg Higgs potential can then be written as

$$V_{CW}(\phi_c) = \frac{1}{64\pi^2} \sum_{i=h,\chi,t,W,Z} (-1)^{F_i} n_i\, m_i^4(\phi_c) \left(\log\frac{m_i^2(\phi_c)}{\mu^2} - c_i \right) \tag{6.61}$$

where the effective degree of freedom n_i takes into account the spin and color degrees of freedom and $F_i = 1$ for fermions and 0 for bosons. For the standard model fields,

$$n_h = 1, \quad n_\chi = 3, \quad n_W = 6, \quad n_Z = 3 \quad \text{and} \quad n_t = 12 \tag{6.62}$$

and the constants $c_i = \frac{5}{6}$ for W and Z bosons and $c_i = \frac{3}{2}$ for all other fields.

The one loop finite temperature correction to the standard model Higgs potential is

$$\Delta V^\beta(\phi_c) = \frac{T^4}{2\pi^2}\left[\sum_{i=W,Z,h,\chi} n_i\, J_B\left(\frac{m_i^2(\phi_c)}{T^2}\right) - n_t\, J_F\left(\frac{m_t^2(\phi_c)}{T^2}\right) \right] \tag{6.63}$$

where J_B and J_F are given by

$$J_{B/F}(y^2) = \int_0^\infty dx\, x^2 \log\left(1 \mp \exp\left(-\sqrt{x^2 + y^2}\right)\right) \tag{6.64}$$

Using the analytical expressions (6.45) and (6.53) for J_B and J_F for the standard model fields, we have the following expression for the Standard Model Higgs potential at high temperature $(T > m_h)$ [5],

$$V^\beta(\phi_c) = D(T^2 - T_0^2)\phi_c^2 - ET\phi_c^3 + \frac{\lambda(T)}{4}\phi_c^4, \tag{6.65}$$

where

$$D = \frac{1}{8v^2}\left(2m_W^2 + m_Z^2 + 2m_t^2\right),$$

$$E = \frac{1}{4\pi v^3}\left(2m_W^3 + m_Z^3\right),$$

$$T_0^2 = \frac{1}{4D}\left(m_h^2 - 8Bv^2\right),$$

$$B = \frac{3}{64\pi^2 v^4}\left(2m_W^4 + m_Z^4 - 4m_t^4\right),$$

$$\lambda(T) = \lambda - \frac{3}{16\pi v^4}\left[2m_W^4 \ln\left(\frac{M_W^2}{A_B T^2}\right) + m_Z^4 \ln\left(\frac{M_Z^2}{A_B T^2}\right) - 4m_t^4 \ln\left(\frac{M_t^2}{A_F T^2}\right)\right], \tag{6.66}$$

and $\ln A_B = \ln a_b - 3/2 = 3.91$ and $\ln A_F = \ln a_f - 3/2 = 1.135$. The masses m_h, m_W, m_z, m_t in these expressions are evaluated at the level at zero temperature.

6.4.1   Hard Thermal Loops-Ring Diagrams

We have derived the one-loops contribution to scalar potential from scalars, gauge bosons and fermions. If one computes loop corrections to the one loop potential then the corrections to the scalar and gauge boson loops become infrared divergent. This infrared divergence cancels when the all order loops are summed over to all orders in the ring diagrams [7, 8].

The inclusion of the hard thermal loops amounts to replacing the tree level mass m_i^2 in $V_{CW} + \Delta V^\beta$ by the Debye mass $\mathcal{M}_i^2 = m_i^2 + \Pi_i$. The thermal self energy correction of the longitudinal gauge bosons of $SU(2) \times U(1)_Y$ are

$$\Pi_{GB}^L = \frac{11}{6}T^2 \text{ diagonal}(g^2, g^2, g^2, g'^2). \tag{6.67}$$

In the $(W_L^\pm, Z_L, \gamma_L)$ basis, the Debye masses are

$$
\mathcal{M}_{W_L}^2 = = m_W^2 + \frac{11}{6} g^2 T^2 ,
$$

$$
\mathcal{M}_{Z_L}^2 = \frac{1}{2} \left[m_z^2 + \frac{11}{6} \frac{g^2}{\cos^2 \theta_W} T^2 + \Delta \right] ,
$$

$$
\mathcal{M}_{\gamma_L}^2 = \frac{1}{2} \left[m_z^2 + \frac{11}{6} \frac{g^2}{\cos^2 \theta_W} T^2 - \Delta \right] ,
$$

(6.68)

where

$$
\Delta^2 = m_Z^4 + \frac{11}{3} \frac{g^2 \cos^2 2\theta_W}{\cos^2 \theta_W} \left[m_Z^2 + \frac{11}{3} \frac{g^2}{\cos^2 \theta_W} T^2 \right] T^2 .
$$

(6.69)

There is no thermal correction to the transverse gauge boson masses due to gauge symmetry. Similarly chiral symmetry prevents thermal corrections to the fermions masses. The thermal self energy of the Higgs and Goldstone bosons are

$$
\Pi_h = \Pi_\chi = T^2 \left(\frac{3}{16} g^2 + \frac{1}{16} g'^2 + \frac{1}{4} y_t^2 + \frac{1}{2} \lambda \right)
$$

(6.70)

Using these relations in $V_{CW}(\mathcal{M}_i) + \Delta V^\beta(\mathcal{M}_i)$ and taking the high temperature expansions of $J_B(\beta \mathcal{M}_i)$, we see the following Terms of order $\mathcal{M}_i^2$ are $T^2 \Pi_i$ which are independent of the scalar field and can be dropped. The field independent terms arising from the logs in V_{CW} and ΔV^β cancel. The leading corrections of the Debye masses occurs in the cubic order in $\mathcal{M}_i$.

A good approximation which is followed in the calculations is to add these cubic order ring diagram terms to the one-loop potential. With this procedure the thermal potential takes the form

$$
V(\phi_c, T) = V_0 + \sum_i V_{CW}(m_i) + \Delta V^\beta(m_i) - \frac{T}{12\pi} \sum_j n_j' \left[(\mathcal{M}_i^2)^{3/2} - (m_i^2)^{3/2} \right] .
$$

(6.71)

where $n_j' = (2, 1, 1, 1, 3)$ for $j = (W_L, Z_L, \gamma_L, h, \chi)$. The extra cubic mass terms help in making the thermal potential suitable for giving rise to First order phase transitions.

6.5   Renormalisation Group Improvement of Scalar Potential

The scalar potential depends on the couplings of the theory and the couplings change with the renormalisation scale. The observable quantities like the scalar potential should not change with the renormalisation scale μ,

$$\mu \frac{d}{d\mu} V(\phi) = 0 \tag{6.72}$$

which implies that

$$\left\{ \mu \frac{\partial}{\partial \mu} + \beta_\lambda \frac{\partial}{\partial \lambda} + \beta_g \frac{\partial}{\partial g} + \beta_y \frac{\partial}{\partial y} - \gamma \phi \frac{\partial}{\partial \phi} \right\} V(\phi) = 0 \tag{6.73}$$

where

$$\beta_{c_i} = \mu \frac{\partial c_i}{\partial \mu}, \tag{6.74}$$

are the beta-functions of the respective couplings $c_i = \{\lambda, g_i, y_i\}$ and

$$\gamma = \mu \frac{\partial \log \sqrt{Z_\phi}}{\partial \mu}. \tag{6.75}$$

is the anomalous dimension arising from the wave function renormalization of the scalar. The renormalisation group equations (6.74) and (6.75) for the couplings $c_i(\mu)$ with the boundary condition $c_i(\mu = M_z)$ determined from experiments. If one substitutes these couplings in the potential $V(\phi)$ then the μ dependence from the couplings must cancel the μ dependent terms in the Coleman-Weinberg potential at the same order of couplings. In practice it is simpler to solve the RGE for $c_i(\mu)$ and then evaluate the potential at $\mu = \phi$.

The beta functions for the standard model couplings $c_i = \{\lambda, g_1, g_2, g_3, y_t\}$ which are the quartic coupling, the $SU(3) \times SU(2) \times U(1)_Y$ gauge couplings and the top quark Yukawa coupling (the Yukawa couplings of the other fermions to the Higgs are too small and do not affect the Higgs potential) computed at one-loop are as follows:

$$\beta_\lambda^{(1)} = \frac{1}{16\pi^2} \left[\lambda \left(-9g_2^2 - 3g_1^2 + 12y_t^2 \right) + 24\lambda^2 + \frac{3}{4}g_2^4 + \frac{3}{8} \left(g_1^2 + g_2^2 \right)^2 - 6y_t^4 \right],$$

$$\beta_{y_t}^{(1)} = \frac{1}{16\pi^2} \left[\frac{9}{2}y_t^3 + y_t \left(-\frac{17}{12}g_1^2 - \frac{9}{4}g_2^2 - 8g_3^2 \right) \right],$$

$$\beta_{g_1}^{(1)} = \frac{1}{16\pi^2} \left[\frac{41}{3}g_1^3 \right], \quad \beta_{g_2}^{(1)} = \frac{1}{16\pi^2} \left[-\frac{19}{6}g_2^3 \right], \quad \beta_{g_3}^{(1)} = \frac{1}{16\pi^2} \left[-7g_3^3 \right].$$

$$\tag{6.76}$$

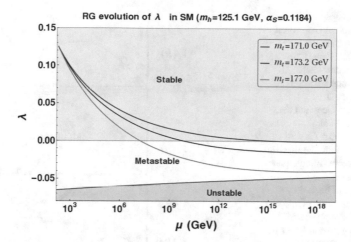

Fig. 6.1 The running of Higgs quartic coupling evaluated with three-loop beta functions of the standard model

The Higgs mass dependence on μ is determined by the γ function which for the standard model is

$$\mu \frac{dm_h^2}{d\mu} = m_h^2 \gamma_h \,,$$

$$\gamma_h^{(1)} = \frac{1}{16\pi^2} \left(12\lambda + 6g_t^2 - \frac{9}{2}g_2^2 - \frac{3}{2}g_1^2 \right) . \qquad (6.77)$$

This group of equations is solved with the initial conditions on the couplings at $\mu = m_t$ given by $\{\lambda = 0.12604,\ y_t = 0.94014,\ g_3 = 1.1666,\ g_2 = 0.64779,\ g_Y = 0.35830\}$ and $v = 246.1\,\text{GeV}$ (compiled in ref. [9]). The beta functions of the standard model have been computed unto three-loops are listed in ref. [10, 11].

In the standard model RG running of the Higgs quartic coupling is such that starting from $\lambda = 1.13$ at $\mu = m_t$ it goes to zero and then becomes negative beyond $\mu = 10^9 - 10^{11}\,\text{GeV}$. This is shown in Fig. 6.1 where the running of *lambda* has been plotted with three-loop beta functions of the standard model. However even taking the one-loop beta functions (6.76) and solving for $\lambda(\mu)$ this feature that λ becomes negative at $\mu > 10^{10}\,\text{GeV}$ can be seen.

The fact that in SM the Higgs quartic coupling $\lambda(\mu)$ can become negative at $\mu \simeq 10^9\,\text{GeV}$ has significant implications for the fate of the universe. The Higgs potential for large field values h (at $T = 0$) is dominated by the quartic term,

$$V(h) = \frac{1}{4} Z_h^2 \lambda(h) h^4 \qquad (6.78)$$

Fig. 6.2 Shape of Higgs potential for the case when $\lambda(\mu) > 0$ for all values μ potential (top panel), for the standard model where tunneling time between the false vacuum at $h = \upsilon \sim 246\,\mathrm{GeV}$ and the global minima at $h \simeq 10^{16} - 10^{18}\,\mathrm{GeV}$ is much larger than the life-time of the universe (middle panel) and the case if say the top-quark had been very heavy $m_t > 178.08\,\mathrm{GeV}$ and the quartic coupling had become negative at lower μ. The universe would have tunneled into the true vacuum state in a time lower than the lifetime of the present universe (top panel)

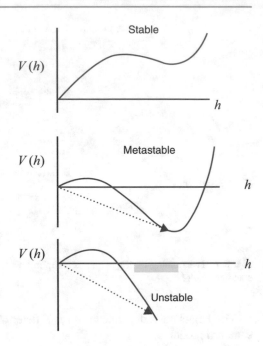

where the wave function renormalization Z_h is determined by solving (6.75) using the γ_h of SM (6.77),

$$Z_h^2 = \exp\left[4 \int_{m_t}^{h} \gamma_h(\mu)\, d \ln \mu\right]. \tag{6.79}$$

A plot of the Higgs potential can have the schematic appearance of one of the three types shown in Fig. 6.2. The shape of the potential depends crucially on two parameters λ (or equivalently the Higgs mass $m_h = \sqrt{2\lambda}\upsilon$) and the top-Yukawa coupling y_t. From an examination of β_λ (6.76) it is clear that larger values of Higgs mass will ensure that the instability scale (where λ becomes negative) increases. On the other hand the top-Yukawa y_t contributes to β_λ with a negative sign which means that large top=mass would make the Higgs potential unstable at a lower μ.

For the best fit values of Higgs and top mass, $m_h = 125.7 \pm 0.03$ and $m_t = 173.34 \pm 0.76$ the shape of the potential is as shown in the middle panel of Fig. 6.2.

In SM λ becomes negative at $\mu = 1.9 \times 10^{10}\,\mathrm{GeV}$ [9, 12]. The tunneling time from the electroweak vacuum of our universe to the minima at $h \sim 10^{17}\,\mathrm{GeV}$ is much larger than the lifetime of the universe which makes the electroweak vacuum metastable [13–15].

For a large top mass $m_t > 178.04\,\mathrm{GeV}$, the universe would be in an unstable vacuum where the tunneling time to the lower vacuum state is less than the lifetime of the universe (Fig. 6.2 lowest panel). On the other hand for a lower top mass $m_t < 171.43\,\mathrm{GeV}$ the universe would be in a stable configuration where the electroweak vacuum is the global minima (Fig. 6.2 lowest panel) [14].

The fact that the Higgs potential is at a metastable vacuum in the present universe also holds up after taking into account the finite temperature corrections to the potential and tunneling rate [16].

6.6 Tunnelling Rate from False Vacuum to True Vacuum

The transition rate per volume from a local minima (false vacuum) to the global minima by quantum tunnelling through a potential barrier (as shown in Fig. 6.3) is given by Callan and Coleman[17, 18],

$$\frac{\Gamma}{V} = A e^{-B} \tag{6.80}$$

where B is the action in Euclidean time, evaluated for a classical solution for the scalar field rolling between the false vacuum and true vacuum through the potential well (the potential barrier in real time changes sign in the eom in Euclidean time). The pre-factor A is taken to be the fourth power of the field value at the true vacuum. The main contribution to the transition rate comes from the value of the bounce action B.

For the standard model which at large h has the form

$$V(h) = \frac{1}{4} \lambda_{eff}(\mu) h^4 \tag{6.81}$$

where $\lambda_{eff}(\mu)$ is effective coupling which on solving the beta functions say to three loop order is

$$\lambda_{eff}(\mu) = Z_h^2(\mu) \left(\lambda^{(1)}(\mu) + \lambda^{(2)}(\mu) + \lambda^{(3)}(\mu) + \cdots \right) \tag{6.82}$$

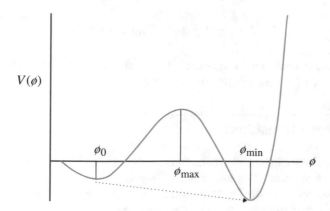

Fig. 6.3 Quantum tunnelling from the false vacuum at $\phi = \phi_0$ the true vacuum at $\phi = \phi_{min}$ through the potential barrier located at $\phi = \phi_{max}$

The bounce solution for a $(\lambda/4)\phi^4$ theory has been computed by Lee and Weinberg [19] and the Euclidean action for the bounce is given by

$$B = \frac{8\pi^2}{3|\lambda|} \tag{6.83}$$

In the standard model the false vacuum is at $h_0 = 246\,\text{GeV}$, the barrier is at $h_{max} \simeq 10^9\,\text{GeV}$, λ_{eff} becomes negative at $\mu = 10^{10}\,\text{GeV}$ and the true vacuum is at $h_{min} = 2.79 \times 10^{17}\,\text{GeV}$. The decay rate per volume of the standard model vacuum for tunnelling to true vacuum is

$$\frac{\Gamma}{V} = h_{min}^4\, \exp\left(\frac{-8\pi^2}{3|\lambda_{eff}(\mu = h_{min})|}\right) \tag{6.84}$$

The quartic coupling changes from $\lambda_{eff}(m_t) = 0.126$ to $\lambda_{eff}(10^{17}\,\text{GeV}) = -0.01408$. The bounce action for the standard model is therefore $B_{SM} = 8\pi^2/(3 \times 0.01408) = 1869.24$ and the decay rate per volume of the standard model vacuum is

$$\frac{\Gamma}{V} = \left(10^{17}\text{GeV}\right)^4 \times 10^{-812}. \tag{6.85}$$

The decay probability in the life-time of the universe is the decay rate times the $(VT)_{light-cone}$, which is the space-time volume of our past-light cone. This is given by

$$(VT)_{light-cone} = \frac{0.15}{H_0} = 3.4 \times 10^{166}\,\text{GeV}^{-4} \tag{6.86}$$

where $H_0 = 67\,\text{km/s mpc} = 1.44 \times 10^{-42}\,\text{GeV}$.
For the standard model (6.85) the probability of vacuum decay is

$$P = \frac{\Gamma}{V}(VT)_{light-cone} \simeq 10^{-578} \tag{6.87}$$

Therefore the standard model vacuum is metastable with a very low probability for vacuum decay within the lifetime of the universe.

6.7   Phase Transitions

The system at equilibrium where the free energy or the finite temperature potential $V(\phi, T)$ is a minima. The minimum of the potential is the vacuum state and quantum excitations of the fields around the minimum corresponds to the particles states. As the universe cools the potential can develop a new minima $\phi \neq 0$ where the potential has a lower value (true vacuum). The universe can shift to the new minima in two different ways, (1) by a continuous change in ϕ as it rolls from the old minima to the new minima or (2) by thermal tunnelling and a discontinuous change from the false vacuum at $\phi = 0$ to the true vacuum at non-zero ϕ. This change in the location of the field globally is called a phase transition. What type of phase transition occurs, depends on the parameters of the potential as we will see below.

Consider a general temperature dependent potential

$$V(\phi, T) = D(T^2 - T_0^2)\phi^2 - E T \phi^3 + \frac{\lambda}{4}\phi^4, \tag{6.88}$$

At high temperatures the $DT^2\phi^2$ term will dominate and there is one minimum of the potential at $\phi_{min} = 0$. As we lower the temperature we see the following changes in the shape of the potential at different temperatures..

At some intermediate temperature $T = T_1$ given by

$$T_1^2 = \frac{8\lambda D T_0^2}{8\lambda D - 9E^2} \tag{6.89}$$

the potential will have an inflection point

$$\phi_{infl} = \frac{3E T_1}{2\lambda}. \tag{6.90}$$

At temperatures $T < T_1$ the inflection point splits into a maxima at

$$\phi_{max} = \frac{3ET}{2\lambda} - \frac{1}{2\lambda}\left(9E^2 T^2 - 8\lambda D(T^2 - T_0^2)\right)^{1/2}, \tag{6.91}$$

and a local minima at

$$\phi_{min} = \frac{3ET}{2\lambda} + \frac{1}{2\lambda}\left(9E^2 T^2 - 8\lambda D(T^2 - T_0^2)\right)^{1/2}. \tag{6.92}$$

As the temperature decreases further, at a particular temperature called the critical temperature, $T = T_c$,

$$T_c^2 = \frac{T_0^2}{1 - E^2/(\lambda D)}. \tag{6.93}$$

the potential at the local minima becomes degenerate with the potential at the origin $\phi_c = 0$,

$$V(\phi = 0, T_c) = V(\phi_{min}, T_c). \tag{6.94}$$

At $T = T_c$ the maxima of the potential is at

$$\phi_{max}(T_c) = E\,T_c\,, \qquad (6.95)$$

and the minima is at

$$\phi_{min}(T_c) = \frac{2ET_c}{\lambda}\,. \qquad (6.96)$$

At temperatures $T < T_c$ the minima at $\phi = 0$ becomes metastable and the global minima is at $\phi > \frac{2ET_c}{\lambda}$. Finally at $T < T_0$ the barrier disappears and the origin $\phi_{max} = 0$ becomes the local minima while the global minima is at

$$\phi_{min} = \frac{3ET_0}{\lambda}\,. \qquad (6.97)$$

If the barrier is large then at temperatures $T_c > T > T_0$ there will be a thermal tunneling of the universe from the false vacuum at $\phi = 0$ to the true vacuum at $\phi_{min} = 3ET_c/\lambda$. This happens by the process of bubble nucleation where in each causal horizon the bubble of the true vacuum ($\phi \neq 0$) is surrounded by the false vacuum ($\phi = 0$). The true vacuum bubbles grow and swallow the region of true vacuum around it. The energy difference between the false vacuum and the true vacuum

$$\epsilon = V(\phi = 0) - V(\phi = \phi_{min}) \qquad (6.98)$$

is released and a sizable fraction of this latent heat can get converted to gravitational waves. This process of tunneling from the false vacuum to true vacuum by bubble nucleation and expansion and release of latent heat is called a First Order Phase Transition (FOPT). The occurrence of a FOPT depends on the presence of a potential barrier as illustrated in Fig. 6.4.

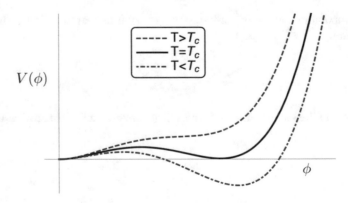

Fig. 6.4 Effective potential as a function of temperature first order phase transition

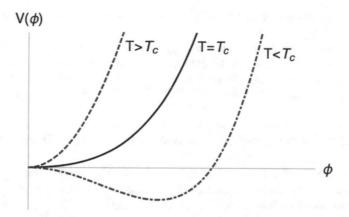

Fig. 6.5 Effective potential as a function of temperature for a second order phase transition

A criterion which is used to describe the strength of FOPT is that for a FOPT we must have

$$\frac{\phi_{min}(T_c)}{T_c} = \frac{2E}{\lambda} > 1 \qquad (6.99)$$

When the cubic term in the potential is absent there is no barrier in the potential and as the universe cools it rolls down adiabatically from the false minima to the try minima. This is called the Second Order Phase Transition (SOPT). The evolution of the thermal potential for the case of a second order phase transition is shown in Fig. 6.5. Second order phase transitions are not accompanied by a release of the latent heat and there is no gravitational wave generation.

Now consider the standard model thermal potential (6.65) and (6.66). We can check the criterion for FOPT,

$$\frac{\phi_{min}(T_c)}{T_c} = \frac{2E}{\lambda} = \frac{2m_W^3 + m_z^3}{\pi v m_h^2} = \left(\frac{42\,\text{GeV}}{m_h}\right)^2. \qquad (6.100)$$

After taking into account the contribution from the ring diagrams this changes to [20],

$$\frac{\phi_{min}(T_c)}{T_c} = \left(\frac{67\,\text{GeV}}{m_h}\right)^2. \qquad (6.101)$$

The measured Higgs mass of 125 GeV therefore implies that the electroweak phase transition was a weak First Order Phase Transition where the barrier arising from the cubic terms in the potential is present but the barrier height is too small for tunneling. The transition between the $\phi_c = 0$ at $T \gg 100\,\text{GeV}$ to $\phi_c = v = 246\,\text{GeV}$ in the present universe was not accompanied by bubble nucleation and release of latent heat.

In the standard model the value of $\lambda(\phi)$ can become negative at very large ϕ ($\lambda(\phi) < 0$ at $\phi > 10^{10}$ GeV). In this case there is a second minima and there may be a tunneling from the metastable vacuum ϕ_{min} Eq. (6.97) to the new global minima as we have discussed in Sect. 6.5.

6.8   Stochastic Gravitational Waves from Phase Transitions

Stochastic gravitational waves signal are characterised by their statistical properties. A stochastic GW signal can be detected by statistical cross-correlation of the strain outputs two laser interferometric GW detectors [21, 22].

Stochastic GW will arise when we have a large number of unresolved sources like compact stars or black hole mergers. They can also arise from early universe processes like inflation and cosmological phase transitions [23].

To predict the energy density and the spectrum of the stochastic GW from phase transitions we need to determine the time $t*$ and temperature T^* of the phase transition.

Gravitational waves produced in the early universe at time t_* and temperature T_* will have their frequency redshifted with scale factor as $f \propto a^{-1}$ and energy density red-shifted as $\rho_{GW} \propto a^{-4}$. The frequency of gravitational waves in the present epoch as a function of their frequency at production f_* is Hubble rate at temperature T_* is $H_* = \left(\frac{4\pi^2 g_*}{45}\right)^{1/2} \frac{T_*^2}{M_P}$.

$$
f = \frac{a_*}{a_0} f_* = \frac{a_*}{a_0} H_* \frac{f_*}{H_*} = \left(\frac{g_{0s}}{g_{*s}}\right)^{1/3} \left(\frac{T_0}{T_*}\right) H_* \frac{f_*}{H_*}
$$

$$
= \left(\frac{g_{0s}}{g_{*s}}\right)^{1/3} \left(\frac{T_0}{T_*}\right) \left(\frac{4\pi^2 g_*}{45}\right)^{1/2} \frac{T_*^2}{M_P} \frac{f_*}{H_*}
$$

$$
= 1.65 \times 10^{-5} \text{ Hz} \left(\frac{g_*}{100}\right)^{1/6} \left(\frac{T_*}{100 \text{ GeV}}\right) \frac{f_*}{H_*} \tag{6.102}
$$

where we have used the entropy conservation $a_*^3 g_{*s} T_*^3 = a_0^3 g_{0s} T_0^3$ and we have taken $T_0 = 2.725 K = 2.348 \times 10^{-13}$ GeV and $g_{*0} = 3.91$ and assumed that all species are in thermal equilibrium and $g_* = g_{*s}$. If the first order phase transition lasts for a duration β^{-1} and takes place in the conformal time interval $(t_*, t_* + \beta^{-1})$. The time scale β^{-1} will be associated with a correlation length R by a velocity $v = R \beta^{-1}$. In the case of GR waves from bubble wall collisions R will be the size of the bubble and v is expansion velocity of the bubble wall. The frequency is related to the comoving wavenumber $k = 2\pi a f$ (where the comoving wavenumber is equal to the physical wavenumber in the present epoch). Writing the factor f_*/H_*

in (6.102) in terms of the conformal wave number we have

$$\frac{f_*}{H_*} = \frac{k}{2\pi \mathcal{H}_*} = \frac{\beta}{2\pi \mathcal{H}_*} \frac{k}{\beta} \tag{6.103}$$

Substituting (6.103) in (6.102) we can estimate the peak velocity of gravitational waves from bubble wall collisions as

$$f_p = 2.62 \times 10^{-6} \, \text{Hz} \left(\frac{g_*}{100}\right)^{1/6} \left(\frac{T_*}{100 \, \text{GeV}}\right) \left(\frac{\beta}{\mathcal{H}_*}\right) \left(\frac{k_p}{\beta}\right) \tag{6.104}$$

For the case of bubble wall collisions $\beta \simeq 10\mathcal{H}_*$ and $k_p \simeq \beta$ [23]. For GW waves generated from MHD turbulence following phase transitions $k_p = (\pi^2/v)\beta$ and it turns out that $f_p^{MHD} \sim 10 f_p^{Bubble}$ [24].

The fractional energy density of the gravitational waves in the present epoch is

$$\Omega_{GW} = \frac{\rho_{GW}}{\rho_{crit}} = \frac{\rho_{*GW}}{\rho_{*crit}} \left(\frac{a_*}{a_0}\right)^4 \frac{\rho_{*crit}}{\rho_{crit}}$$

$$= \Omega_{*GW} \left(\frac{a_*}{a_0}\right)^4 \left(\frac{H_*}{H_0}\right)$$

$$= 1.64 \times 10^{-5} h^{-2} \left(\frac{100}{g_*}\right)^{1/3} \Omega_{*GW} \tag{6.105}$$

where we have taken $H_0 = 2.13 \times 10^{-42} h \, \text{GeV}$. The GW density in the present epoch is independent of temperature (other that the weak dependence on the number of species at T_*) while the shape of the spectrum and the peak frequency depends on the temperature during the phase transition T_*.

The time evolution of the effective entropy degrees of freedom must be taken into account while evolving the frequency and energy density of gravitational waves through the QCD phase transition [25].

The spectral shape of GW radiation is represented by the quantity

$$\frac{\Omega_{GW}}{d \log k} = \frac{1}{\rho_{crit}} \int \frac{dk}{k} \frac{\langle \dot{h}_i(\mathbf{k}, t) \dot{h}_i(\mathbf{k}, t) \rangle}{8\pi G a(t)^2}, \qquad \text{for each polarization } i = +, \times \tag{6.106}$$

The GW spectrum generated by bubble wall collision following a phase transition is given by [24],

$$\frac{d\Omega_{GW}^B}{d \log k} \simeq \frac{2h^2}{3\pi^2} \Omega_{r0} \left(\frac{\mathcal{H}_*}{\beta}\right)^2 \Omega_{S*}^2 v^3 \frac{(k/\beta)^3}{1 + (k/\beta)^4} \tag{6.107}$$

where Ω_{S*} is the fractional energy in the source at the time of phase transition. The GW spectrum generated by MHD turbulence following a phase transition is given by [24],

$$\frac{d\Omega_{GW}^{MHD}}{d\log k} \simeq \frac{8h^2}{\pi^6}\, \Omega_{r0}\left(\frac{\mathcal{H}_*}{\beta}\right)\Omega_{S*}^{3/2}\, v^4\, \frac{(k/\beta)^3}{(1+4k/\mathcal{H}_*)\left[1+(v/\pi^2)(k/\beta)\right]^{11/4}}$$

(6.108)

The era phase transition in the early universe is related to the peak frequency of gravitational waves generated and will be accessible to different gravitational wave detectors [26]. QCD phase transition gives rise to nano-Hertz frequency gravitational waves [27] and will be accessible to pulsar timing array measurements [28]. If there are phase transitions at PeV temperatures they have gravitational wave signal at aLigo frequencies. Search for gravitational waves at different frequency ranges will provide a window into the phase transitions in the early universe [29].

References

1. S. Coleman, E. Weinberg, Radiative corrections as the origin of spontaneous symmetry breaking. Phys. Rev. D **7**, 1888 (1973)
2. G. 't Hooft, M.J.G. Veltman, Regularization and renormalization of gauge fields. Nucl. Phys. B **44**, 189 (1972)
3. L. Dolan, R. Jackiw, Symmetry behavior at finite temperature. Phys. Rev. **D9**, 3320 (1974)
4. J.I. Kapusta, C. Gale, Finite-Temperature Field Theory: Principles and Applications. Cambridge University Press, Cambridge (2009)
5. M. Quiros, Field theory at finite temperature and phase transitions. Acta Phys. Polon. B **38**, 3661 (2007)
6. M. Laine, A. Vuorinen, Basics of thermal field theory. Lect. Notes Phys. **925**, 1 (2016)
7. P.B. Arnold, O. Espinosa, The effective potential and first order phase transitions: Beyond leading-order. Phys. Rev. D **47**, 3546 (1993). Erratum: [Phys. Rev. D **50**, 6662 (1994)]
8. D. Curtin, P. Meade, H. Ramani, Thermal resummation and phase transitions. Eur. Phys. J. C **78**(9), 787 (2018)
9. D. Buttazzo, G. Degrassi, P.P. Giardino, G.F. Giudice, F. Sala, A. Salvio, A. Strumia, Investigating the near-criticality of the Higgs boson. J. High Energy Phys. **1312**, 089 (2013)
10. A.V. Bednyakov, A.F. Pikelner, V.N. Velizhanin, Higgs self-coupling beta-function in the standard model at three loops. Nucl. Phys. B **875**, 552 (2013)
11. K.G. Chetyrkin, M.F. Zoller, β-function for the Higgs self-interaction in the Standard Model at three-loop level. J. High Energy Phys. **1304**, 091 (2013). Erratum: [JHEP **1309**, 155 (2013)]
12. G. Degrassi, S. Di Vita, J. Elias-Miro, J.R. Espinosa, G.F. Giudice, G. Isidori, A. Strumia, Higgs mass and vacuum stability in the standard model at NNLO. J. High Energy Phys. **1208**, 098 (2012)
13. G. Isidori, G. Ridolfi, A. Strumia, On the metastability of the standard model vacuum. Nucl. Phys. B **609**, 387 (2001)
14. V. Branchina, E. Messina, A. Platania, Top mass determination, Higgs inflation, and vacuum stability. J. High Energy Phys. **1409**, 182 (2014)
15. T. Markkanen, A. Rajantie, S. Stopyra, Cosmological aspects of Higgs vacuum metastability. Front. Astron. Space Sci. **5**, 40 (2018)

16. L. Delle Rose, C. Marzo, A. Urbano, On the fate of the standard model at finite temperature. J. High Energy Phys. **1605**, 050 (2016)
17. S.R. Coleman, The fate of the false vacuum. 1. Semiclassical theory. Phys. Rev. D **15**, 2929 (1977). Erratum: [Phys. Rev. D **16**, 1248 (1977)]
18. C.G. Callan Jr., S.R. Coleman, The fate of the false vacuum. 2. First quantum corrections. Phys. Rev. D **16**, 1762 (1977)
19. K.M. Lee, E.J. Weinberg, Tunneling without barriers. Nucl. Phys. B **267**, 181 (1986)
20. A.I. Bochkarev, S.V. Kuzmin, M.E. Shaposhnikov, Electroweak baryogenesis and the Higgs boson mass problem. Phys. Lett. B **244**, 275 (1990)
21. B. Allen, J.D. Romano, Detecting a stochastic background of gravitational radiation: signal processing strategies and sensitivities. Phys. Rev. D **59**, 102001 (1999)
22. E. Thrane, J.D. Romano, Sensitivity curves for searches for gravitational-wave backgrounds. Phys. Rev. D **88**(12), 124032 (2013)
23. C. Caprini, D.G. Figueroa, Cosmological backgrounds of gravitational waves. Class. Quant. Grav. **35**(16), 163001 (2018)
24. C. Caprini, R. Durrer, T. Konstandin, G. Servant, General properties of the gravitational wave spectrum from phase transitions. Phys. Rev. D **79**, 083519 (2009)
25. S. Anand, U.K. Dey, S. Mohanty, Effects of QCD equation of state on the stochastic gravitational wave background. J. Cosmol. Astropar. Phys. **1703**(03), 018 (2017)
26. C. Caprini, et al., Science with the space-based interferometer eLISA. II: gravitational waves from cosmological phase transitions. J. Cosmol. Astropar. Phys. **1604**(04), 001 (2016)
27. C. Caprini, R. Durrer, X. Siemens, Detection of gravitational waves from the QCD phase transition with pulsar timing arrays. Phys. Rev. D **82**, 063511 (2010)
28. F.A. Jenet, G.B. Hobbs, K. Lee, R.N. Manchester, Detecting the stochastic gravitational wave background using pulsar timing. Astrophys. J. Lett. **625**, L123–L126 (2005)
29. P.S.B. Dev, A. Mazumdar, Probing the scale of new physics by advanced LIGO/VIRGO. Phys. Rev. D **93**(10), 104001 (2016). [arXiv:1602.04203 [hep-ph]]
30. G. Isidori, G. Ridolfi, A. Strumia, On the metastability of the standard model vacuum. Nucl. Phys. B **609**, 387 (2001)

Gravitational Waves

7

Abstract

The existence of gravitational waves were first confirmed by the measurement of the time period loss of the Hulse–Taylor binary pulsar which agreed with the calculation of Peters and Mathews of the energy loss of the compact binaries. Since then many more binary pulsars have been observed and the gravitational wave formula from Einstein's gravity tested to 1% accuracy. By monitoring the time period loss of compct stars we can also test for the envy loss due to other light fields like ultra light scalars or gauge bosons which can couple to neutron stars. In this chapter we derive the Peters–Mathews formula in an effective field theory approach as graviton emission from a classical source. We take due care of including the eccentricity of the binary orbits as that accounts for about a factor 10 enhancement in the energy loss rate from compact binaries. We then also use the same technique to derive the expressions for the energy loss due to ultra light scalars and vector bosons from binary stars. The direct observation of gravitational waves by Ligo and Virgo from black hole and neutrons star merger was a landmark evert which has opened the area of multi-messenger astronomy. In this chapter we also derive the analytical expressions for the gravitational was waveform, which can be expected from binary mergers in Einsteins gravity and beyond Einsteins gravity and which can be tested with Ligo and Virgo observations or in the future gravitational wave experiments like Advanced Ligo, LISA, BBO and DECIGO.

7.1 Introduction

The direct observation of gravitational waves from the black hole binary merger GW150914 was a landmark event which confirmed the existence of black holes and ushered in the era of gravitational wave astronomy. The observation of the neutron star merger trough gravitational waves in the GW170817 event which was also seen

© Springer Nature Switzerland AG 2020   191
S. Mohanty, *Astroparticle Physics and Cosmology*,
Lecture Notes in Physics 975, https://doi.org/10.1007/978-3-030-56201-4_7

as the GRB 170817A by Fermi and Integral from the same host galaxy which was later on flowed up by observations by Chandra in X-rays, VLA in radio and Hubble in optical ushered in the era of multi-messenger astronomy.

The gravitational force has a geometrical interpretation based on the Equivalence Principle, which is different from all other fundamental forces which are interpreted in terms of exchange of elementary particles (photon, gluon, $W^{\pm}$, Z). In the regions where the gravitational field is weak (for example well outside the black hole event horizon) the metric can be expanded in the Minkowski background and the field perturbations of the metric interpreted as gravitons. This step was done as a first step in achieving a quantum theory of gravity along the lines of the other quantum field theories of Electroweak and Strong interactions [1–3]. The linearised gravity treatment also makes the calculation of gravitational wave emission [4] conceptually similar to other elementary particle processes and easy to compute compared with the full metric treatment [5]. This method is also used to calculate the emission of scalar particles like axions from neutrons stars which can be observed in pulsar binaries to put constraints on axion-nucleon couplings. The emission rate of light vectors is also calculated using the field theoretic method [6]. There are two types of signals to test the radiation of scalar or vector particles. The compact binaries will lose energy in the process of graviton, scalar and vector emission therefore time period will decrease with time. Observations of binaries [7–12] with atleast one star is a pulsar or a white dwarf will put constraints on the particle emissions. The energy loss due to scalars and vector emissions will also increase the chirp frequency of the observed gravitational signal which can be see in the infall stage by LIGO or VIRGO.

7.2 Linearised Gravity

The linearised gravitonal field $h_{\mu\nu}$ field is the expansion of the metric around the Minkowski background,

$$g_{\mu\nu}(x) = \eta_{\mu\nu} + h_{\mu\nu}(x) \tag{7.1}$$

and we consider the weak field situations $|h_{\mu\nu}| \ll 1$ and keep $h_{\mu\nu}$ to the leading order in the field expansions. The raising and lowering of indices is done by the Minkowski space metric $\eta_{\mu\nu} = \eta^{\mu\nu} = diagonal(1, -1, -1, -1)$. The Christoffel connection and Ricci tensor to the linear order in $h_{\mu\nu}$ are

$$\Gamma^{\lambda}_{\mu\nu} = \frac{1}{2}\eta^{\lambda\rho}\left(\partial_{\mu}h_{\rho\nu} + \partial_{\nu}h_{\rho\mu} - \partial_{\rho}h_{\mu\nu}\right) + O(h^2). \tag{7.2}$$

and

$$R_{\mu\nu} = \partial_\nu \Gamma^\lambda_{\lambda\mu} - \partial_\lambda \Gamma^\lambda_{\mu\nu} + O(h^2)$$

$$= \frac{1}{2}\left(\partial_\alpha \partial^\alpha h_{\mu\nu} - \partial_\lambda \partial_\mu h^\lambda{}_\nu - \partial_\lambda \partial_\nu h^\lambda{}_\mu + \partial_\mu \partial_\nu h^\lambda{}_\lambda\right) \tag{7.3}$$

Einstein's equation $R_{\mu\nu} = 8\pi G(T_{\mu\nu} - \frac{1}{2}g_{\mu\nu}T^\alpha{}_\alpha)$ to the linear order in $h_{\mu\nu}$ reads

$$\Box h_{\mu\nu} - \partial_\lambda \partial_\mu h^\lambda{}_\nu - \partial_\lambda \partial_\nu h^\lambda{}_\mu + \partial_\mu \partial_\nu h^\lambda{}_\lambda = 16\pi G\tilde{T}_{\mu\nu} \tag{7.4}$$

where $\Box \equiv \eta^{\mu\nu}\partial_\mu \partial_\nu$ and

$$\tilde{T}_{\mu\nu} = T_{\mu\nu} - \frac{1}{2}T^\alpha{}_\alpha \eta_{\mu\nu}. \tag{7.5}$$

We can use the general covariance symmetry of general relativity

$$x^\mu \to x'^\mu = x^\mu + \xi^\mu(x),$$

$$\Rightarrow \quad g'^{\mu\nu}(x') = \frac{\partial x'^\mu}{\partial x^\rho}\frac{\partial x'^\nu}{\partial x^\sigma} g^{\rho\sigma}(x),$$

$$\Rightarrow \quad h'_{\mu\nu}(x') = h_{\mu\nu}(x) - \left(\partial_\mu \xi_\nu(x) + \partial_\nu \xi_\mu(x)\right) \tag{7.6}$$

to choose a gauge where (7.4) will be in the simplest form. We choose the harmonic gauge

$$g^{\mu\nu}\Gamma^\lambda_{\mu\nu} = 0$$

$$\Rightarrow \quad \partial_\mu h^\mu{}_\nu = \frac{1}{2}\partial_\nu h^\mu{}_\mu, \tag{7.7}$$

and the e.o.m for the gravitational waves (7.4) reduces to the form

$$\Box h_{\mu\nu} = 16\pi G\tilde{T}_{\mu\nu} \tag{7.8}$$

The harmonic gauge condition (7.7) (four equations) reduces the 10 degrees of freedom of $h^{\mu\nu}$ to 6. If the gravitational wave is propagating in empty space then its e.o.m is $\Box h_{\mu\nu} = 0$ and one can choose a harmonic function ξ^μ (such that $\Box \xi^\mu = 0$) to make one more gauge transformation (7.6) and still retain the form of e.o.m $\Box h'_{\mu\nu} = 0$. This reduces the propagating degrees of the free graviton (polarization states) of gravitational waves to 2 which are also interpreted as the two helicity states of the spin-2 graviton. The conservation of the stress tensor (to the leading order in $h_{\mu\nu}$), $\partial_\mu T^{\mu\nu} = 0$ implies

$$\partial_\mu \tilde{T}^\mu{}_\nu = \frac{1}{2}\partial_\nu \tilde{T}^\mu{}_\mu. \tag{7.9}$$

The plane wave solutions of gravitational waves in free space are of the form

$$h_{\mu\nu}(x) = \mathcal{A}\epsilon_{\mu\nu}\, e^{ik\cdot x} + \mathcal{A}^*\epsilon_{\mu\nu}^*\, e^{-ik\cdot x}\,, \tag{7.10}$$

where $\epsilon_{\mu\nu}$ is a symmetric tensor ($\epsilon_{\mu\nu} = \epsilon_{\nu\mu}$) that represents the polarization states of the gravitons. The plane wave satisfies the equation of motion (7.8) in free space $\Box h_{\mu\nu} = 0$, if

$$k_\alpha k^\alpha = 0\,, \tag{7.11}$$

while the harmonic gauge condition (7.7) is satisfied if

$$k_\mu\, \epsilon^\mu{}_\nu = \frac{1}{2}k_\nu\, \epsilon^\mu{}_\mu\,. \tag{7.12}$$

The harmonic gauge condition reduces the free parameters in $\epsilon_{\mu\nu}$ from 10 to 6. Gravitational waves propagating in empty space has 2 degrees of polarization. We impose one more gauge restriction, the transversality condition,

$$k^\mu\epsilon_{\mu\nu} = k^\nu\epsilon_{\nu\mu} = 0\,. \tag{7.13}$$

The transversality condition (7.13) reduces the free graviton degrees of freedom from 6 to 2. The transversality condition must be consistent with the harmonic gauge condition (7.12) which implies $\epsilon^\mu{}_\mu = 0$. Therefore the free graviton polarization tensor is transverse and traceless and represents two polarization states. The polarization states of a graviton in free space propagating with momentum $k = (\omega, 0, 0, \omega)$ have the following two transverse-traceless polarization states,

$$\epsilon_{\mu\nu}^{(+)TT} = \begin{pmatrix} 0 & 0 & 0 & 0 \\ 0 & 1 & 0 & 0 \\ 0 & 0 & -1 & 0 \\ 0 & 0 & 0 & 0 \end{pmatrix}\,, \qquad \epsilon_{\mu\nu}^{(\times)TT} = \begin{pmatrix} 0 & 0 & 0 & 0 \\ 0 & 0 & 1 & 0 \\ 0 & 1 & 0 & 0 \\ 0 & 0 & 0 & 0 \end{pmatrix}\,.$$

The free graviton propagating in a general direction $\hat{n}$ is given by the solution of $\Box h_{ij} = 0$,

$$h_{ij}(x) = \epsilon_{ij}\, e^{i\omega(\hat{n}\cdot x - t)} + \epsilon_{ij}^*\, e^{-i\omega(\hat{n}\cdot x - t)} \tag{7.14}$$

where ϵ_{ij} is any symmetric 3×3 matrix. In presence of sources there can be other dynamical components of the gravitational field $h_{\mu\nu}$, for example h_{00} is the Newtonian potential sourced by masses and the h_{0i} components are non-zero when there is a rotation of the source. Of the 6 degrees of freedom of the gravitational wave h_{ij} (7.14) only the two transverse ($\hat{n}^i h_{ij} = 0$) and traceless ($h^i{}_i = 0$) modes are the propagating degrees of freedom and the remaining are removed by a choice of gauge. The projector that projects out the gauge modes of the general tensor h_{ij} and retains only the transverse and transverse and traceless part can be constructed from the transverse projection operator $P_{ij} = \delta_{ij} - \hat{n}_i \hat{n}_j$ as follows [2, 13],

$$\Lambda_{ij,kl}^{TT} \equiv P_{ik} P_{jl} - \frac{1}{2} P_{ij} P_{kl} , \qquad (7.15)$$

such that

$$h_{ij}^{TT} = \Lambda_{ij,kl}^{TT} h_{kl} = \Lambda_{ij,kl}^{TT} \tilde{h}_{kl} . \qquad (7.16)$$

where $\tilde{h}_{ij} \equiv h_{ij} - (1/2)h^i{}_i$. The e.o.m (7.8 can also be written as

$$\Box \tilde{h}_{\mu\nu} = 16\pi G T_{\mu\nu} . \qquad (7.17)$$

7.3   Energy Loss by Gravitational Radiation from Binary Neutron Stars or Black Holes

We use the Feynman rules of linearised gravity [3] to compute the gravitational radiation from inspiralling binary black holes [4].

We write a Lagrangian for the graviton field so that we can reproduce the e.o.m from Einsteins gravity. The field $h_{\mu\nu}$ is dimensionless and we define the canonical graviton field with mass dimension-one as $h'_{\mu\nu} = \kappa^{-1} h_{\mu\nu}$ (where $\kappa = \sqrt{32\pi G}$ and the e.o.m of the canonical graviton is from (7.8) given by

$$\Box h'_{\mu\nu} = \frac{\kappa}{2} \tilde{T}_{\mu\nu} \qquad (7.18)$$

The effective Lagrangian from which the equation of motion (7.18) follows can be written as

$$\mathcal{L} = \frac{1}{2} \partial_\alpha h'^{\mu\nu} \partial^\alpha h'_{\mu\nu} + \frac{1}{2} \kappa \, h'^{\mu\nu} \tilde{T}_{\mu\nu} \qquad (7.19)$$

We have made the harmonic gauge choice, $\partial_\mu h'^\mu{}_\nu = (1/2)\partial_\nu h'^\mu{}_\mu$. The universal coupling $\kappa = \sqrt{32\pi G}$ of $h'^{\mu\nu}$ with matter is a reflection of the equivalence principle of classical gravity.

We will treat the binary BH or NS as a classical current T^{ij} determined from the Kepler orbits and calculate the emission of gravitons assuming the interaction

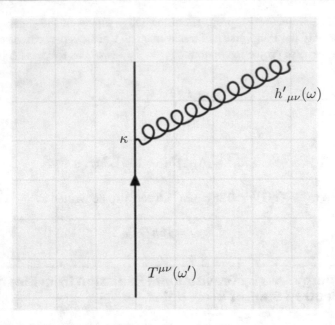

Fig. 7.1 Graviton emission from a classical source

vertex $\kappa h_{ij}^{TT} T^{ij}$. The Feynman diagram of the process is show in Fig. 7.1. The rate
of graviton emission is given by

$$dΓ = \kappa^2 \sum_{\lambda=1}^{2} |T_{ij}(k')\, \epsilon^{TT}{}^{ij}_{(\lambda)}(k)|^2\, 2\pi\, \delta(\omega - \omega')\, \frac{d^3k}{(2\pi)^3}\, \frac{1}{2\omega} \tag{7.20}$$

where $T_{ij}(k')$ is the Fourier transforms of $T_{ij}(x)$. Expanding the modulus squared
in (7.20) we have

$$dΓ = \frac{1}{2}\, \frac{\kappa^2}{(2\pi)^2} \sum_{\lambda=1}^{2} \epsilon^{TT}{}^{ij}_{(\lambda)}{}^{*}(k)\, \epsilon^{TT}{}^{lm}_{(\lambda)}(k)\, T_{ij}^{*}(\omega')\, T_{lm}(\omega')\, \omega\, \delta(\omega - \omega')\, d\omega\, d\Omega_k \tag{7.21}$$

The completeness relation of the graviton polarization tensor is

$$\sum_{\lambda=1}^{2} \epsilon^{TT}{}^{ij}_{(\lambda)}{}^{*}(k)\, \epsilon^{TT}{}^{lm}_{(\lambda)}(k) = 2\Lambda_{ij,lm}^{TT}. \tag{7.22}$$

Substituting (7.22) in (7.21), and doing the angular integrals[1] in the projection operator [2],

$$
\int d\Omega_k \Lambda_{ij,lm}^{TT} T_{ij}^* T_{lm} = \frac{2\pi}{15} \left[11\delta_{il}\delta_{jm} - 4\delta_{ij}\delta_{lm} + \delta_{im}\delta_{jl} \right] T_{ij}^* T_{lm}
$$

$$
= \frac{8\pi}{5} \left[T_{ij}(\omega') T_{ji}^*(\omega') - \frac{1}{3}|T^i{}_i(\omega')|^2 \right] \tag{7.23}
$$

we obtain

$$
d\Gamma = \frac{\kappa^2}{5\pi} \left[T_{ij}(\omega') T_{ji}^*(\omega') - \frac{1}{3}|T^i{}_i(\omega')|^2 \right] \omega \, \delta(\omega - \omega') \, d\omega \tag{7.24}
$$

and the rate of energy loss by the graviton radiation is

$$
\frac{dE_{gw}}{dt} = \int \frac{\kappa^2}{5\pi} \omega^2 \left[T_{ij}(\omega') T_{ji}^*(\omega') - \frac{1}{3}|T^i{}_i(\omega')|^2 \right] \delta(\omega - \omega') \, d\omega \tag{7.25}
$$

To obtain the rate of energy loss by the binary system in Kepler orbit due to graviton emission, we next compute the stress tensor T_{ij} for this system.

In a binary system of stars with masses m_1, m_2 separated by distance R in an elliptical orbit around the centre of mass, the stress tensor is

$$
T_{\mu\nu}(x') = \mu \delta^3(\mathbf{x}' - \mathbf{x}(t)) U_\mu U_\nu \tag{7.26}
$$

where $\mu = m_1 m_2/(m_1 + m_2)$ is the reduced mass and $U_\mu = \gamma(1, \dot{x}, \dot{y}, \dot{z})$ is the four velocity of the reduced mass in the elliptical orbit. Assuming an orbit in the $x - y$ plane the Kepler orbit can be written in the parametric form [14]

$$
x = a(\cos\xi - e), \quad y = a\sqrt{1 - e^2}\sin\xi, \quad \Omega t = \xi - e\sin\xi, \tag{7.27}
$$

where a is the semi-major axis and e the eccentricity of the elliptical orbit. Here $\Omega \equiv (G(m_1 + m_2)/a^3)^{1/2}$ is the fundamental frequency and the Kepler orbit in

---

1

$$
\Lambda_{ij,kl}^{TT}(\hat{n}) = \delta_{ik}\delta_{jl} - \frac{1}{2}\delta_{ij}\delta_{kl} - n_j n_l \delta_{ik} - n_i n_k \delta_{jl} + \frac{1}{2}n_k n_l \delta_{ij} + \frac{1}{2}n_i n_j \delta_{kl} + \frac{1}{2}n_i n_j n_k n_l ,
$$

$$
\int d\Omega \hat{n}_i \hat{n}_j = \frac{4\pi}{3}\delta_{ij}, \quad \int d\Omega \hat{n}_i \hat{n}_j \hat{n}_l \hat{n}_m = \frac{4\pi}{15}(\delta_{il}\delta_{lm} + \delta_{il}\delta_{jm} + \delta_{im}\delta_{jl}).
$$

Fourier space can be expressed in terms of harmonics of Ω. The Fourier transform of the velocity $(\dot{x}, \dot{y})$ in the Kepler orbit can be evaluated as follows

$$\dot{x}_n \equiv \frac{1}{T} \int_0^T e^{i\Omega n t} \, \dot{x} \, dt$$

$$= \frac{-a}{2\pi} \int_0^{2\pi} e^{in(\xi - e \sin \xi)} \, \sin \xi \, d\xi \qquad (7.28)$$

where $T = 2\pi/\Omega$ and we have used the relation $\dot{x} dt = -a \sin \xi d\xi$.

Similarly we can write

$$\dot{y}_n = \frac{a\sqrt{1-e^2}}{2\pi} \int_0^{2\pi} e^{in(\xi - e \sin \xi)} \, \cos \xi \, d\xi$$

$$= \frac{a\sqrt{1-e^2}}{2\pi e} \int_0^{2\pi} e^{in(\xi - e \sin \xi)} \, d\xi$$

$$(7.29)$$

We can express $\dot{x}_n$ and $\dot{y}_n$ in terms of Bessel's functions by using the identity

$$J_n(z) = \frac{1}{2\pi} \int_0^{2\pi} e^{i(n\xi - z \sin \xi)} d\xi. \qquad (7.30)$$

Using (7.30) in (7.28) and (7.29) we obtain the velocities in Fourier space

$$\dot{x}_n = -ia\Omega J'(ne), \quad \dot{y}_n = \frac{a\Omega\sqrt{1-e^2}}{e} J(ne) \qquad (7.31)$$

where prime over the Bessel function denotes derivative w.r.t to the argument. From the Fourier components of the velocities $\dot{x}_n$ and $\dot{y}_n$ we can find the orbit equation in Fourier space,

$$x_n = \frac{\dot{x}_n}{-i\Omega n} = \frac{a}{n} J'(ne), \quad y_n = \frac{\dot{y}_n}{-i\Omega n} = \frac{ia\sqrt{1-e^2}}{ne} J(ne) \qquad (7.32)$$

We need Fourier transform of (7.26) with $\omega' = n\Omega$,

$$T_{ij}(\omega') = \mu \int d^3\mathbf{x}' e^{\mathbf{k}' \cdot \mathbf{x}'} \delta^3(\mathbf{x}' - \mathbf{x}(t)) \frac{1}{T} \int_0^T dt \, e^{in\Omega t} \dot{x}_i(t) \dot{x}_j(t)$$

$$= \mu \frac{1}{T} \int_0^T dt \, e^{in\Omega t} \dot{x}_i(t) \dot{x}_j(t) \qquad (7.33)$$

where we have expanded the $e^{\mathbf{k}'\cdot\mathbf{x}} = 1 + \mathbf{k}' \cdot \mathbf{x} + \cdots$ and retained the leading term as $\mathbf{k}' \cdot \mathbf{x} \sim \Omega a \ll 1$ for binary star orbits. We can change the time integral to integral over ξ, $dt = \Omega^{-1}(1 - e\cos\xi)d\xi$ and express $\dot{x} = -a\sin\xi(d\xi/dt) = -a\sin\xi/(1 - e\cos\xi)$ and $\dot{y} = a\sqrt{1 - e^2}\cos\xi(d\xi/dt) = a\sqrt{1 - e^2}/(1 - e\cos\xi)$ and express the components of $T_{ij}(\omega')$ as

$$T_{xx}(\omega') = \frac{\mu\Omega^2 a^2}{2\pi} \int_0^{2\pi} d\xi\, e^{in(\xi - e\sin\xi)} \frac{\sin^2 \xi}{(1 - e\cos\xi)},$$

$$T_{yy}(\omega') = \frac{\mu\Omega^2 a^2(1 - e^2)}{2\pi} \int_0^{2\pi} d\xi\, e^{in(\xi - e\sin\xi)} \frac{\cos^2 \xi}{(1 - e\cos\xi)},$$

$$T_{xy}(\omega') = \frac{\mu\Omega^2 a^2\sqrt{1 - e^2}}{2\pi} \int_0^{2\pi} d\xi\, e^{in(\xi - e\sin\xi)} \frac{(-\sin\xi\,\cos\xi)}{(1 - e\cos\xi)}. \tag{7.34}$$

Integrating by parts and using the Bessel function identities[2] we can write the stress tensor components (7.34) in terms of Bessel's functions as

$$T_{xx}(\omega') = -\frac{\mu\,\omega'^2 a^2}{2n}\left[J_{n-2}(ne) - 2e\,J_{n-1}(ne) + 2e\,J_{n+1}(ne) - J_{n+2}(ne)\right],$$

$$T_{yy}(\omega') = \frac{\mu\,\omega'^2 a^2}{2n}\left[J_{n-2}(ne) - 2e\,J_{n-1}(ne) + \frac{4}{n}\,J_n(ne)\right.$$

$$\left. +2e\,J_{n+1}(ne) - J_{n+2}(ne)\right],$$

$$T_{xy}(\omega') = \frac{-i\,\mu\omega'^2 a^2}{2n}\,(1 - e^2)^{1/2}\left[J_{n-2}(ne) - 2\,J_n(ne) + J_{n+2}(ne)\right]. \tag{7.35}$$

From (7.35) we obtain

$$\left[T_{ij}(\omega')T_{ji}^*(\omega') - \frac{1}{3}|T^i{}_i(\omega')|^2\right] = \mu^2\,a^4\,\omega'^4\,\tilde{g}(n, e) \tag{7.36}$$

where

$$\tilde{g}(n, e) = \frac{1}{32\,n^2}\left\{[J_{n-2}(ne) - 2e\,J_{n-1}(ne) + 2e\,J_{n+1}(ne) - J_{n+2}(ne)]^2\right.$$

$$\left. +(1 - e^2)[J_{n-2}(ne) - 2\,J_n(ne) + J_{n+2}(ne)]^2 + \frac{4}{3n^2}[J_n(ne)]^2\right\} \tag{7.37}$$

---

[2]

$$J_{n-1}(z) - J_{n+1}(z) = 2J_n'(z),\quad J_{n-1}(z) + J_{n+1}(z) = \frac{2n}{z}J_n(z).$$

Substituting (7.36) into (7.25) we have the rate of energy loss of a binary system in an elliptical orbit as sum of radiation in the n harmonics of the fundamental frequency Ω,

$$\frac{dE_{gw}}{dt} = \frac{32G}{5} \sum_{n=1}^{\infty} (n\,\Omega)^2 \cdot \mu^2\, a^4\, (n\,\Omega)^4\, \tilde{g}(n, e). \qquad (7.38)$$

Using the Bessel function identity $\sum_{n=1}^{\infty} n^6\, \tilde{g}(n, e) = (1 - e^2)^{-7/2} \left(1 + \frac{73}{24}e^2 + \frac{37}{96}e^4\right)$ [5] we find that the energy loss by gravitational wave emission for binary stars,

$$\boxed{\frac{dE_{gw}}{dt} = \frac{32}{5} G\,\Omega^6 \left(\frac{m_1 m_2}{m_1 + m_2}\right)^2 a^4\, (1 - e^2)^{-7/2} \left(1 + \frac{73}{24}e^2 + \frac{37}{96}e^4\right).}$$

$$(7.39)$$

The time period of the elliptical orbit depends upon the energy E, so energy loss leads to a change of the time period of the orbit at the rate

$$\frac{dP_b}{dt} = -6\pi\, G^{-3/2}(m_1 m_2)^{-1}(m_1 + m_2)^{-1/2} a^{5/2} \left(\frac{dE_{gw}}{dt}\right) \qquad (7.40)$$

gives us the Peters and Mathews formula[5] for the classical gravitational radiation from binary systems,

$$\boxed{\frac{dP_b}{dt} = -\frac{192\pi}{5} G^{5/3} \Omega^{5/3} \frac{m_1 m_2}{(m_1 + m_2)^{1/3}} (1 - e^2)^{-7/2} \left(1 + \frac{73}{24}e^2 + \frac{37}{96}e^4\right).}$$

$$(7.41)$$

As expected tree level quantum calculation agrees with the classical result.

In a circular orbit the angular velocity of the binary is a constant over the orbital period and the Fourier expansion of the orbit $x(\omega)$ and $y(\omega)$ would be only term, the $\omega = \Omega$. In a eccentric orbit the angular velocity is not constant and this means that the Fourier expansion must sum over the harmonics $n\Omega$ of the fundamental. This makes the calculation tedious, but for binary pulsars with large eccentricity the enhancement factor $f(e)$ to the energy loss can be large and must be retained. For the Hulse–Taylor binary where $e = 0.617$ the enhancement factor $f(e) = 12$.

For the Hulse–Taylor binary, the expression (7.41) we have $\Omega = 0.2251 \times 10^{-3}$ sec^{-1}, the pulsar mass $m_1 = 1.4414 \pm 0.0002 M_\odot$, mass of the companion $m_2 = 1.3867 \pm 0.0002 M_\odot$, $e = 0.6171338(4)$ and for the parameters of the H-T binary, yields the energy loss is $dE/dt = 3.2 \times 10^{33}$ erg/sec and orbital period loss due to the gravitational radiation $\dot{P}_b = -2.40263 \pm 0.0005 \times 10^{-12}$ which can be compared with the observed value from the Hulse–Taylor binary [7,9] (after correcting for the

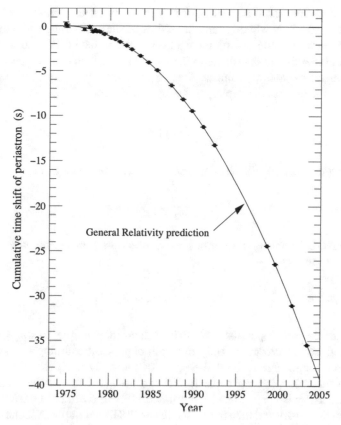

Fig. 7.2 Cumulative period loss over observation time of Hulse–Taylor binary pulsar shows agreement with GR prediction of gravitational wave radiation. Figure taken from PDG [15]

galactic motion) $\dot{P}_b(observed) = -2.40262 \pm 0.00005 \times 10^{-12}$. The cumulative time period loss over a period of 40 years of observations is shown in Fig. 7.2 with the predictions from general relativity. The period loss by gravitational emission has been seen in many more [10–12] neutron star and neuron star -white dwarf binaries since the first observation from the Hulse–Taylor binary.

7.4   Ultra-Light Scalar Radiation from Compact Binary Stars

Ultra light scalars with mass $m \sim 10^{-20} - 10^{-22}$ eV will have De Broglie wavelengths $\lambda = \hbar/(mv)$ is of the scale of dwarf galaxies ~ 2 kpc and can be candidates for dark matter [16, 17]. To prevent the light mass from receiving large quantum corrections these scalars can be Pseudo–Nambu–Golstone–Bosons (PNGB) which have a shift symmetry $\phi \to \phi + 2\pi f$. Axion like particles (ALPS) in this mass range can serve as the ultra light dark matter. The mass of these particles is smaller than the orbital frequency of neutron stars or neutron star- white dwarf binaries $\Omega \sim 10^{-19}$ eV which means that if they forma a long range field the exterior of neutron stars or white dwarfs then they will be radiated like the gravitational waves from binaries and contribute to the period loss which can be observed if one of the compact binaries is a pulsar. Due to the shift symmetry PNGB couple to fermions by the derivative coupling to the pseudo-vector[3]

$$\mathcal{L} = \frac{1}{2f} (\partial_\mu \phi) \, \bar{\psi} \gamma_\mu \gamma_5 \psi \tag{7.43}$$

which using the e.o.m of the fermion $\not{\partial} \psi = m \psi$ can also be written as

$$\mathcal{L} = -\frac{m}{f} \phi \, \bar{\psi} \gamma_5 \psi \tag{7.44}$$

and the PNGB couples to fermions as a pseudo-sclar. In the non-relativistic limit (7.44) reduces to

$$\mathcal{L} = -\frac{m}{f} \phi \psi^\dagger \Sigma \, \psi \tag{7.45}$$

where Σ is the spin operator. The PNGB therefore couple to the spin of the fermion and in a macroscopic body this spin dependent coupling can give rise to macroscopic long range force if the spins of the body are polarized. Such spin dependent macroscopic long range forces can be tested in spin-polarized torsion balance experiments [18]. It has been pointed [19, 20] out that in case there is CP violation in the theory and then it is possible for PNGB to couple as scalars $g_s \phi \bar{\psi} \psi$ and generate long range forces in unpolarized macroscopic bodies like neutron stars.

Neutron stars or white dwarfs immersed in the periodic potential of PNGB dark matter can also have a long range field which comes from matching the field value inside and outside the NS or WD [21].

In this section we derive the expression for the energy and period loss of binary orbits by emission of ultra-light scalars ($m < \Omega \sim 10^{-19}$ eV). Consider a coupling between scalar fields ϕ and the baryons of the form

$$\mathcal{L}_s = g_s \, \phi \, \bar{\psi} \, \psi \tag{7.46}$$

---

[3]Derivative coupling to the vector current vanishes when the fermion number is conserved

$$\mathcal{L} = \frac{1}{2f} (\partial_\mu \phi) \, \bar{\psi} \gamma_\mu \psi = -\frac{1}{2f} \phi \partial_\mu (\bar{\psi} \gamma_\mu \psi) = 0 \,. \tag{7.42}$$

which for a macroscopic baryon source can be written as

$$\mathcal{L}_s = g_s \, \phi \, q(x) \tag{7.47}$$

where $q(x) = \psi^\dagger \psi$ is the baryon number density. A neutron star with radius $\sim 10 \, \text{km}$ can be regarded as a point source since the Compton wavelength of the radiation $\sim \Omega^{-1} = 10^9 \, \text{km}$ is much larger than the dimension of the source. The baryon number density $q(x)$ for the binary stars (denoted by $a = 1, 2$) may be written as

$$q(x) = \sum_{a=1,2} N_a \, \delta^3(\mathbf{x} - \mathbf{x}_a(t)) \tag{7.48}$$

where $N_a \sim 10^{57}$ is the total number of baryons in the neutron star and $\mathbf{x}_a(t)$ represents the Keplerian orbit of the binary stars. For the coupling (7.47) and the source (7.48) the rate of scalar particles emitted from the neutron star in orbit with frequency Ω is

$$d\Gamma = g_s^2 |q(\omega)|^2 (2\pi) \, \delta(\omega - \omega') \frac{d^3\omega'}{(2\pi)^3 \, 2\omega'} \tag{7.49}$$

the rate of energy loss by massless scalar radiation is

$$\frac{dE_s}{dt} = \int g_s^2 |q(\omega)|^2 \, \omega' \, (2\pi) \, \delta(\omega - \omega') \frac{d^3\omega'}{(2\pi)^3 \, 2\omega'} \tag{7.50}$$

where $q(\omega)$ is the Fourier expansion of the source density (7.48)

$$q(\omega) = \frac{1}{2\pi} \int e^{i\mathbf{k}\cdot\mathbf{x}} e^{-i\omega t} \sum_{a=1,2} N_a \delta^3(\mathbf{x} - \mathbf{x}_a(t)) d^3x \, dt \tag{7.51}$$

with $\omega = n\Omega$. Going over to the c.m. coordinates $\mathbf{r} = (x, y)$ by substituting $\mathbf{x}_1 = m_2 \mathbf{r}/(m_1 + m_2)$, $\mathbf{x}_2 = -m_1 \mathbf{r}/(m_1 + m_2)$ we have

$$q(\omega) = (N_1 + N_2) \, \delta(\omega) + \left(\frac{N_1}{m_1} - \frac{N_2}{m_2} \right) M \, (\, i k_x \, x(\omega) + i k_y \, y(\omega))$$

$$+ O(\mathbf{k} \cdot \mathbf{r})^2 \tag{7.52}$$

where $(x(\omega), y(\omega))$ are the Fourier components of the Kepler orbit of the reduced mass in the c.m. frame given by (7.32). The first term in $n(\omega)$ is a delta function which has vanishing contribution to (7.52). The leading non-zero contribution comes from the second term in (7.52). Substituting for the Kepler orbit (7.32),

$$x(\omega) = \frac{a}{n} J_n'(ne), \quad y(\omega) = \frac{i a \sqrt{1 - e^2}}{ne} J_n(ne) \tag{7.53}$$

in (7.52) we obtain the expression for $|q(\omega)|^2$ given by

$$|q(\omega)|^2 = \frac{1}{12} \left(\frac{N_1}{m_1} - \frac{N_2}{m_2} \right)^2 M^2 a^2 \Omega^2 \left[J_n'^2(ne) + \frac{(1 - e^2)}{e^2} J_n^2(ne) \right] \tag{7.54}$$

where we have used the angular average $\langle k_x^2 \rangle = \langle k_y^2 \rangle = \frac{1}{3}(n\Omega)^2$. Substituting (7.54) in (7.50) we have the rate of energy loss by massless scalars

$$\frac{dE_s}{dt} = \frac{1}{6\pi} \left(\frac{N_1}{m_1} - \frac{N_2}{m_2} \right)^2 M^2 g_s^2 a^2 \Omega^4 \sum_n n^2 \left[J_n'^2(ne) + \frac{(1 - e^2)}{e^2} J_n^2(ne) \right] \tag{7.55}$$

The mode sum can be carried out using the Bessels function series formulas [5] to obtain $\sum_n n^2 \left[J_n'^2(ne) + (1 - e^2)e^{-2} J_n^2(ne) \right] = (1/4) (2 + e^2)(1 - e^2)^{-5/2}$. The energy loss (7.55) in terms of the orbital parameters Ω, a and e is given b

$$\boxed{\frac{dE_s}{dt} = \frac{1}{24\pi} \left(\frac{N_1}{m_1} - \frac{N_2}{m_2} \right)^2 M^2 g_s^2 \Omega^4 a^2 \frac{(1 + e^2/2)}{(1 - e^2)^{5/2}}.} \tag{7.56}$$

Since $N_a m_n = m_a - \epsilon_a$ where $\epsilon_a = \frac{G m_a^2}{R_a}$ is the gravitational binding energy and m_n the neutron mass, the factor $\left(\frac{N_1}{m_1} - \frac{N_2}{m_2} \right) = G \left(\frac{m_1}{R_1} - \frac{m_2}{R_2} \right)$. For the H-T binary $m_1 - m_2 \simeq 0.02 M_\odot$ and $R_a \sim 10$ km, therefore $\left(\frac{N_1}{m_1} - \frac{N_2}{m_2} \right) \simeq 3 \times 10^{-3}$ GeV^{-1}.

The rate of change of the orbital period due to energy loss of scalar bosons as well as gravitons is

$$\frac{dP_b}{dt} = -6\pi G^{-3/2} (m_1 m_2)^{-1} (m_1 + m_2)^{-1/2} a^{5/2} \left(\frac{dE_s}{dt} + \frac{dE_{GW}}{dt} \right), \tag{7.57}$$

where $\frac{dE_{GW}}{dt}$ is the rate of energy loss due to quadrupole formula for the gravitational radiation and is given by (7.39).

For the H-T binary the rate of energy loss turns out to be

$$\frac{dE_s}{dt} = g_s^2 \times 9.62 \times 10^{67} \text{ ergs/sec} \tag{7.58}$$

Assuming that this is less than 1% of the gravitational energy loss i.e. $dE/dt \leq 10^{31}$ ergs/sec, we obtain gives an upper bound on scalar nucleon coupling $g_s < 3 \times 10^{-19}$. This bound holds for models where the mass of the Nambu–Goldstone boson is smaller than the frequency $\Omega \simeq 10^{-19}$ eV of the binary orbit.

The period loss in the H-T system has been determined by measuring the time of periastron over a period of almost 19 years. The accuracy of the measured value of period loss increases quadratically with time. If in the course of observation one finds a significant discrepancy between the observed value of period loss and the prediction of the gravitational quadrupole formula, it would be a compelling signal of physics beyond standard model.

7.5 Vector Boson Radiation by Neutron Star Binaries

Ultra light vector bosons can also be radiated from neutron star binaries if the gauge boson mass is lower than the angular frequency of the binaries. Gauge bosons which couple to baryon number can be constrained by fifth force experiments. It is more difficult to constrain vector bosons which couple only to lepton number. In the standard model the any one of the combination of leptonic charges $L_e - L_\mu$, $L_e - L_\tau$ and $L_\mu - L_\tau$ can be gauged and is anomaly free [22,23]. Of these the $L_e - L_\mu$ and $L_e - L_\tau$ long range gauge bosons which are generated by astrophysical bodies like the earth and the sun can be probed by neutrino oscillations experiment [24]. In case the gauge bosons couple to the $L_\mu - L_\tau$ quantum number, they can couple to the muon charge of neutron stars and if their mass $M_{Z'} < \Omega$ the angular frequency of binary neutron stars, they can be radiated over time and result in a loss of the time period of the binary stars and hence can be probed in the binary pulsar timings. In this section we derive the formula for energy loss of binary neutron stars due to emission of ultra light vector bosons [6].

The interaction between the ultra-light gauge boson Z' and fermions is described by the Lagrangian

$$\mathcal{L}_I = g J^\mu Z'_\mu + \frac{M_{Z'}^2}{2} Z'_\mu Z'^\mu \tag{7.59}$$

The Z' boson is a U(1) gauge boson and gauge invariance demands that the current is conserved $k_\mu J^\mu = 0$. The mass term arises from a spontaneous symmetry breaking.

The rate of energy loss by vector boson emission by a classical current is

$$\frac{dE_V}{dt} = g^2 \sum_{\lambda=1}^{3} [J^\mu(k') J^{\nu*}(k') \epsilon_\mu^\lambda(k) \epsilon_\nu^{\lambda*}(k)] \, 2\pi \, \delta(\omega - \omega') \, \omega \, \frac{d^3k}{(2\pi)^3 2\omega} \tag{7.60}$$

where $J^\mu(k')$ is the Fourier transform of $J^\mu(x)$ and $\epsilon_\mu^\lambda(k)$ is the polarization vector of massive vector boson. The polarization sum is given as,

$$\sum_{\lambda=1}^{3} \epsilon_\mu^\lambda(k)\epsilon_\nu^{\lambda *}(k) = -g_{\mu\nu} + \frac{k_\mu k_\nu}{M_{Z'}^2}. \tag{7.61}$$

If Z' boson is a gauge field then from gauge invariance $k_\mu J^\mu = 0$ and the second term in the polarization sum Eq. (7.61) will not contribute to the energy loss formula. The expression for the energy loss reduces to

$$\frac{dE_V}{dt} = \frac{g^2}{2(2\pi)^2} \int |-(J^0(\omega'))^2 + (J^i(\omega'))^2| \delta(\omega - \omega')\omega \left(1 - \frac{M_{Z'}^2}{\omega^2}\right)^{\frac{1}{2}} d\omega d\Omega_k. \tag{7.62}$$

The current density for the binary stars denoted by $a = 1, 2$ may be written as,

$$J^\mu(x) = \sum_{a=1,2} Q_a \delta^3(\mathbf{x} - \mathbf{x}_a(t))u_a^\mu, \tag{7.63}$$

where Q_a is the total charge of the neutron star due to muons and $\mathbf{x}_a(t)$ denotes the Kepler orbit of the binaries. $u_a^\mu = (1, \dot{x}_a, \dot{y}_a, 0)$ is the non relativistic four velocity in the x–y plane of the neutron stars. The Fourier transform of Eq. (7.63) for the spatial part with $\omega' = n\Omega$ is,

$$J^i(\omega') = \int \frac{1}{T} \int_0^T dt e^{in\Omega t} \dot{x}_a^i(t) \sum_{a=1,2} Q_a d^3\mathbf{x}' e^{-i\mathbf{k}'.\mathbf{x}'} \delta^3(\mathbf{x}' - \mathbf{x}_a(t)). \tag{7.64}$$

We expand the $e^{i\mathbf{k}'.\mathbf{x}'} = 1 + i\mathbf{k}'.\mathbf{x}' + \dots$ and retained the leading order term as $\mathbf{k}'.\mathbf{x}' \sim \Omega a \ll 1$ for binary star orbits. Hence Eq. (7.64) becomes

$$J^i(\omega') = \frac{Q_1}{T} \int_0^T dt e^{in\Omega t} \dot{x}_1^i(t) + \frac{Q_2}{T} \int_0^T dt e^{in\Omega t} \dot{x}_2^i(t). \tag{7.65}$$

In the c.o.m coordinates we have $x_1^i = \frac{m_2 x^i}{m_1 + m_2} = \frac{M}{m_1}x^i$ and $x_2^i = -\frac{m_1 x^i}{m_1 + m_2} = -\frac{M}{m_2}x^i$. $M = m_1 m_2/(m_1 + m_2)$ is the reduced mass of the compact binary system. Hence we rewrite the current density as

$$J^i(\omega') = \frac{1}{T}\left(\frac{Q_1}{m_1} - \frac{Q_2}{m_2}\right)M \int_0^T dt e^{in\Omega t} \ddot{x}^i(t). \tag{7.66}$$

For the Kepler orbit using (7.31) in (7.66) we obtain

$$J^x(\omega') = \Omega\left(\frac{Q_1}{m_1} - \frac{Q_2}{m_2}\right)M\frac{1}{2\pi}\int_0^T dt e^{in\Omega t} \dot{x}^i(t)$$

$$= -ia\Omega\left(\frac{Q_1}{m_1} - \frac{Q_2}{m_2}\right)M J_n'(ne). \tag{7.67}$$

and

$$J^y(\omega') = \Omega\left(\frac{Q_1}{m_1} - \frac{Q_2}{m_2}\right)M\frac{a\sqrt{1-e^2}}{e}J_n(ne). \tag{7.68}$$

Therefore

$$(J^i)^2 = (J^x)^2 + (J^y)^2 = (a^2\Omega^2 M^2\left(\frac{Q_1}{m_1} - \frac{Q_2}{m_2}\right)^2\left[J_n'^2(ne) + \frac{(1-e^2)}{e^2}J_n^2(ne)\right]. \tag{7.69}$$

From Eq. (7.63) we have

$$J^0(\omega) = \frac{1}{2\pi}\int e^{i\mathbf{k}'\cdot\mathbf{x}'}e^{-i\omega t}\sum_{a=1,2} Q_a\delta^3(\mathbf{x}' - \mathbf{x}_a(t))d^3\mathbf{x}'dt. \tag{7.70}$$

Going to the c.o.m frame the integral results in

$$J^0(\omega) = (Q_1 + Q_2)\delta(\omega) + iM\left(\frac{Q_1}{m_1} - \frac{Q_2}{m_2}\right)(k_x x(\omega) + k_y y(\omega)) + O((\mathbf{k}.\mathbf{r})^2), \tag{7.71}$$

where $x(\omega) = a J_n'(ne)/n$ and $y(\omega) = ia\sqrt{1-e^2}J_n(ne)/ne$ are the Fourier transform of the orbital coordinates. The first term in Eq. (7.71) does not contribute due to the delta function $\delta(\omega)$. Therefore considering the second term as the leading order contribution we obtain

$$(J^0(\omega))^2 = -\frac{1}{3}a^2 M^2\Omega^2\left(1 - \frac{M_{Z'}^2}{n^2\Omega^2}\right)\left(\frac{Q_1}{m_1} - \frac{Q_2}{m_2}\right)^2$$

$$\times \left(J_n'^2(ne) + \frac{1-e^2}{e^2}J_n^2(ne)\right), \tag{7.72}$$

where we have used $< k_x^2 >=< k_y^2 >= k^2/3$ and $\omega = n\Omega$. Using Eqs. (7.69) and (7.72) in Eq. (7.62) we obtain the rate of energy loss

$$\frac{dE_V}{dt} = \frac{g^2}{3\pi} a^2 M^2 \left(\frac{Q_1}{m_1} - \frac{Q_2}{m_2} \right)^2 \Omega^4$$

$$\times \sum_{n=1} n^2 \left[J_n'^2 (ne) + \frac{(1 - e^2)}{e^2} J_n^2 (ne) \right] \left(1 - \frac{n_0^2}{n^2} \right)^{\frac{1}{2}} \left(1 + \frac{1}{2} \frac{n_0^2}{n^2} \right). \quad (7.73)$$

where $n_0 = (M_{Z'}/\Omega)$. In the limit $M_{Z'} \ll \Omega$ energy loss from Eq. (7.73) can be written as a closed form expression

$$\boxed{\frac{dE_V}{dt} = \frac{g^2}{6\pi} a^2 M^2 \left(\frac{Q_1}{m_1} - \frac{Q_2}{m_2} \right)^2 \Omega^4 \frac{(1 + \frac{e^2}{2})}{(1 - e^2)^{\frac{5}{2}}}.} \quad (7.74)$$

The rate of change of the orbital period due to energy loss of vector bosons is

$$\frac{dP_b}{dt} = -6\pi G^{-3/2} (m_1 m_2)^{-1} (m_1 + m_2)^{-1/2} a^{5/2} \left(\frac{dE_V}{dt} + \frac{dE_{GW}}{dt} \right), \quad (7.75)$$

where $\frac{dE_{GW}}{dt}$ is the rate of energy loss due to quadrupole formula for the gravitational radiation and is given by (7.39).

7.6   Waveform of Gravitational Waves

Gravitational waves waves whose existence was proved by the energy loss of Hulse–Taylor binary Sect. 7.3 and were finally directly observed by the LIGO collaboration in 2016 [25] by laser interferometry. When a gravitational waves $h_{\mu\nu}(t)$ passes through the interferometer the arm lengths changes as a function of time. The proper length (along say the x-axis) of an interferometer arm is related to its coordinate length L_c as

$$L = \int_0^{L_c} dx \sqrt{g_{xx}} = \int_0^{L_c} dx \left(1 + h_{xx}^{TT} \right)^{1/2} \quad (7.76)$$

which implies that the strain along the x-axis is $\Delta L/L = (1/2) h_{xx}^{TT}$. For a h_+ polarisation gravitational wave propagating along the z-axis, $h_+(t, z_0) = h_{xx}^{TT} = -h_{yy}^{TT}$ which means that when the x-axis elongates the y-axis is contracts as the gravitational wave passes through the interferometer. For a $h_\times$ polarized gravitational wave $h_\times(t, z_0) = h_{xy}^{TT} = h_{yx}^{TT}$ and the squeezing and stretching is rotated by $45°$ compared to the h_+ polarization case.

The waveform of the gravitational wave is therefore directly observable and it can tell us about the distance and nature of the source. In this section we derive the form of gravitational waves from mergers of binary black-holes or neutron stars.

The gravitational wave from a source described by the stress tensor T_{ij} is given by

$$\Box \tilde{h}_{ij} = -16\pi G \, T_{ij} \tag{7.77}$$

This equation can be solved by the Green's function method. The Green's function $G(t, \mathbf{x}; t', \mathbf{x}')$ is the solution of the delta-function source equation

$$\Box G(t, \mathbf{x}; t', \mathbf{x}') = \delta^3(\mathbf{x} - \mathbf{x}') \, \delta(t - t') \tag{7.78}$$

The Green's function solution of (7.78) is

$$G(t, \mathbf{x}; t', \mathbf{x}') = -\frac{\delta(t' - [t - |\mathbf{x} - \mathbf{x}'|])}{4\pi \, |\mathbf{x} - \mathbf{x}'|}. \tag{7.79}$$

The signal emitted at time t' from a source located at $\mathbf{x}'$ reaches the detector at $\mathbf{x}$ at retarded time $t = [t' + |\mathbf{x} - \mathbf{x}'|]$.

With the Green's function (7.79) we can solve for the gravitational wave $h_{ij}(t, \mathbf{x})$ generated by the source $T_{ij}(t', \mathbf{x}')$ as,

$$\tilde{h}_{ij}(t, \mathbf{x}) = 4G \int d^3x' \, \frac{T_{ij}(t - |\mathbf{x} - \mathbf{x}'|, \mathbf{x}')}{|\mathbf{x} - \mathbf{x}'|} \tag{7.80}$$

We choose the origin of the coordinate system at the source, The location of the detector is $\mathbf{x} = \mathbf{n}\,r$. When the distance of the detector $|\mathbf{x}| = r$ is large compared to the size of the source R_s we can expand

$$|\mathbf{x} - \mathbf{x}'| = r - \mathbf{n} \cdot \mathbf{x}' + O\left(\frac{R_s^2}{r}\right) \tag{7.81}$$

and we can write (7.80) (to the leading order in $1/r$) as,

$$\tilde{h}_{ij}(t, \mathbf{x}) = \frac{4G}{r} \int d^3x' \, T_{ij}\left(t - r - \mathbf{n} \cdot \mathbf{x}', \mathbf{x}'\right) \tag{7.82}$$

The transverse-traceless components of the gravitational waves are obtained by the use of the TT-projector (7.15),

$$h_{ij}^{TT}(t, \mathbf{x}) = \frac{4G}{r} \Lambda_{ij,kl}^{TT} \int d^3x' \, T^{kl}\left(t - r - \mathbf{n} \cdot \mathbf{x}', \mathbf{x}'\right) \tag{7.83}$$

We Taylor expand T^{kl} in (7.83) around the retarded time $t - r$ as,

$$T^{kl}\left(t - r - \mathbf{n} \cdot \mathbf{x}', \mathbf{x}'\right) = T^{kl} + n_i x'^i \, \dot{T}^{kl} + \frac{1}{2} n_i x'^i \, n_j x'^j \, \ddot{T}^{kl} + \cdots \Big|_{(t-r,\mathbf{x}')} \tag{7.84}$$

where overdots on T^{kl} represent derivative w.r.t time t. We can define the moments of the stress tensor of the source as

$$S^{kl} \equiv \int d^4\mathbf{x}' T^{kl}, \quad S^{kl,i} \equiv \int d^4\mathbf{x}' T^{kl} x'^i, \quad S^{kl,ij} \equiv \int d^4\mathbf{x}' T^{kl} x'^i x'^j. \tag{7.85}$$

The expression for the gravitational wave-form at the detector (7.83) can therefore be expressed as a function of the time-derivatives of the moments of the stress tensor evaluated at time $t - r$,

$$h_{ij}^{TT}(t, \mathbf{x}) = \frac{4G}{r} \Lambda_{ij,kl}^{TT} \left[S^{kl} + n_i \dot{S}^{kl,i} + \frac{1}{2} n_i n_j \ddot{S}^{kl,ij} + \cdots \right]_{t-r} \tag{7.86}$$

One can relate the moments of T^{ij} components of the stress tensor (7.85) with the moments of the T^{00} and T^{0i} components by using the conservation equation $\partial_\mu T^{\mu\nu} = 0$. We define the moments of the T^{00} which are the mass distributions of the source

$$M \equiv \int d^4\mathbf{x}' T^{00}, \quad M^i \equiv \int d^4\mathbf{x}' T^{00} x'^i, \quad M^{ij} \equiv \int d^4\mathbf{x}' T^{00} x'^i x'^j. \tag{7.87}$$

We also define the moments of the pressure as

$$P^i \equiv \int d^4\mathbf{x}' T^{0i}, \quad P^{ij} \equiv \int d^4\mathbf{x}' T^{0i} x'^j, \quad P^{ijk} \equiv \int d^4\mathbf{x}' T^{0i} x'^j x'^k. \tag{7.88}$$

Using the conservation equation $\partial_0 T^{00} + \partial_i T^{0i} = 0$ we see that

$$\dot{M} = \int d^4\mathbf{x}' \, \partial_0 T^{00} = -\int d^4\mathbf{x}' \, \partial_i T^{0i} = -\int dS_i \, T^{0i} = 0 \tag{7.89}$$

where we have applied Stokes Theorem and assumed that T^{oi} vanishes on the surface $|\mathbf{x}'| \to \infty$. We can similarly show that

$$\dot{M}^{jk} = \int d^4\mathbf{x}' \, \partial_0 T^{00} x'^j x'^k = -\int d^4\mathbf{x}' \, \partial_i T^{0i} x'^j x'^k$$

$$= \int d^4\mathbf{x}' \, T^{0i} x'^j \delta_i^k + \int d^4\mathbf{x}' \, T^{0i} \delta_i^j x'^k = P^{jk} + P^{kj}. \tag{7.90}$$

Taking one more time derivative of (7.90) we obtain

$$\ddot{M}^{jk} = \int d^4\mathbf{x}' \, T^{0j} x'^k + (j \leftrightarrow k) = 2S^{jk} \tag{7.91}$$

The moments of the stress tensor can in this way be related to time derivatives of mass distribution of the source[4]

$$S^{kl} = \frac{1}{2}\ddot{M}^{kl}, \quad \dot{S}^{kl,m} = \frac{1}{6}\dddot{M}^{jkl} \quad \cdots \tag{7.92}$$

Using these relations we can write the gravitational waveform (7.86) in terms of the time derivatives of the mass distribution of the source as

$$h_{ij}^{TT}(t,\mathbf{x}) = \frac{4G}{r}\Lambda_{ij,kl}^{TT}\left[\frac{1}{2}\ddot{M}^{kl} + \frac{1}{6}\dddot{M}^{rkl} n_r + \cdots\right]_{t-r} \tag{7.93}$$

The terms in the series expansion in (7.93) are the time derivatives of multipole moments of mass distribution with the leading term which are suppressed by powers of $\frac{d}{dt}|\mathbf{x}'| \sim v'/c$ the velocity of the source constituents. Consider the leading order quadruple term in (7.93). We can split the tensor $\ddot{M}^{kl}$ as a sum of the traceless part and the trace,

$$\ddot{M}^{kl} = \left(\ddot{M}^{kl} - \frac{1}{3}\eta^{kl}\ddot{M}^r{}_r\right) + \frac{1}{3}\eta^{kl}\ddot{M}^r{}_r \tag{7.94}$$

The TT projection operator $\Lambda_{ij,kl}^{TT}$ operating on the second term gives zero and retaining only the traceless quadrupole distribution

$$\ddot{Q} \equiv \ddot{M}^{kl} - \frac{1}{3}\eta^{kl}\ddot{M}^r{}_r = \frac{d^2}{dt^2}\left[\int d^4\mathbf{x}'\rho(t',\mathbf{x}')\left(x'^k x'^l - \frac{1}{3}\eta^{kl} x'^i x'_i\right)\right]_{t'=t-r} \tag{7.95}$$

(where $\rho = T^{00}$) we can write the expression (7.93) as

$$h_{ij}^{TT}(t,\mathbf{x}) = \frac{2G}{r}P_{ik}(\mathbf{n})P_{jl}(\mathbf{n})\ddot{Q}^{kl} \tag{7.96}$$

4

$$\dot{M} = 0, \quad \dot{M}^k = P^k, \quad \dot{M}^{ij} = P^{ij} + P^{ji}, \quad \dot{M}^{ijk} = P^{ijk} + P^{kij} + P^{jki},$$
$$\dot{P}^j = 0, \quad \dot{P}^{jk} = S^{jk}, \quad \dot{P}^{ijk} = S^{ij,k} + S^{ji,k}.$$

where we have retained only the the transverse-projection part $P_{ik}(\mathbf{n}) P_{jl}(\mathbf{n})$ of the operator $\Lambda_{ij,kl}^{TT}$ as the source term $\ddot{Q}^{kl}$ is already traceless. The projection operators contain the information about the direction of the source in the sky. Extracting the direction of the source fro the signal is crucial in multi messenger searches for optical or neutrino signal searches associated with the gravitational wave signal.

We will now derive the gravitational wave forms for the h_+ and $h_\times$ polarization waves from binary star mergers events.

7.7   Waveforms from Compact Binaries

Consider a binary system of compact stars with masses m_1 and m_2 in a circular Kepler orbit separated by distance R. Choosing the origin of the coordinate at the center of mass, the two stars revolve around the c.m at radii $r_1 = m_2 R/(m_1 + m_2)$ and $r_2 = m_1 R/(m_1 + m_2)$ respectively with angular velocity given by the Keplers Law

$$\Omega^2 = \frac{G(m_1 + m_2)}{R^3}. \tag{7.97}$$

Choose the plane of orbit of the stars to be the $x - y$ plane. The orbit equations of the two stars are $\mathbf{x}_1 = (r_1 \cos \Omega t, r_1 \sin \Omega t, 0)$ and $\mathbf{x}_2 = (r_2 \cos(\Omega t + \pi), r_2 \sin(\Omega t + \pi), 0)$. The components of the quadrupole moment of the binary system is therefore

$$
\begin{aligned}
M_{xx} &= m_1 \, (r_1 \cos \Omega t)^2 + m_2 \, (-r_2 \cos \Omega t)^2 \\
&= \mu R^2 \cos^2 \Omega t \\
&= \frac{1}{2} \mu R^2 \, (1 - \cos(2\Omega t))
\end{aligned}
\tag{7.98}
$$

where $\mu = m_1 m_2/(m_1 + m_2)$ is the reduced mass and we can see that the fundamental frequency of the quadruple oscillation which is the frequency of the gravitational waves generated is twice the orbital frequency of the binary, $\Omega_{gw} = 2\Omega$. Similarly calculating the other components of the quadrupole distribution, we find that the non-zero components are

$$
\begin{aligned}
M_{yy} &= m_1 \, (r_1 \sin \Omega t)^2 + m_2 \, (-r_2 \sin \Omega t)^2 \\
&= \frac{1}{2} \mu R^2 \, (1 + \cos(2\Omega t)) , \\
M_{xy} &= m_1 \left(r_1^2 \cos \Omega t \, \sin \Omega t \right) + m_2 \left(r_2^2 \cos \Omega t \sin \Omega t \right) \\
&= \frac{-1}{2} \mu R^2 \sin(2\Omega t)
\end{aligned}
\tag{7.99}
$$

Dropping the time independent constants we see that the tensor M_{ij} is traceless and therefore $Q_{ij} = M_{ij}$. Now consider the detector to be on the z-axis, $\mathbf{n} = (0, 0, 1)$. The projector operators $P_{ik}(\mathbf{n}) P_{jl}(\mathbf{n}) Q_{kl} = Q_{ij}$ as Q_{ij} as the non-zero components of Q_{ij} are orthogonal to $\mathbf{n}$. The gravitational waveform observed along the z-axis at a distance r will be

$$h_{xx}^{TT} = \frac{2G}{r} \ddot{M}_{xx} = \frac{4G}{r} \mu^2 R^2 \Omega^2 \cos(2\Omega t_r)$$

$$h_{yy}^{TT} = \frac{2G}{r} \ddot{M}_{yy} = -\frac{4G}{r} \mu^2 R^2 \Omega^2 \cos(2\Omega t_r)$$

$$h_{xy}^{TT} = \frac{2G}{r} \ddot{M}_{xy} = \frac{4G}{r} \mu^2 R^2 \Omega^2 \sin(2\Omega t_r) \tag{7.100}$$

where $t_r = t - r$ is the retarded time and the gravitational wave angular frequency is $\Omega_{gw} = 2\Omega$. The waveform of $h_+ = h_{xx}^{TT} = -h_{yy}^{TT}$ and $h_\times = h_{xy}^{TT}$ are therefore

$$h_+(t) = \frac{G}{r} \mu^2 R^2 \Omega_{gw}^2 \cos(\Omega_{gw} t + \phi_0)$$

$$h_\times(t) = \frac{G}{r} \mu^2 R^2 \Omega_{gw}^2 \sin(\Omega_{gw} t + \phi_0) \tag{7.101}$$

we can replace R is terms of Kepler frequency $\Omega = \Omega_{gw}/2$ using (7.97), and write $h+$ in (7.101) as

$$h_+(t) = \frac{G^{5/3}}{r} 2^{1/3} \mu (m_1 + m_2)^{2/3} \Omega_{gw}^{2/3} \cos(\Omega_{gw} + \psi_0). \tag{7.102}$$

And similarly for $h_\times$. We can combine the mass factors by defining a 'chirp mass'

$$\mathcal{M}_c = \mu^{3/5} (m_1 + m_2)^{2/5} \tag{7.103}$$

and write the expressions for the gravitational wave signals from binaries as

$$h_+(t) = h \cos(\Omega_{gw} t + \phi_0) ,$$

$$h_\times(t) = h \sin(\Omega_{gw} t + \phi_0) , \tag{7.104}$$

where the amplitude of the GW of both the polarizations is

$$\boxed{h = \frac{G^{5/3}}{r} 2^{1/3} \mathcal{M}_c^{5/3} \Omega_{gw}^{2/3} .} \tag{7.105}$$

Numerically the amplitude of gravitational waves from binary black-holes in the last seconds before coalescence is

$$h \simeq 10^{-21} \left(\frac{\mathcal{M}_c}{10 M_\odot} \right)^{5/3} \left(\frac{f}{100 \, \text{Hz}} \right)^{2/3} \left(\frac{100 \, \text{Mpc}}{r} \right) \qquad (7.106)$$

where $f = \Omega_{gw}/(2\pi)$ is the frequency of GW.

Angular Distribution of Gravitational Waves
Now we generalize the geometry such that the plane of binary orbit is tilted by an angle i. The quadrupole moment tensor M'_{ij} of in the tilted plane can be computed from the quadupole moment tensor of the orbit in x–z plane (7.98) and (7.99) by a rotation around the y-axis by an angle i,

$$M'_{ij} = R^y(i)_{ik} R^y(i)_{lj} M_{kl} = \left(R^y(i) \, M \, R^y(i)^T \right)_{ij} \qquad (7.107)$$

To obtain the TT projection of a wave along $\mathbf{n} = (0, 0, 1)$ we use the projection operator P_{ij}

$$P_{ij}(\mathbf{n}) = \delta_{ij} - n_i n_j = \begin{pmatrix} 1 & 0 & 0 \\ 0 & 1 & 0 \\ 0 & 0 & 0 \end{pmatrix} \qquad (7.108)$$

to construct the TT projection operator $\Lambda^{TT}_{ij,kl}$ (7.15). The TT components of the quadrupole tensor are

$$\Lambda^{TT}_{ij,kl} M'_{kl} = \begin{pmatrix} \frac{1}{2}(M'_{xx} - M'_{yy}) & M'_{xy} & 0 \\ M'_{yx} & -\frac{1}{2}(M'_{xx} - M'_{yy}) & 0 \\ 0 & 0 & 0 \end{pmatrix}$$

$$= \begin{pmatrix} \frac{1}{2}(M_{xx} \cos^2 i - M_{yy}) & M_{xy} \cos i & 0 \\ M_{yx} \cos i & -\frac{1}{2}(M_{xx} \cos^2 i - M_{yy}) & 0 \\ 0 & 0 & 0 \end{pmatrix} \qquad (7.109)$$

The waveforms generated in this configuration of the source will be

$$h^{TT}_{ij} = \frac{2G}{r} \Lambda^{TT}_{ij,kl} \ddot{M}'_{kl} \qquad (7.110)$$

and the two modes of gravitational waves from an binary orbit inclined at an angle i with respect to the observer (along the z-axis) will be of the form

$$h_+(t, z) = h \left(\frac{1 + \cos^2 i}{2} \right) \cos(\Omega_{gw} t + \phi_0),$$

$$h_\times(t, z) = h \, (\cos i) \, \sin(\Omega_{gw} t + \phi_0). \tag{7.111}$$

where h is given in (7.105). If from the earth the binary orbit plane is edge-on ($i = \pi/2$) then there will be no $h_\times$ polarisation gravitational waves from that source. The geometry can be further generalized considering a rotation by ϕ around the z-axis of the inclined orbit. The waveforms for this geometry of the source can be calculated by taking the rotation matrix in (7.107) to be $R = R^z(\phi) R^y(i)$ and then taking the TT projection the resulting matrix M'_{ij} as in (7.109) [13].

Chirp Signal of Coalescing Binaries Coalescing binaries in the last 15 min or so of their orbits are the main source of the signals seen at LOGO and VIRGO detectors whose detection efficiency peaks at $f_{GW} = 2\pi\Omega_{gw} \sim 100$ Hz range. The during the coalescing phase of black-hole or neutron star mergers the orbits become circular and the orbital frequency increases as the separation decreases (maintaining the Kepler relation (7.97). The gravitational binding energy is radiated as gravitational waves. The change of frequency of the GW with time can be written from the change of time period (7.41) ($f = \Omega_{gw}/(2\pi) = \Omega/\pi = 1/(\pi P_b)$) due to energy loss in any process (emission of gravitational, scalar or vector waves) is,

$$\boxed{\frac{df}{dt} = 6\pi^{1/3} \, G^{-2/3} \, f^{1/3} (m_1 m_2)^{-1} (m_1 + m_2)^{1/3} \left(\frac{dE}{dt} \right).} \tag{7.112}$$

Using the expression for energy loss by gravitational waves using (7.39) we obtain[5]

$$\frac{df}{dt} = \frac{96}{5} \pi^{8/3} \, (G M_c)^{5/3} \, f^{11/3}. \tag{7.113}$$

The effect on the gravitational wave signal due to emission of ultralight scalar or vector particles can be determined by adding to dE_{gw}/dt for gravitational emission in (7.112) the energy loss rate for scalars from Eq. (7.56) and for vectors from (7.74). The chirp frequency will rise faster when there are other particles emitted besides gravitons. If both stars are pulsars and there is a long range Coulomb force between

---

[5]We write a in (7.39) for Kepler orbit in terms of gravitational wave frequency using $\Omega = \pi f = G^{1/2} M^{1/2} a^{-3/2}$ and we have dropped the eccentricity factors in dE/dt for simplicity.

them due to light scalars or vector exchange then the effective gravitational constant will change as

$$G \to G\left[1 \pm \frac{\alpha' Q_1 Q_2}{G M_1 M_2}\right] \tag{7.114}$$

and this will also change the rate of gravitational wave frequency [26].

We can integrate (7.113) to give us the time dependence of the frequency of the gravitational waves from in-spiraling binaries

$$f(t) = \left(\frac{1}{f_c^{8/3}} + \frac{5}{36\,\pi^{8/3}}\,(G\mathcal{M}_c)^{-5/3}\,(t-t_c)\right)^{-3/8}$$

$$\simeq \left(\frac{36}{5}\right)^{3/8} \pi \,(G\mathcal{M}_c)^{5/8}(t-t_c)^{-3/8} \tag{7.115}$$

We see that frequency increases with time till the coalesce time t_c which is regarded as the last stable circular orbit $R_c = 6GM$ ($M = m_1 + m_2$) at which point the frequency reaches the value $f_c = (6^{3/2}\pi M)^{-1}$. After $R \sim 6GM$ in a BH-BH binary the two black-holes will coalesce within one orbit. For NS-NS or NS-BH in-spirals the NS will be tidally disrupted after the binary separation reaches $R = 6GM$. In the in-spiraling phase the GRW amplitude (7.105) rises with along with frequency till the coalesce time $t = t_c$ and this is the characteristic 'chirp signal' of the in-spiraling binaries. In Fig. 7.3 we show the observation of the gravitational wave signal of binary black-hole merger event GW150914 [27]. The chirp phase is followed by the merger to form the final black-hole. The perturbations of this final black-hole produces GRW which are seen in the signal. The black-hole perturbations are damped by GW emission in the ring down phase.

From (7.113) we see that by measuring the frequency f and $\dot{f}$ one can measure the chirp mass of the binary. One must take into account for the cosmological redshift in frequency $f(z) = (1 + z)f$ (where f is the frequency at the source). The equations for amplitude (7.105) and rate of change of frequency (7.113) are invariant under the transformations [28],

$$\{f, t, \mathcal{M}_c, r\} \to \left\{\frac{f}{(1+z)},\, t(1+z),\, \mathcal{M}_c(1+z),\, r(1+z)\right\} \tag{7.116}$$

The luminosity distance in a spatially flat universe the luminosity distance is in terms of the luminosity of the source $\mathcal{L}$ and observed flux $\mathcal{F}$ is,

$$d_L(z) = \left(\frac{\mathcal{L}}{4\pi\mathcal{F}}\right)^{1/2} = \frac{1+z}{H_0} \int_0^z \frac{dz'}{\left(\Omega_m(1+z')^3 + \Omega_r(1+z')^4 + \Omega_\Lambda\right)^{1/2}}$$

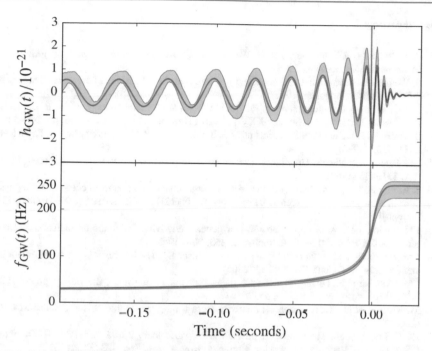

Fig. 7.3 Gravitational wave amplitude and frequency at 90% C.L for the black hole binary in spiral event GW150914. Plot taken from [27]

and the angular distance is $d_A = d_L/(1+z)$. The extra suppression by factor $(1+z)$ of the luminosity distance is due to time dilation of the flux rate.

The distance to the source $r(z) = d_L(z)$ needs to be determined independently as the amplitude (7.105) and frequency (7.113) equations have a degeneracy $(\mathcal{M}_c(z), d_L(z)) \rightarrow (\mathcal{M}_c(z)(1+z), d_L(z)(1+z))$. If the in-spiraling binary is from NS-NS mergers then one may locate the associated gamma ray bursts (GRB) from gamma ray, x-ray and radio observations. If from the optical counterpart the red-shift z of the source can be measured then the measurement of the gravitational wave amplitude will provide the "standard sirens" and cosmological parameters like H_0 can be determined by the multi-messenger analysis [29].

When the final state of the merger is a black hole there is an opportunity to study strong field general relativity from the perturbations of the black hole horizon called 'ringing down' which leaves a damped oscillatory gravitational wave signal [27].

In multi-messenger observations as observation of GW170817 by Ligo and Virgo [30] and GRB 170817A by Fermi [31], the gamma-rays arrived after 2 s delay from the arrival of the gravitational waves. The delay of photons is due to the delay of the photons in the dense envelop of matter blown off from the merged neutron stars. This timing of the gravitational waves provides an opportunity for testing theories of GR like Born-Infeld gravity where the velocity of gravitational waves differs from velocity of light [32].

References

1. R.P. Feynman, F.B. Morinigo, W.G. Wagner, B. Hatfield, *Feynman lectures on gravitation* (Addison-Wesley, Reading 1995)
2. S. Weinberg, *Gravitation and cosmology: principles and applications of the general theory of relativity*. Wiley, Hoboken (1972)
3. M.J.G. Veltman, Quantum theory of gravitation, in *Methods in field theory*, ed. by R. Balian, J. Zinn-Justin. Les Houches, Session XXVIII (North Holland, Amsterdam, 1976), pp. 265–328
4. S. Mohanty, P. Kumar Panda, Particle physics bounds from the Hulse-Taylor binary. Phys. Rev. D **53**, 5723 (1996)
5. P.C. Peters, J. Mathews, Gravitational radiation from point masses in a Keplerian orbit. Phys. Rev. **131**, 435 (1963)
6. T. Kumar Poddar, S. Mohanty, S. Jana, Vector gauge boson radiation from compact binary systems in a gauged $L_\mu - L_\tau$ scenario. Phys. Rev. D **100**(12), 123023 (2019). [arXiv:1908.09732 [hep-ph]]
7. J.H. Taylor, J.M. Weisberg, A new test of general relativity: gravitational radiation and the binary pulsar PS R 1913+16. Astrophys. J. **253**, 908 (1982)
8. J.M. Weisberg, J.H. Taylor, Relativistic binary pulsar B1913+16: thirty years of observations and analysis. ASP Conf. Ser. **328**, 25 (2005)
9. J.M. Weisberg, Y. Huang, Relativistic measurements from timing the binary pulsar PSR B1913+16. Astrophys. J. **829**(1), 55 (2016)
10. M. Kramer, et al., Tests of general relativity from timing the double pulsar. Science **314**, 97 (2006)
11. P.C.C. Freire, et al., The relativistic pulsar-white dwarf binary PSR J1738+0333 II. The most stringent test of scalar-tensor gravity. Mon. Not. Roy. Astron. Soc. **423**, 3328 (2012)
12. J. Antoniadis, et al., A massive pulsar in a compact relativistic binary. Science **340**, 6131 (2013)
13. M. Maggiore, *Gravitational waves. Vol. 1: theory and experiments* (Oxford University Press, Oxford, 2007)
14. L.D. Landau, E.M. Lifshitz, *Classical theory of fields* (Pergamon Press, Oxford, 1987)
15. C. Amsler et al., (Particle Data Group), Phys. Lett. B **667**, 1 (2008)
16. W. Hu, R. Barkana, A. Gruzinov, Cold and fuzzy dark matter. Phys. Rev. Lett. **85**, 1158 (2000)
17. L. Hui, J.P. Ostriker, S. Tremaine, E. Witten, Phys. Rev. D **95**(4), 043541 (2017)
18. G. Raffelt, Limits on a CP-violating scalar axion-nucleon interaction. Phys. Rev. D **86**, 015001 (2012)
19. J.E. Moody, F. Wilczek, New macroscopic forces? Phys. Rev. D **30**, 130 (1984)
20. D. Chang, R.N. Mohapatra, S. Nussinov, Could goldstone bosons generate an observable 1/r potential? Phys. Rev. Lett. **55**, 2835 (1985). https://doi.org/10.1103/PhysRevLett.55.2835
21. A. Hook, J. Huang, Probing axions with neutron star inspirals and other stellar processes. JHEP **1806**, 036 (2018)
22. R. Foot, Charge quantization in the standard model and some of its extensions. Mod. Phys. Lett. A**(6)**, 527 (1991)
23. X.-G. He, G.C. Joshi, H. Lew, R.R. Volkas, New Z' phenomenology. Phys. Rev. D **44**, 2118 (1991)
24. A.S. Joshipura, S. Mohanty, Constraints on flavor dependent long range forces from atmospheric neutrino observations at super-Kamiokande. Phys. Lett. B **584**, 103–108 (2004)
25. B.P. Abbott, et al. [LIGO Scientific and Virgo Collaborations], Observation of gravitational waves from a binary black hole merger. Phys. Rev. Lett. **116**(6), 061102 (2016)
26. J. Kopp, R. Laha, T. Opferkuch, W. Shepherd, Cuckoo?s eggs in neutron stars: can LIGO hear chirps from the dark sector? JHEP **1811**, 096 (2018)
27. B.P. Abbott, et al. [LIGO Scientific and Virgo Collaborations], Tests of general relativity with GW150914. Phys. Rev. Lett. **116**(22), 221101 (2016). Erratum: [Phys. Rev. Lett. **121**(12), 129902 (2018)]

28. C. Cutler, E.E. Flanagan, Gravitational waves from merging compact binaries: how accurately can one extract the binary's parameters from the inspiral wave form? Phys. Rev. D **49**, 2658 (1994)
29. B. Abbott, et al. [LIGO Scientific and Virgo], A gravitational-wave measurement of the Hubble constant following the second observing run of Advanced LIGO and Virgo. [arXiv:1908.06060 [astro-ph.CO]]
30. B. Abbott, et al. [LIGO Scientific and Virgo], GW170817: observation of gravitational waves from a binary neutron star inspiral. Phys. Rev. Lett. **119**(16), 161101 (2017)
31. G.P. Lamb, S. Kobayashi, GRB 170817A as a jet counterpart to gravitational wave triggerGW 170817. Mon. Not. Roy. Astron. Soc. **478**(1), 733–740 (2018)
32. S. Jana, G.K. Chakravarty, S. Mohanty, Constraints on Born-Infeld gravity from the speed of gravitational waves after GW170817 and GRB 170817A. Phys. Rev. D **97**(8), 084011 (2018)

Black Holes

8

Abstract

Black holes mergers have been observed through their gravitational signals and the ring down of the black holes have been seen in the LIGO observations (2016). From the gravitational wave observations the mass and spin of the black holes can be estimated. Moreover recently there is a direct observation of the black hole shadow by the Event Horizon Telescope team of the super massive black hole in the center of M87 galaxy. These observations are a good motivation for studying the details of the rotating black holes as well as more general types of black holes like the dilation-axion black holes predicted from string theory. The deviation from the Kerr metric may be observable in gravitational wave signals or more directly from the shape and size of the photon shadows of the super-massive galactic center black holes. Kerr black holes may also generate high energy particles by the mechanism of super-radiance which we discuss.

8.1 Introduction

The Karl Schwarzschild gave the metric that goes by his name in 1916 to describe the vacuum solution of Einstein's equation which described space-time in the exterior of stars [1]. He as well as Einstein noticed that his solution had a singularity at the edius $r = 2GM$ but since this "Schwarzschild" radius was well within the stars (for solar mass stars $2GM_{\odot} \sim 3\,\text{km}$) where the vacuum solution was not applicable it did not have any physical significance.

Eddington in 1924 suggested that white dwarfs are protected from further collapsing into smaller and denser objects due to the degeneracy pressure of electrons which balanced the gravitational force of primarily the protons.

In 1931 Chandrasekhar found that there was a limiting mass of white dwarf's which is $1.4\,M_{\odot}$ beyond which the degeneracy pressure of electrons cannot support

© Springer Nature Switzerland AG 2020
S. Mohanty, *Astroparticle Physics and Cosmology*,
Lecture Notes in Physics 975, https://doi.org/10.1007/978-3-030-56201-4_8

the gravitational force of the protons and concluded that stars with masses more than the Chandrasekhar limit must collapse into infinite density singularities [2].

Landau suggested that the core of all stars were supported by the degeneracy pressure of neutrons which would prevent formation of singularity on stellar collapse. Zwicky proposed that such neutron cores are produced in supernova.

Oppenheimer and Snyder [3] examined the stellar collapse in general relativity and concluded that in the frame of reference of attached to a collapsing shell of dust in a star the Schwarzschild singularity at $r = 2GM$ had no significance and the endpoint of a collapse of a very massive star is a singularity at the center.

The Kerr metric of the spinning black hole was discovered in 1963 [4]. The name "black hole" to describe the gravitational objects with a singularity was coined by Wheeler in 1967.

8.2   Schwarzschild Black Hole

The Schwarzschild metric is a vacuum solution of Einstein equations and describes the metric outside a specially symmetric mass M [1],

$$ds^2 = \left(1 - \frac{2GM}{r}\right) dt^2 - \left(1 - \frac{2GM}{r}\right)^{-1} dr^2 - r^2 \left(d\theta^2 + \sin^2\theta d\phi^2\right) \quad (8.1)$$

The value of r where the $g_{tt} \to 0$ and $g_{rr} \to \infty$ is the Schwarzschild radius

$$r_s = 2GM . \quad (8.2)$$

The singularity of the metric at $r = r_s$ can be removed by transforming to a new set of coordinates as we shall see below. The Riemann tensors which represent the tidal acceleration however is finite at $r = r_s$ and the surface $r = r_s$ is a singularity due to our choice of coordinates. The scalar constructed from the Riemann tensor

$$R^{\mu\nu\alpha\beta} R_{\mu\nu\alpha\beta} = \frac{48G^2M^2}{r^6} \quad (8.3)$$

diverges at $r = 0$ and is a real singularity of the Schwarzschild space-time.

If a star collapses such that its radius becomes smaller than r_s such the horizon is outside its physical radius then it is called a black hole. Before we study in detail the trajectories of particles in Schwarzschild metric we introduce Killing vectors and Killing tensors as they are useful for identifying the conserved quantities in particles trajectories around black holes.

8.3   Killing Vectors and Conserved Quantities

Killing vectors describe the symmetries of the metric. If under a coordinate transformation along the direction ξ^μ

$$x^\mu \to x'^\mu = x^\mu + \epsilon \xi^\mu, \quad |\epsilon| \ll 1. \tag{8.4}$$

and the form of the metric remains the same in the new coordinates

$$g'_{\mu\nu}(x') = g_{\mu\nu}(x'). \tag{8.5}$$

The symmetry transformation is then called an isometry of the metric and the set of vectors ξ^μ are the Killing vectors. Under the transformation (8.4) the metric transformation is an isometry when

$$
\begin{aligned}
g_{\rho\sigma}(x') = g'_{\rho\sigma}(x') &= g_{\mu\nu}(x)\frac{\partial x^\mu}{\partial x'^\rho}\frac{\partial x^\nu}{\partial x'^\sigma}\\
&= g_{\mu\nu}(x)\left[\delta^\mu_\rho - \epsilon\partial_\rho\xi^\mu\right]\left[\delta^\mu_\rho - \epsilon\partial_\rho\xi^\mu\right]\\
\Rightarrow \left(g_{\rho\sigma}(x) + \epsilon\xi^\alpha\partial_\alpha g_{\rho\sigma}(x)\right) &= g_{\mu\nu}(x)\left[\delta^\mu_\rho - \epsilon\partial_\rho\xi^\mu\right]\left[\delta^\mu_\rho - \epsilon\partial_\rho\xi^\mu\right]
\end{aligned}
\tag{8.6}
$$

where the first equality is the condition of isometry and the second equality follows from the fact that under (8.4) the metric transforms as a tensor. Simplifying (8.6) we obtain the condition on the vectors ξ^μ in order that the shift (8.4) is a symmetry,

$$\xi^\alpha\partial_\alpha g_{\rho\sigma} + g_{\rho\nu}\partial_\sigma\xi^\nu + g_{\mu\sigma}\partial_\rho\xi^\mu = 0 \tag{8.7}$$

The **Lie derivative** of any tensor along any vector ξ^μ is

$$
\begin{aligned}
\mathcal{L}_\xi T^{\mu_1\mu_2...\nu_k}{}_{\nu_1\nu_2...\nu_l} = &\,(\xi^\lambda\partial_\lambda)T^{\mu_1\mu_2...\nu_k}{}_{\nu_1\nu_2...\nu_l}\\
&- (\partial_\lambda\xi^{\mu_1})T^{\lambda\mu_2...\nu_k}{}_{\nu_1\nu_2...\nu_l} - (\partial_\lambda\xi^{\mu_2})T^{\mu_1\lambda...\nu_k}{}_{\nu_1\nu_2...\nu_l}\\
&+ (\partial_{\nu_1}\xi^\lambda)T^{\mu_1\mu_2...\nu_k}{}_{\lambda\nu_2...\nu_l} + (\partial_{\nu_2}\xi^\lambda)T^{\mu_1\mu_2...\nu_k}{}_{\nu_1\lambda...\nu_l}
\end{aligned}
\tag{8.8}
$$

The Lie derivative along a vector is a tensor and it can be written in a manifestly tensor form as

$$
\begin{aligned}
\mathcal{L}_\xi T^{\mu_1\mu_2...\nu_k}{}_{\nu_1\nu_2...\nu_l} = &\,(\xi^\lambda\nabla_\lambda)T^{\mu_1\mu_2...\nu_k}{}_{\nu_1\nu_2...\nu_l}\\
&- (\nabla_\lambda\xi^{\mu_1})T^{\lambda\mu_2...\nu_k}{}_{\nu_1\nu_2...\nu_l} - (\nabla_\lambda\xi^{\mu_2})T^{\mu_1\lambda...\nu_k}{}_{\nu_1\nu_2...\nu_l}\\
&+ (\nabla_{\nu_1}\xi^\lambda)T^{\mu_1\mu_2...\nu_k}{}_{\lambda\nu_2...\nu_l} + (\nabla_{\nu_2}\xi^\lambda)T^{\mu_1\mu_2...\nu_k}{}_{\nu_1\lambda...\nu_l}
\end{aligned}
\tag{8.9}
$$

By expanding the covariant derivatives in (8.9) the Chirtoffel connection terms cancel and we obtain (8.8).

Consider the Lie derivative of the metric tensor in the direction some vector ξ^μ we see that

$$\mathcal{L}_\xi g_{\rho\sigma} = \xi^\alpha \, \partial_\alpha g_{\rho\sigma} + g_{\rho\nu} \, \partial_\sigma \xi^\nu + g_{\mu\sigma} \, \partial_\rho \xi^\mu \tag{8.10}$$

From (8.7) and (8.10) we see that a set of vectors ξ^μ are Killing vectors if they obey the conditions

$$\mathcal{L}_\xi g_{\mu\nu} = 0 \tag{8.11}$$

Equation (8.10) can also be written in the manifestly tensor form

$$\mathcal{L}_\xi g_{\rho\sigma} = \xi^\alpha \, \nabla_\alpha g_{\rho\sigma} + g_{\rho\nu} \, \nabla_\sigma \xi^\nu + g_{\mu\sigma} \, \nabla_\rho \xi^\mu$$
$$= g_{\rho\nu} \, \nabla_\sigma \xi^\nu + g_{\mu\sigma} \, \nabla_\rho \xi^\mu \tag{8.12}$$

Therefore a vector ξ^μ is a Killing vector if it satisfies the condition

$$\nabla_\nu \xi_\mu + \nabla_\mu \xi_\nu = 0. \tag{8.13}$$

From (8.7) we see that if in the coordinate system $(x_0, x_1, x_2, x_3) = \{x_i\}$ the metric is independent of any x_i then one of $\xi^\mu = (\ldots 0, 1, 0 \ldots)$ where 1 is in the 'i' th position is a Killing vector.

The Schwarzschild metric (8.1) is independent of t and r so we can identify two of the Killing vectors (in (t, r, θ, ϕ) coordinate system) as $\xi^\mu_{(t)} = (1, 0, 0, 0)$ and $\xi^\mu_{(\phi)} = (0, 0, 0, 1)$. But these are not the only Killing vectors. The Schwarzschild metric has a spherical symmetry so the other two components of angular momentum L_x and L_y are also constants for a particle in a geodesic orbit around the black hole. The other two Killing vectors which can be determined from symmetry arguments or by using (8.7) are $\xi^\mu_3 = (0, 0, -y, z)$ and $\xi^\mu_4 = (0, x, 0, -z)$ (in t, x, y, z) coordinates) which correspond to the conserved angular momentum components L_x and L_y respectively.

8.3.1   Killing Vectors and Conserved Energy-Momentum

Identifying the Killing vectors of a metric are of great help in computing the trajectories of particles in the gravitational field described by that metric as they tell us what quantities remain constant along the geodesics. Given a Killing vector ξ^μ and the four velocity $V^\mu = \frac{dx^\mu}{d\tau}$, the quantity $V^\mu \xi_\mu$ remains constant along the geodesic,

$$\frac{D}{D\lambda}\left(V^{\mu}\xi_{\mu}\right) \equiv V^{\alpha}\nabla_{\alpha}\left(V^{\mu}\xi_{\mu}\right)$$

$$= V^{\alpha}V^{\mu}\left(\nabla_{\alpha}\xi_{\mu}\right) + V^{\alpha}\xi_{\mu}\left(\nabla_{\alpha}V^{\mu}\right)$$

$$= 0 \tag{8.14}$$

where the first term on the r.h.s of (8.14) is zero due to the property of the Killing vectors $(\nabla_{\alpha}\xi_{\mu}) = -(\nabla_{\mu}\xi_{\alpha})$ and the second term is zero from the geodesic equation $\nabla_{\alpha}V^{\mu} = 0$.

8.3.1.1 Killing Tensors

A Killing tensor is a symmetric tensor of with n indices, $\xi_{\mu_1\mu_2\cdots\mu_n}$ with the property

$$\nabla_{(\lambda}\xi_{\mu_1\mu_2\cdots\mu_n)} = \nabla_{\lambda}\xi_{\mu_1\mu_2\cdots\mu_n} + \nabla_{\mu_1}\xi_{\lambda\mu_2\cdots\mu_n} + \cdots + \nabla_{\mu_n}\xi_{\mu_1\mu_2\cdots\lambda} = 0 \tag{8.15}$$

If $\xi_{\mu_1\mu_2\cdots\mu_n}$ is a Killing tensor then the quantity $\xi_{\mu_1\mu_2\cdots\mu_n}V^{\mu_1}V^{\mu_2}\cdots V^{\mu_n}$ is conserved along a geodesic,

$$V^{\alpha}\nabla_{\alpha}\left(\xi_{\mu_1\mu_2\cdots\mu_n}V^{\mu_1}V^{\mu_2}\cdots V^{\mu_n}\right) = 0 \tag{8.16}$$

This follows from the observation that

$$V^{\alpha}\nabla_{\alpha}\left(\xi_{\mu_1\mu_2\cdots\mu_n}V^{\mu_1}V^{\mu_2}\cdots V^{\mu_n}\right) = V^{\alpha}\left(\nabla_{(\alpha}\xi_{\mu_1\mu_2\cdots\mu_n)}\right)V^{\mu_1}V^{\mu_2}\cdots V^{\mu_n}$$

$$+ \xi_{\mu_1\mu_2\cdots\mu_n}V^{\alpha}\nabla_{\alpha}\left(V^{\mu_1}V^{\mu_2}\cdots V^{\mu_n}\right)$$

$$= 0 \tag{8.17}$$

where the first term in the r.h.s is zero by definition of Killing tensor and the second term is zero by using the geodesic equation.

The metric $g_{\mu\nu}(x)$ is a Killing tensor which implies that the quantity $g_{\mu\nu}V^{\mu}V^{\nu}$ is a constant along the geodesic. The particle four-momenta is $P^{\mu} = mV^{\mu}$ for particles with non-zero mass and the conserved quantity is $g_{\mu\nu}P^{\mu}P^{\nu} = m^2$. For massless particles the four momentum is $P^{\mu} = V^{\mu}$ (by a suitable choice of the affine parameter λ) and the conserved quantity along the massless particle geodesic is $g_{\mu\nu}P^{\mu}P^{\nu} = 0$. For computing particle trajectories in curved space it is often more convenient to use these constraint equations instead of solving the geodesic equations.

8.3.2 Horizons

Particles which fall inside the surface $r = 2M$ cannot escape the black hole. Outward directed light rays at all points $r > 2M$ can escape falling into the black hole whereas at $r < 2M$ both out and in directed light rays fall towards $r = 0$. The 2-sphere described by $r = 2M$ is the horizon of the Schwarzschild black hole. For other types of black holes which are not static (metric component are time dependent) or stationary (line element not invariant under the transformation $dt \rightarrow -dt$ and spherically symmetric there can be several types of horizons defined and which may not coincide as in the case of Schwarzschild black hole.

A hyper-surface is defined by a scalar function $\Sigma(x) = 0$ of coordinates. The normal to the surface at any point x is defined as

$$n_\mu(x) \equiv \partial_\mu \Sigma(x) \tag{8.18}$$

and the tangent vectors t^μ at a given point is the set of four vector orthogonal to the normal $t^\mu n_\mu \equiv 0$.

Consider the hyper-surface of the Schwarzschild metric (8.1) defined by the function $r = $ constant. The normal vector to this surface is

$$n_\mu = \partial_\mu r = (0, 1, 0, 0) \tag{8.19}$$

The *apparent horizon* is defined as a null hyper-surface i.e. a hyper-surface whose normal is a null four-vector. For the Schwarzschild metric the null surface is given by the relation

$$n^\mu n_\mu = g^{\mu\nu} n_\mu n_\nu = g^{rr} = \left(1 - \frac{2GM}{r}\right) = 0 \tag{8.20}$$

and the apparent horizon of the is the 2-sphere described by $r = 2M$. On the null surface the tangent vector is related to the normal vector as $t^\mu = g^{\mu\nu} n_\mu$. The *Killing horizon* is the a surface where the Killing vector (Sect. 8.3) $\xi^\mu(x)$ becomes an null vector. Consider the $r = $ constant surface. The two Killing vectors of the Schwarzschild metric (8.1) (in (t, r, θ, ϕ) coordinate system) are $\xi^\mu_{(t)} = (1, 0, 0, 0)$ and $\xi^\mu_{(\phi)} = (0, 0, 0, 1)$. Consider the norm of the Killing vector $\xi^\mu_{(t)} = (1, 0, 0, 0)$ on a $r = $ constant 2-sphere, it becomes a null vector when

$$\xi^\mu_{(t)} \xi_{\mu(t)} = \xi^\mu_{(t)} g_{\mu\nu} \xi^\nu_{(t)} = g_{tt} = \left(1 - \frac{2GM}{r}\right) = 0. \tag{8.21}$$

For the Schwarzschild metric the Killing horizon is identical to the apparent horizon $r = 2M$ hyper-surface.

The locus of the $g_{tt} = 0$ is the *infinite redshift surface*. If a source at r_{em} near the black-hole emits a light pulse of frequency ν_{em} it will be observed at a distant point $r_{obs} \gg 2M$ with frequency

$$\nu_{obs} = \left(\frac{g_{tt}(r_{em})}{g_{tt}(r_{obs})}\right)^{1/2} \nu_{em} \simeq \left(1 - \frac{2GM}{r_{em}}\right)^{1/2} \nu_{em}. \tag{8.22}$$

When $r_{em} \to 2M$ for any finite value of ν_{em} the observed frequency is infinitely red-shifted, $\nu_{obs} \to 0$. In the Schwarzshild metric the infinite redshift surface coincides with the horizon. In the Kerr metric these are different as we shall see.

8.3.3 Particle Orbits in Schwarzschild Metric

To describe particle orbits around a black hole, it is necessary to identify the conserved quantities. The Schwarzschild metric (8.1) is independent of t and r so we can identify two of the Killing vectors (in (t, r, θ, ϕ) coordinate system) as $\xi^{\mu}_{(t)} = (1, 0, 0, 0)$ and $\xi^{\mu}_{(\phi)} = (0, 0, 0, 1)$. These correspond to the two conserved quantities (from this point onwards we set $G = 1$)

$$E = \xi^{\mu}_{(t)} V_{\mu} = \xi^{\mu}_{(t)} g_{\mu\nu} V^{\nu}$$

$$= \left(1 - \frac{2M}{r}\right) \frac{dt}{d\lambda} \tag{8.23}$$

and

$$L_z = \xi^{\mu}_{(\phi)} V_{\mu} = \xi^{\mu}_{(\phi)} g_{\mu\nu} V^{\nu}$$

$$= r^2 \sin^2 \theta \frac{d\phi}{d\lambda} \tag{8.24}$$

the energy and the z-component of the angular momentum respectively.

8.3.3.1 Massless Particle Trajectories
Massless particles trajectories are the null geodesics

$$g_{\mu\nu} \frac{dx^{\mu}}{d\lambda} \frac{dx^{\nu}}{d\lambda} = 0. \tag{8.25}$$

Since L_x and L_y are conserved quantities, if choose the initial condition $L_x = L_y = 0$, which implies and $\frac{d\theta}{d\lambda} = 0$, then the orbit of massless particles will remain equatorial in the plane, $\theta = \pi/2$. The eom for massless particles in equatorial orbit is from (8.33),

$$\left(1 - \frac{2M}{r}\right) \dot{t}^2 - \left(1 - \frac{2M}{r}\right)^{-1} \dot{r}^2 + r^2 \dot{\phi}^2 = 0, \tag{8.26}$$

where dot stands for $\frac{d}{d\lambda}$. This equation can be written in terms of the conserved energy E (8.23) and angular momentum $L = L_z$ (8.24) as

$$E^2 = \dot{r}^2 + \frac{L^2}{r^2} \left(1 - \frac{2M}{r}\right) \tag{8.27}$$

This can be cast in the form total energy written as a sum of kinetic and potential energy,

$$E_{eff} = \frac{1}{2}\dot{r}^2 + V_{eff}(r),$$

(8.28)

where $E_{eff} = E^2/2$ is the conserved effective energy and the effective potential is

$$V_{eff}(r) = \frac{L^2}{2r^2} - \frac{ML^2}{r^3},$$ (8.29)

Here the first term is the centrifugal term which exists in the Newtonian theory and the second term is the GR correction which can give bound orbits even of massless particles- something that is not there in Newtonian theory. In Fig. 8.1 we have a the plot of V_{eff} for different L.

The potential (8.29) is flat, $\frac{\partial V_{eft}}{\partial r} = 0$, at the radius

$$r_\gamma = 3M = \frac{3}{2}r_s$$ (8.30)

massless particles can form closed orbits. This means that around a black hole photons from the surrounding accretion disc can pile up at $r = 3M$ as a photon ring. Since the photon ring is outside the horizon $r_s = 2M$ it can also occur in any ultra compact object (for example neutron star) whose radius is between $2M$ and $3M$. Photons in the photon ring do not have stable orbits as any perturbation

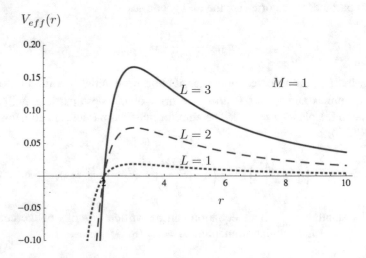

Fig. 8.1 Effective potential for massless particles with different angular momenta L. There is an unstable closed orbit at $r = 3M$ for all L

(say by photon-photon collision) will cause them to either fall into the black-hole or compact-star or escape to $r \to \infty$.

The type of trajectory of a photon (see Fig. 8.2) will depend on the ratio of its energy by angular momentum. For the closed photon orbit $\dot{r} = 0$ from (8.27) we have

$$\frac{E^2}{L^2} = \frac{1}{r^2} - \frac{2M}{r^3} \tag{8.31}$$

Closed circular orbits occur at the maxima of the potential is at $r = 3M$, evaluating (8.31) at $r = 3M$ we see that the critical ratio of E/L for getting closed photon orbits is

$$\left.\frac{E^2}{L^2}\right|_c = \frac{1}{27M^2} \tag{8.32}$$

For photons moving from a distance towards the black hole the angular momentum and energy are related to impact parameter, $L = \mathbf{r} \times \mathbf{P} = bE$. Since E and L are conserved from (8.32) we see that for the photon to be trapped in the photon ring the impact parameter must be $b_c = 3\sqrt{3}M$. Photons with impact parameter $b > b_c$ will escape the black hole, while photons with $b < b_c$ will fall into the event horizon. To the distant observer the image of the black hole will therefore be a shadow region surrounded by a ring of light with radius $r_c \sim 3\sqrt{3}M$.

The mass of the Milky Way galactic center black hole candidate Saggitarius A* (determined by monitoring the Kepler orbits of the stars around it) is $M = 4 \times 10^6 M_\odot$.

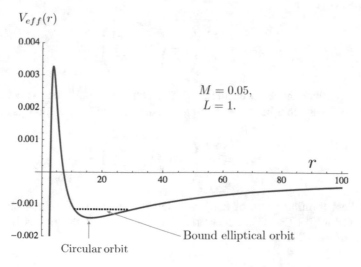

Fig. 8.2 Effective potential for massive particles in a Schwarzschild metric

8.3.3.2 Trajectories of Massive Particles

Massive particle trajectories are given by

$$g_{\mu\nu}\frac{dx^\mu}{d\lambda}\frac{dx^\nu}{d\lambda} = 1. \tag{8.33}$$

which for equatorial orbits is

$$\left(1 - \frac{2M}{r}\right)\dot{t}^2 - \left(1 - \frac{2M}{r}\right)^{-1}\dot{r}^2 + r^2\dot{\phi}^2 = 1. \tag{8.34}$$

Again replacing $\dot{t}$ and $\dot{\phi}$ with the conserved quantities E (8.23) and angular momentum $L = L_z$ (8.24), which now are the energy per unit mass and angular momentum per mass of the orbiting particle, we obtain,

$$\frac{1}{2}(E^2 - 1) = \frac{1}{2}\dot{r}^2 + \frac{L^2}{2r^2} - \frac{M}{r} - \frac{L^2M}{r^3} \tag{8.35}$$

which can again be written in the form (8.28) with the effective energy $E_{eff} = (E^2 - 1)/2$ and the effective potential for radial motion given by

$$V_{eff} = \frac{L^2}{2r^2} - \frac{M}{r} - \frac{L^2M}{r^3} \tag{8.36}$$

where the first term is the centrifugal term, the second term is the Newtonian potential and the last term is the GR correction.

The potential (8.36) (shown in Fig. 8.2) has two extrema where $V'_{eff} = 0$. The maxima is at

$$r_{max} = \frac{L^2}{r_s}\left(1 - \sqrt{1 - \frac{3r_s^2}{L^2}}\right) \tag{8.37}$$

where the particles have an unstable orbit which when perturbed can fall to the center or to the stable orbit at the minima. The minima of the potential is at

$$r_{min} = \frac{L^2}{r_s}\left(1 + \sqrt{1 - \frac{3r_s^2}{L^2}}\right) \tag{8.38}$$

The smallest limiting value of r_{min} is called the radius of the Innermost Stable Circular Orbit (ISCO). This occurs for $L = \sqrt{3}r_s$ and the radius of ISCO is $r_{ISCO} = 3r_s = 6GM$.

For circular orbits, the Kepler angular frequency is

$$\Omega \equiv \frac{d\phi}{dt} = \frac{d\phi/d\lambda}{dt/d\lambda} = \frac{L}{r^2} \frac{(1 - r_s/r)}{E} = \left(\frac{r_s}{2r^3}\right)^{1/2} = \frac{\sqrt{GM}}{r^{3/2}} . \tag{8.39}$$

where we have used (8.23), (8.24) and (8.35). Kepler's law relation between the time period and radius remain the same in Einstein's theory as in Newtonian theory.

8.4 Charged Black Hole

8.4.1 Astrophysical Charge of Black Holes

A black hole swallows up matter from the accretion discs and in this process can an uncharged black hole can acquire a net charge over time. The electrons and protons in a neutral plasma in the accretion disc are rested from falling towards the black hole by the radiation pressure from the photons in the plasma. The radiation pressure on the electrons is larger and as a result the black holes acquire a surplus of protons over electrons and acquire a net charge. An estimate of the charge acquired by black holes from this differential accretion of protons and electrons can be estimated as follows [5]. The equation of motion for protons and electrons accreting into a charged black hole are respectively,

$$m_p \dot{v}_p = -\frac{m_p GM}{r^2} + \frac{\sigma_{\gamma p} L}{4\pi r^2} + \frac{eQ}{r^2} , \tag{8.40}$$

$$m_e \dot{v}_e = -\frac{m_e GM}{r^2} + \frac{\sigma_{\gamma e} L}{4\pi r^2} - \frac{eQ}{r^2} . \tag{8.41}$$

The black hole charge will build up due to the net accumulation of protons over electrons and the equilibrium charge will be reached when $\dot{v}_p = \dot{v}_e$. Using this relation in (8.41) and using $m_p \gg m_e$ and the photon scattering cross section $\sigma_{\gamma e} \gg \sigma_{\gamma p}$ we find that the equilibrium charge is reached when photon pressure on electrons is equal to the electrostatic attraction by the charge in the black holes

$$Q_{eq} = \frac{\sigma_{\gamma e} L}{4\pi e} = 0.96 \times 10^{21} \left(\frac{M}{M_\odot}\right) \left(\frac{L}{L_{Edd}}\right) \tag{8.42}$$

in terms of the mass of the black hole and luminosity of the accretion disc where the limiting Eddington luminosity L_{Edd} is when the accretion of matter is balanced by the photon pressure and is given by the expression

$$L_{Edd} = \frac{4\pi G m_p}{\sigma_{\gamma e}} = 1.3 \times 10^{38} \left(\frac{M}{M_\odot}\right) \frac{erg}{sec} . \tag{8.43}$$

8.4.2   Reissner-Nordstrom Metric of the Charged Black Holes

A charged black hole has the metric [6] given by

$$ds^2 = \left(1 - \frac{2GM}{r} + \frac{GQ^2}{r^2}\right) dt^2$$

$$- \left(1 - \frac{2GM}{r} + \frac{GQ^2}{r^2}\right)^{-1} dr^2 - r^2 \left(d\theta^2 + \sin^2\theta d\phi^2\right) \qquad (8.44)$$

The characteristic length scale associated with the charge of the black hole (8.44) is

$$r_Q = (GQ^2)^{1/2} = 1.38 \times 10^{-39} \left(\frac{|Q|}{e}\right) \text{ km} \qquad (8.45)$$

This is small compared to with the Schwarzschild radius $r_s = 2GM = 2.96(M/M_\odot)$km for black holes with astrophysical charge (8.42), so the charge accumulated from the accretion of matter in a black hole has negligible effect on the space-time curvature compared to the effect of its mass.

8.5   Kerr Black Holes

The space-time metric of a rotating black hole is described by the Kerr metric [4, 7] which in Boyer-Lindquist [8] coordinates (t, r, θ, ϕ) is described by the line element

$$ds^2 = -dt^2 + \Sigma \left(\frac{dr^2}{\Delta} + d\theta^2\right) + (r^2 + a^2)\sin^2\theta d\phi^2 + \frac{2Mr}{\Sigma}\left(a\sin^2\theta d\phi - dt\right)^2,$$

$$\Delta(r) = r^2 - 2Mr + a^2, \quad \Sigma(r, \theta) = r^2 + a^2\cos^2\theta. \qquad (8.46)$$

where a is the rotation parameter which is related to magnitude of the angular momentum as $a = J/M$.

The non-zero components of the Kerr metric are

$$g_{tt} = -\left(1 - \frac{2Mr}{\Sigma}\right), \quad g_{rr} = \frac{\Sigma}{\Delta}, \quad g_{\theta\theta} = \Sigma,$$

$$g_{t\phi} = -\frac{2Mr}{\Sigma} a\sin^2\theta, \quad g_{\phi\phi} = \frac{\sin^2\theta}{\Sigma}\left((r^2 + a^2)^2 - a^2\Delta\sin^2\theta\right)$$

The non-zero components of the inverse metric are

$$g^{tt} = -\frac{1}{\Delta}\left(r^2 + a^2 + \frac{2Mra^2\sin^2\theta}{\Sigma}\right), \quad g^{rr} = \frac{\Delta}{\Sigma}, \quad g^{\theta\theta} = \frac{1}{\Sigma},$$

$$g^{t\phi} = -\frac{2Mra}{\Sigma\Delta}, \quad g^{\phi\phi} = \frac{\Delta - a^2\sin^2\theta}{\Sigma\Delta\sin^2\theta}. \tag{8.47}$$

The Kerr metric is stationary (metric components do not spend upon time) but not static as it is not invariant under $dt \to -dt$. It is axisymmetric (no ϕ dependence) and is invariant under the simultaneous transformations $dt \to -dt$ and $d\phi \to -d\phi$.

A rotating black hole with non-zero electric charge Q and magnetic charge P is described by the Kerr-Newman metric [9, 10] which has the same form as (8.47) with $\Delta(r)$ given in (8.46) generalized to $\Delta(r) = r^2 - 2Mr + a^2 + Q^2 + P^2$.

Birkoff's theorem states that the exterior metric of any spherically matter density is the Schwarzschild metric. A similar theorem does not exist for the Kerr metric. The exterior metric of a rotating star does is not the Kerr metric. The Kerr metric and the rotating star metric match only asymptotically. The asymptotic limit of the Kerr metric (leading order in M/r and a/r^2) which is the metric outside rotating stars is

$$ds^2 = -\left(1 - \frac{2M}{r}\right)dt^2 - \frac{4aM}{r^2}\sin^2\theta \, d\phi dt$$

$$+ \left(1 + \frac{2M}{r}\right)dr^2 + r^2\left(d\theta^2 + \sin^2 d\phi^2\right). \tag{8.48}$$

The identification of the coordinates can be done by taking the $M = 0, J = 0$ limit (keeping a fixed) of (8.46). The result is a flat metric not in spherical but in ellipsoidal (r, θ, ϕ) coordinates,

$$ds^2 = -dt^2 + \frac{(r^2 + a^2\cos^2\theta)^2}{r^2 + a^2}dr^2 + (r^2 + a^2\cos^2\theta)d\theta^2 + (r^2 + a^2)\sin^2\theta d\phi^2$$

$$= -dt^2 + dx^2 + dy^2 + dz^2 \tag{8.49}$$

where

$$x = (r^2 + a^2)^{1/2}\sin\theta\cos\phi, \quad y = (r^2 + a^2)^{1/2}\sin\theta\sin\phi, \quad z = r\cos\theta. \tag{8.50}$$

In these coordinates $r = constant$ spatial hyper-surfaces are ellipsoids,

$$\frac{x^2 + y^2}{r^2 + a^2} + \frac{z^2}{r^2} = 1. \tag{8.51}$$

8.5.1   Ring Singularity

The metric (8.46) is singular where $\Delta = 0$ and $\Sigma = 0$. An evaluation of the invariant quantity from the Riemann tensor

$$R_{\mu\nu\alpha\beta}R^{\mu\nu\alpha\beta} = \frac{48M^2}{\Sigma^6}\left(r^2 - a^2\cos^2\theta\right)\left[\Sigma^2 - 16r^2a^2\cos^2\theta\right] \qquad (8.52)$$

shows that the real singularity is the locus $\Sigma = r^2 + a^2\cos^2\theta = 0$ which means that the singularity is located at $r = 0$ and $\theta = \pi/2$. This implies that the locus of the singularity ($r = 0, \theta = \pi/2$) is the equatorial ring with radius a given in Cartesian coordinates as $x^2 + y^2 = a^2$ and $z = 0$.

8.5.2   Inner and Outer Horizon

The apparent horizon of a black hole is a surface whose normal is a null-vector. Consider the surface $r = constant$ at a fixed t. These are different ellipsoids at different values of the constant. Consider the vector $n_\mu = (0, 1, 0, 0)$ in the ellipsoidal coordinates. Then

$$n_\mu n^\mu = n_\mu n_\nu g^{\mu\nu} = g^{rr} = \frac{\Delta}{\Sigma} \qquad (8.53)$$

The zeroes of g^{rr} are by the relation $\Delta = r^2 - 2Mr + a^2 = 0$ are the apparent horizons since the normal to these surfaces, n_μ is a null vector. There are two solutions. The outer horizon $r = r_+$ given by

$$r_+ = M + \sqrt{M^2 - a^2} \qquad (8.54)$$

and an inner horizon located at $r = r_-$ given by

$$r_- = M - \sqrt{M^2 - a^2}\,. \qquad (8.55)$$

The two horizons described by the coordinates $(r_\pm, \theta, \phi)$ are ellipsoids.

In the region $r > r_+$, the $r = $ constant hyper surfaces are time-like. In the region $r_- < r < r_+$ the $r = $ constant surfaces are space-like. A particle in this region will have a trajectory such that it falls towards r_- and continues inside the black hole. In $r < r_-$ region $r-$constant surfaces are again time-like. This region contains the ring singularity at $r = 0$.

When $a < M$ the singularity is covered by a horizon. The case $a = M$ when the horizon coincides with the singularity is called extremal.

The horizons of the Kerr metric and the ring singularity are shown in Fig. 8.3.

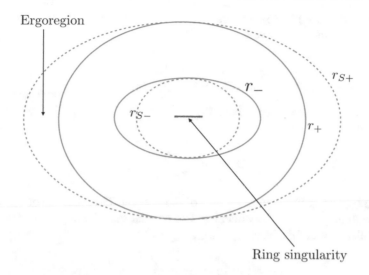

Fig. 8.3 Regions of the Kerr metric. r_+ is the outer horizon and r_- the inner horizon. The regions between the outer infinite redshift surface r_{S+} and the outer horizon r_+ is the ergo-region. The curvature singularity is the ring $r = 0, \theta = \pi/2$

To estimate the typical values of a expected of a solar mass black hole consider the case of a black hole formed when a sun like star collapses without shedding its angular momentum. The mass of the sun is $M_\odot = 1.99 \times 10^{30}$ kg and its angular momentum is $J_\odot = 1.92 \times 10^{41}$ kg m^2 s^{-1}. The value of the parameter a for the sun is $a_\odot = J_\odot/M_\odot = 322mc$. The value of $GM_\odot = 1.48c^2$km. Therefore if the sun were to collapse into a Kerr black hole it will have $a_\odot = 0.22GM_\odot/c$.

8.5.3 Frame Dragging and ZAMO

If a particle with zero angular momentum (ZAMO) at infinity falls towards a black hole, Its conserved angular momentum along as it fall towards r_+ is

$$L = \xi^\mu_{(\phi)} P_\mu = g_{t\phi}p^0 + g_{\phi\phi}p^\phi = 0 \qquad (8.56)$$

This zero angular momentum particle acquires an angular velocity as it falls towards r_+ given by

$$\Omega = \frac{d\phi}{dt} = \frac{p^\phi}{p^0} = \frac{-g_{\phi t}}{g_{\phi\phi}} \qquad (8.57)$$

where we have used (8.56). Substituting the metric components (8.47) we find the expression for the angular velocity for a ZAMO

$$\Omega = \frac{2Mar}{(r^2 + a^2)^2 - a^2 \Delta \sin^2\theta}. \tag{8.58}$$

At the horizons $r = r_\pm$, we have $\Delta = 0$ and the angular velocity (8.58) of a ZAMO is

$$\Omega(r_\pm) = \frac{2Mar_\pm}{(r_\pm{}^2 + a^2)^2} \tag{8.59}$$

This is regarded as the angular velocity of the horizon. The inner and outer horizons rotate with different angular velocities. The angular velocity of the outer horizon is the angular speed of the black hole and is given by

$$\Omega_H = \Omega(r_+) = \frac{a}{2Mr_+} = \frac{a}{(r_+{}^2 + a^2)} \tag{8.60}$$

8.5.4   Ergosphere

In the Kerr metric the infinite redshift surface g_{tt} is given by the relation

$$g_{tt} = -\left(1 - \frac{2Mr}{\Sigma}\right) = -\frac{1}{\Sigma}(r - r_{S+})(r - r_{S-}) = 0 \tag{8.61}$$

where the two roots $r_{S\pm} = M \pm \sqrt{M^2 - a^2 \cos^2\theta}$. The outer surface $r = r_{S+}$ is called the *ergo-sphere*. The horizons lie between region bounded by r_{S+} and r_{S-}, $r_{S-} \le r_- \le r_+ \le r_{S+}$. The region between the ergo-sphere r_{S+} and the outer horizon r_+

$$r_+ < r < r_{S+} \tag{8.62}$$

is the *ergo-region*.

The Kerr metric has a Killing vector $\xi_{(t)}^\mu = (1, 0, 0, 0)$ which is asymptotically time-like and is associated with the conserved energy $E = -\xi_{(t)}^\mu p_\mu$. As the norm of $\xi_{(t)}^\mu$ is

$$\xi_{(t)}^\mu \xi_{\mu(t)} = g_{tt} = -\frac{1}{\Sigma}(r - r_{S+})(r - r_{S-}), \tag{8.63}$$

which is null on r_{S+} and becomes space-like in the ergo-region. Since the ergo-region is outside the horizon, particles which fall in the ergo-region have the possibility of escaping falling in the black hole. The fact that $\xi^\mu_{(t)}$ becomes space-like in the ergo-region gives rise to some exotic processes like the Penrose process and super-radiance.

8.5.5 Killing Horizon

The linear combination of the Killing vectors

$$\xi^\mu_H = \xi^\mu_{(t)} + \Omega_H \, \xi^\mu_{(\phi)} = (1, 0, 0, \Omega_H) \tag{8.64}$$

is null at $r = r_+$,

$$g_{\mu\nu}\xi^\mu_H\xi^\nu_H = g_{tt} + 2\Omega_H g_{t\phi} + \Omega^2_H g_{\phi\phi} = 0 \quad \text{at } r = r_+. \tag{8.65}$$

The surface $r = r_+$ is the *Killing horizon* associated with the Killing vector ξ^μ_H.

Since ξ^μ_H is a null vector on the horizons $r = r_+$, its tangent vector $t^\nu = \xi^\mu_H \nabla_\mu \xi^\nu_H \propto \xi^\nu_H$. The surface gravity κ on the horizon is defined as the proportionality constant

$$\xi^\mu_H \nabla_\mu \xi^\nu_H \Big|_{r=r_+} = \kappa \xi^\nu_H \tag{8.66}$$

8.6 The Penrose process

It is possible to extract energy and angular momentum from rotating black by classical processes, known collectively as the Penrose process [11–16]. Consider a process where a particle splits into two in the ergo-region after which one of the secondary particles falls into the black hole and the second one escapes. The conserved energy of the initial particle which falls from an asymptotically flat distant is

$$E = -\xi^\mu_{(t)} p_\mu = -g_{tt} p^0 - g_{t\phi} p^\phi$$

$$= \frac{1}{\Sigma} (r - r_{S+}) (r - r_{S-}) \, p^0 + \frac{2Mr}{\Sigma} a \sin^2 \theta \, p^\phi, \tag{8.67}$$

and its angular momentum is

$$L = \xi^\mu_{(\phi)} p_\mu = g_{t\phi} p^0 + g_{\phi\phi} p^\phi$$

$$= -\frac{2Mr}{\Sigma} a \sin^2 \theta \, p^0 + \frac{\Delta - a^2 \sin^2 \theta}{\Sigma \Delta \sin^2 \theta} p^\phi \tag{8.68}$$

the minus sign in the definition of energy is so that asymptotically far from the black hole ($r \rightarrow \infty$) the energy is p^0 and angular momentum is p^ϕ.

The particle splits into two in the ergosphere and in the local inertial frame of the particle we have the conservation of energy

$$p^0 = p_1^0 + p_2^0 \tag{8.69}$$

Contracting this with $\xi_{(t)}^\mu$ we get the relation

$$E = E_1 + E_2 \tag{8.70}$$

Similarly from the conservation of the angular momentum in the local inertial frame we get the angular momentum conservation relation

$$L = L_1 + L_2 \tag{8.71}$$

In the ergo-sphere $g_{tt} > 0$. If one of the decay products say particle '1" plunges into the horizon and particle '2" escapes to infinity then $E_1 < 0$. Then the energy of the particle that escapes will be $E_2 = E - E_1 = E + |E_1|$ and the final particle energy in the asymptotic region is larger than the initial particle energy. This process of energy extraction from the black hole is called the *Penrose process* and is entirely classical. The energy extraction by the Penrose process results in a net reduction of the black hole mass and angular momentum.

8.7   Superradiance

The phenomenon whereby in-falling quantum fields is reflected from the black hole horizon with larger amplitude is called superradiance [17–19]. Consider a rotating black hole with a cloud of light bosons. The bosons fields in the black hole gravitational potential exist in quantized energy levels. A light boson with frequency ω is has an superradiance instability if

$$\omega < m \, \Omega_H = \frac{ma}{2Mr_+} \tag{8.72}$$

where m is the azimuthal quantum number of the scalar boson quantum state, This can be shown as follows [20]. The null Killing vector which is orthogonal to the horizon is $\xi_H^\mu = \xi_{(t)}^\mu + \Omega_H \xi_{(\phi)}^\mu = (1, 0, 0, \Omega_H)$. Now consider the a scalar field incident on the horizon from infinity

$$\phi(x) = f(r, \theta) \, e^{-i\omega t + im\varphi} + \text{hc} \,. \tag{8.73}$$

The conserved energy of the field is

$$P_\mu = -T_{\mu\nu}\,\xi_t^\nu$$

$$= -\partial_\mu\phi\,\partial_t\phi + \frac{1}{2}g_{\mu t}\mathcal{L}, \tag{8.74}$$

where $T_{\mu\nu}$ is the energy-momentum tensor and $\mathcal{L}$ is the Lagrangian density of the scalar field. The time-averaged energy flux going through the horizon is

$$\langle P_\mu\,\xi_H^\mu\rangle = -\langle\,(\partial_t\phi + \Omega_H\,\partial_\varphi\phi)\,\partial_t\phi\,\rangle$$

$$= \omega\,(\omega - m\,\Omega_H)\,|f|^2 \tag{8.75}$$

and the energy flux in-falling through the horizon is negative when $\omega < m\,\Omega_H$. This is interpreted as the reflection of incident waves with amplification or superradiance.

An important aspect of superradiance is that while bosonic fields can have super radiant instability if they fulfill the condition (8.72), neither massless [21] nor massive [22] fermonic fields have superradiance.

The superradiant instability follows from the area theorem of blackholes [23]. Since the proof of the theorem assumes the weak energy condition (that the energy momentum tensor should the condition $T_{\mu\nu}\xi^\mu\xi^\nu > 0$ for any timelike vector ξ^μ) is violated by fermonic fields which explains the difference why bosonic fields are superradiant but fermionic fields are not [21].

8.7.1 Superradiance of Scalar Fields

A massive scalar field can form a cloud around a spinning black hole and in some circumstances the amplitude of the wave function can grow exponentially [24–27]. The energy of the scalar cloud is drained d from the rotation of the black hole. This superradiant instability has a lifetime and a plot of the lifetime of the black holes with rotation parameter may show gaps which depends on the mass of the scalar, signifying the superradiant instability.

To derive the quantized energy levels of the scalar cloud and its growth time scale consider the wave equation of a scalar of mass μ,

$$\Box\Phi + \mu^2\Phi = (-g)^{-1/2}\partial_\mu\left((-g)^{1/2}g^{\mu\nu}\partial_\nu\right)\Phi + \mu^2\Phi = 0 \tag{8.76}$$

For the Kerr metric in Boyer Lindquist coordinates (t, r, θ, ϕ) using the inverse metric components (8.47) and determinant $g = \Sigma^2\sin\theta^2$ the wave equation for the scalar field takes the form

$$\left(\frac{(r^2 + a^2)}{\Delta} - a^2 \sin^2 \theta \right) \partial_t^2 \Phi - \partial_r \left(\Delta \, \partial_r \Phi \right) + \frac{4Mar}{\Delta} \partial_t \partial_\varphi \Phi$$

$$+ \left(\frac{a^2}{\Delta} - \frac{1}{\sin^2 \theta} \right) \partial_\varphi^2 \Phi - \frac{1}{\sin \theta} \partial_\theta \left(\sin \theta \, \partial_\theta \Phi \right) + \mu^2 \Sigma \, \Phi = 0 \qquad (8.77)$$

This equation can be separated in the variables by taking Φ of the form

$$\Phi(t, r, \theta, \phi) = e^{-i\omega t} \, e^{im\phi} \, S(\theta) \, R(r) . \qquad (8.78)$$

The wave equation (8.77) then separates into equations for $R(r)$ and $S(\theta)$ as follows,

$$\Delta \frac{d}{dr} \left(\Delta \frac{d}{dr} R \right) + \left(\omega^2 (r^2 + a^2)^2 - 4Mam\omega r + a^2 m^2 \right.$$

$$\left. - \Delta (a^2 \omega^2 + \mu^2 r^2 + A_{lm}) \right) R = 0 \qquad (8.79)$$

and

$$\frac{1}{\sin \theta} \frac{d}{d\theta} \left(\sin \theta \frac{d}{d\theta} S \right) + \left[a^2 (\omega^2 - \mu^2) \cos^2 \theta - \frac{m^2}{\sin^2 \theta} + A_{lm} \right] S = 0 \qquad (8.80)$$

where the eigenvalue of the angular wave function $e^{im\varphi} S(\theta)$ is $A_{lm} = l(l + 1) + O(a^2 \omega^2)$.

The radial equation can be solved by a change of variables from r to r_* defined by

$$\frac{dr_*}{dr} = \frac{r^2 + a^2}{\Delta} \qquad (8.81)$$

which can be integrated (with the boundary condition $r_* \to \infty$ when $r \to \infty$ and $r_* \to -\infty$ when $r \to r_+$) to give

$$r_* = r + \frac{2M}{r_+ - r_-} \left(r_+ \ln \left| \frac{r - r_+}{2M} \right| - r_- \ln \left| \frac{r - r_-}{2M} \right| \right) \qquad (8.82)$$

By making the transformation

$$R(r) = \frac{X(r_*)}{r^2 + a^2} \qquad (8.83)$$

the radial equation reduces to the form [24]

$$\frac{d^2 X}{dr_*^2} + (\omega^2 - V(r))X = 0 \tag{8.84}$$

with

$$V(r) = \frac{\mu^2 \Delta}{r^2 + a^2} + \frac{4Mram\omega - a^2 m^2 + \Delta(A_{lm} + (\omega^2 - \mu^2)a^2}{(r^2 + a^2)^2}$$

$$+ \frac{\Delta(3r^2 - 4Mr + a^2)}{(r^2 + a^2)^3} - \frac{3\Delta^2 r^2}{(r^2 + a^2)^4}. \tag{8.85}$$

From (8.85) we can check that the momentum of the scalar field $k = (\omega^2 - V)^{1/2}$ has the asymptotic limits $k \to (\omega^2 - \mu^2)^{1/2}$ as $r \to \infty$ and $k \to k_+ = (\omega - m\Omega_H)$ as $r \to r_+$. The asymptotic solutions for the scalar field in far and near regions from the horizon can therefore be written as

$$\Phi \sim \frac{1}{r} e^{-i\sqrt{\omega^2 - \mu^2}\, r_* - i\omega t} e^{im\varphi} S(\theta), \quad r \to \infty,$$

$$\Phi \sim \frac{1}{\sqrt{2Mr_+}} e^{-i(\omega - m\Omega_H)r_* - i\omega t} e^{im\varphi} S(\theta), \quad r \to r_+, \tag{8.86}$$

where the negative sign in the exponent of the radial wave function is chosen such that the wave function corresponds to in-falling particles near the horizon and at asymptotic distance. If the frequency of the waves is such that $\omega < m\Omega_H$ then we see that the near horizon solution will correspond to a flux of outgoing particles. This is the supper radiant instability in scalar fields which arises when $\omega < m\Omega_H$. As the energy ω is positive, the superradiant condition only occurs when the azimuthal quantum number $m > 0$ which implies that the scalar fields which are emitted from the horizon will be co-rotating with the Kerr black hole.

8.7.2 Teukolsky Equation

Teukolsky has shown [28–30] that in the Kerr background the second order wave equation for all particles (bosons and fermions of spin s) is separable in r and θ coordinates. The wave-function of a particle with energy ω and azimuthal quantum number m can be written as,

$$\Psi(t, r, \theta, \varphi) = \int d\omega e^{-i\omega t} e^{im\varphi} R(r) S(\theta), \tag{8.87}$$

and the equations for the radial wave function $R(r)$ and angular wave function $S(\theta)$ can be separated as,

$$\Delta^{-s}\frac{d}{dr}\left(\Delta^{s+1}\frac{dR}{dr}\right)+\left(\frac{K^2-2is(r-M)K}{\Delta}+4is\omega r-\lambda\right)R(r)=0 \qquad (8.88)$$

and

$$\frac{1}{\sin\theta}\frac{d}{d\theta}\left(\sin\theta\frac{dS}{d\theta}\right)+\left(a^2\omega^2\cos^2\theta-\frac{m^2}{\sin^2\theta}\right.$$

$$\left.-2a\omega s\cos\theta-\frac{2ms\cos\theta}{\sin^2\theta}-s^2\cot^2\theta+s+A_{slm}\right)S(\theta)=0,$$

$$(8.89)$$

where

$$K=(r^2+a^2)\omega-am \quad \text{and} \quad \lambda=A_{slm}+a^2\omega^2-2am\omega \qquad (8.90)$$

and the eigenvalues of the spin-weighted spheroidal harmonics $S_{slm}\equiv e^{im\varphi}S(\theta)$, for small $a\omega$, are given by $A_{slm}=l(l+1)-s(s+1)+O(a^2\omega^2)$ [31].

The Teukolsky equation forms the basis for calculating the gravitational radiation by relativistic particles around black holes [32].

8.8   M87* Galactic Center Black Hole

The Event Horizon Telescope (EHT) collaboration has released the first ever picture [33] of the shadow of a black hole, that of the supermassive black hole at the center of the M87 galaxy. The angular diameter of the photon ring measured by EHT [34] is $\alpha_{sh}=3\sqrt{3}r_s/D_A=42.0\pm3\mu as$ where $r_s=2GM/c^2$. By measuring the distance to M87 by standard candle methods [34] which give $D_A=16.8\pm0.8\,\text{Mpc}$, the mass of the black hole M87* was obtained to be $M=(6.5\pm0.2|_{stat}\pm0.7|_{sys})\times10^9 M_\odot$. The mass measurement from the photon shadow is in agreement with the mass measurement from observations of stellar orbits around M87* which gave the mass as $M=(6.6\pm0.4)\times10^9 M_\odot$ [35].

The measurement of the properties of M87* black hole has resulted in several constraints on the black hole parameters as well as cosmological parameters. A constraint on the angular velocity Ω_H which can be parameterized by the dimensionless rotation parameter $a_*=J/(GM^2)$ as

$$\Omega_H=\frac{a}{2Mr_+}=\frac{1}{2r_g}\frac{a_*}{(1+\sqrt{1-a_*^2})} \qquad (8.91)$$

(where $r_g = GM$) is placed by measurement of the orbital angular momentum of the $\lambda = 1.3$ mm radio waves from M87* [36]. This measurement of the twisted light gives a bound on the rotation parameter

$$a_* = 0.9 \pm 0.1 .\tag{8.92}$$

and the angle of inclination of the black hole rotation axis is $i = 17°$.

From the bounds on the angular velocity which can be placed from (8.92) and (8.91) one can place a bound on the mass of scalar field which will cause a superradiant instability (8.72). Superradiance over a large timescale of the lifetime of the black hole will extract the rotational energy of the black hole and reduce the spin a_* to zero [20, 37]. From the fact that the spin parameter is non-zero (8.92) the range of scalar masses ($\mu_s \sim \omega$, $m = 1$) which could have given rise to superradiance can be ruled out. The rate of superradiant emission is given by Detweiler [25]

$$\Gamma_s = \frac{1}{24} a_* r_g^8 \mu_s^9 .\tag{8.93}$$

Assuming that over a lifetime $\tau_{BH} \sim 10^9$ years one can place a upper bound on the mass of the scalar that can be emitted in order to deplete the black hole of its entire angular momentum the range of mass of scalar that is ruled out is [38]

$$2.9 \times 10^{-21} \text{ eV} < \mu_s < 4.6 \times 10^{-21} \text{ eV} .\tag{8.94}$$

8.9   Primordial Black Holes as Dark Matter

Primordial black holes (PBH) are the black holes which do not from collapse of stars but arise when the primordial density fluctuation in a causal horizon $\delta\rho/\rho \sim O(1)$. The existence of PBH arising from primordial density fluctuations was first proposed by Zeldovich and Novikov [39] followed by Hawking [40]. The idea that PBH could be the dark matter was first proposed by Chapline [41]. Primordial black holes with mass $M > 10^{-16} M_\odot$ survive evaporation by Hawking radiation over the lifetime of the universe. The phenomenological consequences of PBH of different mass range as dark matter has been done recently by Carr et al. [42]. It is possible that the 10–100 $M_\odot$ mass black holes observed through gravitational waves at Ligo or the $10^6 - 10^9$ mass super-masive blak holes which are there in the center of galaxies are of primordial origin.

8.9.1   Formation of PBH

A PBH can form when an a perturbation inside the horizon and grows to the extent that a substantial fraction of the mass of the horizon collapses into a black-hole.

The mass inside a horizon in the radiation era is

$$M_H = \frac{4\pi}{3}\rho H^{-3} = \frac{1}{2GH} = 0.6\frac{M_P^3}{\sqrt{g_\rho}\,T^2}. \tag{8.95}$$

All matter inside the horizon does not collapse into a black hole. The mass of the black hole which is formed is smaller than the mass inside the horizon by the collapse efficiency factor $\gamma \simeq 0.2$ [42]. Thus the mass of a primordial black hole formed in the radiation era at temperature T is

$$M = \gamma M_H = 1.5\,M_\odot\left(\frac{\gamma}{0.2}\right)\left(\frac{T}{200\,\text{MeV}}\right)^{-2}\left(\frac{g_\rho}{10.75}\right)^{-1/2}. \tag{8.96}$$

A black hole will be formed if the density perturbation in a horizon exceeds a critical value [43]

$$\delta_c = \frac{3(1+\omega)}{(5+3\omega)}\sin^2\left[\frac{\pi\sqrt{\omega}}{1+\omega}\right] \tag{8.97}$$

where $\omega = p/\rho$ is the equation of state of the collapsing fluid. In the radiation era $\omega = 1/3$, the critical density $\delta_c = 0.4135$.

8.9.2   PBH Formation by Inflation Generated Perturbations

Suppose the perturbation of co-moving wavenumber k enters the horizon (then its wavelength $\lambda_{phy} = a/k = H^{-1}$) then the mass of the blackhole which is formed is

$$M(k) = \gamma M_H(k) = \gamma\frac{4\pi}{3}\rho H^{-3}\big|_{H=k/a} = \gamma\frac{4\pi}{3}\rho\left(\frac{a}{k}\right)^3 \tag{8.98}$$

The ratio of mass inside the horizon at any temperature T to the mass the corresponding mass at matter radiation equality is

$$\frac{M_H(k)}{M_H(k_{eq})} = \left(\frac{g_\rho}{g_{\rho eq}}\right)\left(\frac{T}{T_{eq}}\right)^4\left(\frac{k_{eq}}{k}\right)^3\left(\frac{a}{a_{eq}}\right)^3 \tag{8.99}$$

where we used $\rho = (\pi^2/30)g_\rho T^4$ for density in the radiation era. The temperature ratio can be written in terms cf. ratio of scale factors using the conservation of entropy $S = (\pi^2/30)g_* T^3 a^3$ which implies that $T \propto g_*^{-1/3} a^{-1}$. Taking $g_\rho \simeq g_*$ which is a good assumption prior to $e^+ e^-$ recombination, we have

$$\frac{M_H(k)}{M_H(k_{eq})} = \left(\frac{g_{*eq}}{g_*}\right)^{1/3}\left(\frac{k_{eq}}{k}\right)^3\left(\frac{a_{eq}}{a}\right) \tag{8.100}$$

The ratio of the scale factors can be evaluated using the relation

$$H = \frac{k}{a} = 1.66 M_P^{-1} g_*^{1/2} T^2 \tag{8.101}$$

to give

$$\left(\frac{a_{eq}}{a}\right) = \left(\frac{k_{eq}}{k}\right) \left(\frac{g_*}{g_{*eq}}\right)^{1/2} \left(\frac{T}{T_{eq}}\right)^2 = \left(\frac{k_{eq}}{k}\right) \left(\frac{M_H(k)}{M_H(k_{eq})}\right)^{-1} \tag{8.102}$$

where we have used the relation (8.95), $M_H \propto (g_*^{1/2} T^2)^{-1}$. Using (8.102) in (8.100) we can solve for the mass ratio to obtain

$$\frac{M_H(k)}{M_H(k_{eq})} = \left(\frac{g_{*eq}}{g_*}\right)^{1/6} \left(\frac{k_{eq}}{k}\right)^2 \tag{8.103}$$

The mass of a black hole formed in the radiation era when a perturbation of comoving wavenumber k enters the horizon can be written in terms of the mass of the horizon at matter radiation equality as [43],

$$M(k) = \gamma M_H(k_{eq}) \left(\frac{g_{*eq}}{g_*}\right)^{1/6} \left(\frac{k_{eq}}{k}\right)^2$$

$$= 10^{19} \text{ gm } \left(\frac{\gamma}{0.2}\right) \left(\frac{g_*}{106.75}\right)^{-1/6} \left(\frac{k}{2.2 \times 10^{13} \text{ Mpc}^{-1}}\right)^{-2} \tag{8.104}$$

At matter-radiation equality, $T_{eq} = 9350\,°\text{K}$, $z_{eq} = 3400$, $g_{*eq} = 3.36$, $k_{eq} = a_{eq} H_{eq} = 0.0091 \text{ Mpc}^{-1}$, and mass of the Horizon is $M_H(k_{eq}) = 3.1 \times 10^{17} M_\odot$. This relation between black-hole mass and wave-number of perturbation is different if there is a non-standard cosmological evolution after reheating [44].

Given a curvature power spectrum generated during inflation $P_\xi(k)$, the density perturbation which enters the horizon during an epoch with $\bar{P}/\bar{\rho} = \omega$ is

$$P_\delta(k) = \frac{4(1+\omega)^2}{(5+3\omega)^2} \left(\frac{k}{aH}\right)^4 P_\xi(k) \tag{8.105}$$

The probability distribution of density fluctuation $\delta(M)$ is assumed to be a Gaussian,

$$p(\delta(M(k))) = \frac{1}{\sqrt{2\pi}\sigma(M)} \exp\left(\frac{-\delta^2(M)}{2\sigma^2(M)}\right) \tag{8.106}$$

with the variance given by the power spectrum of δ smoothed over a length scale R with a window function $W(k, R)$,

$$\sigma^2(R) = \int_0^\infty W^2(q, R) P_\delta(q) \frac{dq}{q}. \tag{8.107}$$

We smooth over the horizon scale taking $R = k^{-1}$ and obtain the relation between the varian $\sigma^2(M)$ and the primordial power spectrum P_ξ as

$$\sigma^2(M(k)) = \int \frac{dq}{q} W^2(qk^{-1}) \frac{16}{21} \left(qk^{-1}\right)^4 P_\xi(q). \tag{8.108}$$

The window function can be taken to be a top-hat with width k or a Gaussian with variance k^2. The probability of forming blackholes of mass M in each horizon is therefore

$$\beta(M(k)) = \int_{\delta_c}^1 d\delta p(\delta(M))$$

$$\simeq \frac{\sigma(M)}{\delta_c} \exp\left(\frac{-\delta_c^2}{2\sigma^2(M)}\right) \tag{8.109}$$

The fraction of density fraction PBH of mass M in the present era is

$$f_{PBH}(M) = \frac{\rho_{PBH}(M)}{\rho_c}$$

$$= \left(\frac{\beta(M)}{8 \times 10^{-15}}\right) \left(\frac{0.12}{\Omega_{PBH}h^2}\right) \left(\frac{\gamma}{0.2}\right)^{3/2} \left(\frac{106.75}{g_*(T_M)}\right)^{1/4} \left(\frac{M}{10^{20}\text{gm}}\right)^{-1/2} \tag{8.110}$$

The total PGB fraction of all masses is

$$\Omega_{PBHtot} = \int \frac{dM}{M} f_{PBH}(M) \tag{8.111}$$

In Fig. 8.4 we show the exclusion region for PBH density fraction vs. mass from different observations. Only PBH in the mass range $(10^{-15} - 10^{-10})M_\odot$ can be the full dark matter of the universe. Consider PBH of mass $M = 10^{19}$ gm $= 5 \times 10^{-15} M_\odot$. This mass black hole is formed when the comoving scale $k_{bh} = 2.2 \times 10^{13}$ Mpc^{-1} enters the horizon. At the COBE scale $k_0 = 0.02$ Mpc the power spectrum is $P_\xi(k_0) = 2.43 \times 10^{-9}$. For black hole formation we need the peak density fluctuation $\delta_c = 0.21$ and this corresponds to $P_\xi(k_{bh}) = 0.021$. If the power spectrum is of the form $P_\xi(k) = A(k_0)(k/k_0)^{n_s-1}$ then we require a large running of $n_s(k)$ between the CMB scales and the PBH scales. Supposing

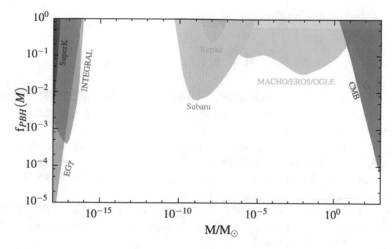

Fig. 8.4 Exclusion regions for PBH density fraction vs. mass from different observations

the COBE scale $k_0 = 0.02\,\mathrm{Mpc}^{-1}$ leaves the inflation horizon N_{60} e-foldings before the end of inflation. Similarly let the black hole scale k_{bh} exit the inflation horizon N_{bh} e-foldings before the end of inflation. Then $k_{bh} = a_{bh}H_{inf-bh}$ and $k_0 = a_{COBE}H_{inf-COBE}$ and the relation between N_{COBE} and N_{bh} is therefore

$$N_{COBE} - N_{bh} = \ln\left(\frac{k_{bh}}{k_0}\right) - \ln\left(\frac{H_{inf-bh}}{H_{inf-COBE}}\right)$$

$$\simeq 35 - \ln\left(\frac{H_{inf-bh}}{H_{inf-COBE}}\right) \tag{8.112}$$

If the inflation potential does not change too much between N_{COBE} and N_{bh} then the black hole perturbations leave the horizon 35 e-foldings after the COBE modes leave the horizon.

In single field inflation models the power spectrum at some hinge scale $k = k_0$ is

$$P_\xi(k_0) = \left(\frac{H_{inf}}{2\pi}\right)^2 \left(\frac{H_{inf}}{\dot{\phi}^2}\right)^2 = \frac{V}{24\pi^2 M_P^2 \,\epsilon_V} \tag{8.113}$$

and the scale dependence is of the form

$$P_\xi(k) = P_\xi(k_0)\left(\frac{k}{k_0}\right)^{n_s-1}, \qquad n_s - 1 = -6\epsilon_V - 2\eta_V \tag{8.114}$$

where ϵ_V and η_V are the slow roll parameters

$$\epsilon_V = \frac{M_P^2}{2}\left(\frac{V'}{V}\right)^2, \qquad \eta_v = M_P^{-2}\frac{V''}{V}. \qquad (8.115)$$

where the primes denote derivative of $V(\phi)$ w.r.t. ϕ. To have PBH production from inflation power spectrum the value of ϵ_V must be made very small at time of N_{bh}. The potential must have an inflection point where $\epsilon_V(N_{bh}) \sim 0$ to get a large increase in P_ξ at N_{bh} [45–47]. Another way is that the inflation goes in two stages, first the CMB perturbations with $P_\xi \sim 10^{-9}$ are generated and in the second stage of inflation $P_\xi \sim 10^{-2}$ is generated which produces PBH at small scales ($k \sim 10^{13}\,\mathrm{Mpc}^{-1}$) [43]. One can generate a large running of n_s by making the inflaton mass and couplings filed dependent [48]. The large scalar power is a source of secondary stochastic gravitational waves which may be observed at LISA frequencies of 0.01 Hz [44, 49].

PBH can also be generated in first order phase transitions where there is a large release of latent heat [50]. They can also be created in collapse of domain walls [51] for example by QCD axions [52].

8.10 Black Hole Temperature

Stephen Hawking's proof [53–55] that black holes have a thermodynamic temperature opened up many new areas of foundations of physics. The unitary evolution of quantum states is violated by the ensemble of thermal outgoing states. Conserved quantities like charge and baryon number can fall into black holes and those will evaporate leading to the paradox—what happened to the information which fall into black holes. A temperature in thermodynamics is associated with a large number of microscopic states. The full description of black hole metric can be given in terms of its charge, mass, angular momentum and a few such macroscopic numbers. Therefore this leaves the question open about what are the macroscopic states which result in the black hole entropy [56].

The Hawking temperature of a Schwarzschild black hole of mass M is

$$T_H = \frac{1}{8\pi G M} \qquad (8.116)$$

which is miniscule for astrophysical black holes but it is of practical observational interest for less massive primordial black holes. For a $M = 10^{19}\,\mathrm{gm}$ PBH the Hawking temperature is 1 keV. The emission spectrum is not a pure black body as the black hole also has an absorption cross section and is given by Hawking [53,54],

$$\frac{d^2N}{dt\,dE} = \frac{1}{2\pi}\frac{\Gamma_s}{e^{E/T_H} - (-1)^{2s}} \qquad (8.117)$$

where Γ_s is the grey body factor which depends upon the absorption cross section and s is the spin of the particle emitted. The grey body factor depends upon the absorption cross section as $\Gamma_s = E^2 \bar{\sigma}/\pi$. For relativistic particles $m \ll T_H$ the absorption cross section is the geometrical cross section $\bar{\sigma} = \pi(3\sqrt{3}GM)^2$ for particles of all spins [57]. For lower energies near threshold grey body factors for the leptons and photons [58–60] and for quarks and gluons [61,62] has been worked out numerically.

PBH of mass 1.5×10^{17} gm will Hawking radiate photons in the energy band $E = 0.02 - 0.6$ MeV [63]. From the observed diffused gamma ray flux measured by INTEGRAL [64] in this energy band, PBH of mass smaller than 1.5×10^{17} gm is ruled out as the entire dark matter [63].

References

1. K. Schwarzschild, On the gravitational field of a mass point according to Einstein's theory. Sitzungsber. Preuss. Akad. Wiss. Berlin (Math. Phys.) **1916**, 189 (1916). [arXiv:physics/9905030]
2. S. Chandrasekhar. The maximum mass of ideal white dwarfs. Astrophys. J. **74**, 81–82 (1931)
3. J.R. Oppenheimer, H. Snyder, On continued gravitational contraction. Phys. Rev. **56**, 455–459 (1939)
4. R.P. Kerr, Gravitational field of a spinning mass as an example of algebraically special metrics. Phys. Rev. Lett. **11**, 237 (1963)
5. Y.B. Zeldovich, I.D. Novikov, Stars and Relativity (University of Chicago Press, Chicago, 1971)
6. S. Chandrasekhar, The Mathematical Theory of Black Holes (Oxford University Press, Oxford, 1998), p. 205
7. S.A. Teukolsky, The Kerr metric. Class. Quant. Grav. **32**(12), 124006 (2015) [arXiv:1410.2130 [gr-qc]]
8. R.H. Boyer, R.W. Lindquist, Maximal analytic extension of the Kerr metric. J. Math. Phys. **8**(2), 265(1967)
9. T. Adamo, E.T. Newman, The Kerr-Newman metric: a review. Scholarpedia **9**, 31791 (2014)
10. B. Carter, Global structure of the Kerr family of gravitational fields. Phys. Rev. **174**, 1559 (1968)
11. R. Penrose, Gravitational collapse: the role of general relativity. Riv. Nuovo Cim. **1**, 252 (1969) [Gen. Rel. Grav. **34**, 1141 (2002)]
12. R. Penrose, G.R. Floyd, Extraction of rotational energy from a black hole. Nature **229**, 177 (1971)
13. J.M. Bardeen, W.H. Press, S.A. Teukolsky, Rotating black holes: locally nonrotating frames, energy extraction, and scalar synchrotron radiation. Astrophys. J. **178**, 347 (1972)
14. R.M. Wald, Energy limits on the penrose process. Astrophys. J. **191**, 231 (1974)
15. M. Banados, J. Silk, S.M. West, Kerr black holes as particle accelerators to arbitrarily high energy. Phys. Rev. Lett. **103**, 111102 (2009). [arXiv:0909.0169 [hep-ph]]
16. T. Harada, M. Kimura, Black holes as particle accelerators: a brief review. Class. Quant. Grav. **31**, 243001 (2014). [arXiv:1409.7502 [gr-qc]]
17. W.H. Press, S.A. Teukolsky, Floating orbits, superradiant scattering and the black-hole bomb. Nature **238**, 211–212 (1972)
18. V. Cardoso, O.J.C. Dias, J.P.S. Lemos, S. Yoshida, The black hole bomb and superradiant instabilities. Phys. Rev. D **70**, 044039 (2004). Erratum: [Phys. Rev. D **70**, 049903 (2004)]. [arXiv:hep-th/0404096]

19. R. Brito, V. Cardoso, P. Pani, Superradiance: energy extraction, black-hole bombs and implications for astrophysics and particle physics. Lect. Notes Phys. **906**, 1 (2015)
20. A. Arvanitaki, S. Dimopoulos, S. Dubovsky, N. Kaloper, J. March-Russell, String axiverse. Phys. Rev. D **81**, 123530 (2010)
21. W. Unruh, Separability of the neutrino equations in a Kerr background. Phys. Rev. Lett. **31**, 1265–1267 (1973)
22. S. Chandrasekhar, The solution of Dirac's equation in Kerr geometry. R. Soc. Lond. Proc. Ser. A **349**, 571–575 (1976)
23. J. Bekenstein, Extraction of energy and charge from a black hole. Phys. Rev. **D7**, 949–953 (1973)
24. T. Zouros, D. Eardley, Instabilities of massive scalar perturbations in a rotating black hole. Annal. Phys. **118**, 139–155 (1979). https://doi.org/10.1016/0003-4916(79)90237-9
25. S.L. Detweiler, Klein-gordon equation and rotating black holes. Phys. Rev. D **22**, 2323 (1980)
26. S.R. Dolan, Instability of the massive Klein-Gordon field on the Kerr spacetime. Phys. Rev. D **76**, 084001 (2007). [arXiv:0705.2880 [gr-qc]]
27. S.R. Dolan, Superradiant instabilities of rotating black holes in the time domain. Phys. Rev. D **87**(12), 124026 (2013)
28. S.A. Teukolsky, Rotating black holes separable wave equations for gravitational and electromagnetic perturbations. Phys. Rev. Lett. **29**, 1114–1118 (1972)
29. S.A. Teukolsky, Perturbations of a rotating black hole. 1. Fundamental equations for gravitational electromagnetic and neutrino field perturbations. Astrophys. J. **185**, 635–647 (1973)
30. W.H. Press, S.A. Teukolsky, Perturbations of a rotating black hole. II. Dynamical stability of the Kerr metric. Astrophys. J. **185**, 649–674 (1973)
31. E. Berti, V. Cardoso, M. Casals, Eigenvalues and eigenfunctions of spin-weighted spheroidal harmonics in four and higher dimensions. Phys. Rev. D **73**, 024013 (2006). Erratum: [Phys. Rev. D **73**, 109902 (2006)]
32. M. Sasaki, H. Tagoshi, Analytic black hole perturbation approach to gravitational radiation. Living Rev. Rel. **6**, 6 (2003). https://doi.org/10.12942/lrr-2003-6. [arXiv:gr-qc/0306120 [gr-qc]]
33. K. Akiyama, et al. [Event Horizon Telescope Collaboration], First M87 event horizon telescope results. I. The shadow of the supermassive black hole. Astrophys. J. **875**(1), L1 (2019)
34. K. Akiyama, et al. [Event Horizon Telescope Collaboration], First M87 event horizon telescope results. VI. The shadow and mass of the central black hole. Astrophys. J. **875**(1), L6 (2019)
35. K. Gebhardt, J. Thomas, The black hole mass, stellar M/L, and Dark Halo in M87. Astrophys. J. **700**, 1690 (2009)
36. F. Tamburini, B. Thide, M. Della Valle, Measurement of the spin of the M87 black hole from its observed twisted light (2019). arXiv:1904.07923 [astro-ph.HE]
37. A. Arvanitaki, S. Dubovsky, Exploring the string axiverse with precision black hole physics. Phys. Rev. D **83**, 044026 (2011)
38. H. Davoudiasl, P.B. Denton, Ultra light boson dark matter and event horizon telescope observations of M87*. Phys. Rev. Lett. **123**, 021102 (2019)
39. Y.B. Zel'dovitch, I.D. Novikov, The hypothesis of cores retarded during expansion and the hot cosmological model. Soviet Astron. **10**(4), 602–603 (1966)
40. S. Hawking, Gravitationally collapsed objects of very low mass. Mon. Not. R. Astron. Soc. **152**, 75 (1971)
41. G.F. Chapline, Cosmological effects of primordial black holes. Nature **253**, 251–252 (1975)
42. B. Carr, F. Kuhnel, M. Sandstad, Primordial black holes as dark matter. Phys. Rev. D **94**(8), 083504 (2016)
43. K. Inomata, M. Kawasaki, K. Mukaida, Y. Tada, T.T. Yanagida, Inflationary primordial black holes as all dark matter. Phys. Rev. D **96**(4), 043504 (2017)
44. S. Bhattacharya, S. Mohanty, P. Parashari, Primordial black holes and gravitational waves in non-standard cosmologies (2019). arXiv:1912.01653
45. J. Garcia-Bellido, E. Ruiz Morales, Primordial black holes from single field models of inflation. Phys. Dark Univ. **18**, 47–54 (2017)

46. C. Germani, T. Prokopec, On primordial black holes from an inflection point. Phys. Dark Univ. **18**, 6–10 (2017)
47. H. Motohashi, W. Hu, Primordial black holes and slow-roll violation. Phys. Rev. D **96**(6), 063503 (2017)
48. M. Drees, E. Erfani, Running-mass inflation model and primordial black holes. J. Cosmol. Astropart. Phys. **04**, 005 (2011)
49. K. Inomata, M. Kawasaki, K. Mukaida, Y. Tada, T.T. Yanagida, Inflationary primordial black holes for the LIGO gravitational wave events and pulsar timing array experiments. Phys. Rev. D **95**(12), 123510 (2017)
50. K. Jedamzik, J.C. Niemeyer, Primordial black hole formation during first order phase transitions. Phys. Rev. D **59**, 124014 (1999)
51. M.Y. Khlopov, S.G. Rubin, A.S. Sakharov, Primordial structure of massive black hole clusters. Astropart. Phys. **23**, 265 (2005)
52. F. Ferrer, E. Masso, G. Panico, O. Pujolas, F. Rompineve, Primordial black holes from the QCD axion. Phys. Rev. Lett. **122**(10), 101301 (2019)
53. S.W. Hawking, Black hole explosions? Nature **248**, 30 (1974)
54. S.W. Hawking, Particle creation by black holes. Commun. Math. Phys. **43**(3), 199–220 (1975)
55. J.B. Hartle, S.W. Hawking, Path integral derivation of black hole radiance. Phys. Rev. D**13**, 2188–2203 (1976)
56. A. Almheiri, T. Hartman, J. Maldacena, E. Shaghoulian, A. Tajdini, The entropy of Hawking radiation (2020). [arXiv:2006.06872 [hep-th]]
57. C. Doran, A. Lasenby, S. Dolan, I. Hinder, Fermion absorption cross section of a Schwarzschild black hole. Phys. Rev. D **71**, 124020 (2005)
58. D.N. Page, Particle emission rates from a black hole: massless particles from an uncharged, non-rotating hole. Phys. Rev. D**13**, 198 (1976)
59. D.N. Page, Particle emission rates from a black hole. 2. Massless particles from a rotating hole. Phys. Rev. D **14**, 3260 (1976)
60. D.N. Page, Particle emission rates from a black hole. 3. Charged leptons from a non rotating hole. Phys. Rev. D **16**, 2402 (1977)
61. J.H. MacGibbon, B.R. Webber, Quark and gluon jet emission from primordial black holes: the instantaneous spectra. Phys. Rev. D **41**, 3052 (1990)
62. J. H. MacGibbon, Quark-and gluon-jet emission from primordial black holes. II. The emission over the black-hole lifetime. Phys. Rev. D **44**, 376 (1991)
63. R. Laha, J.B. Muñoz, T.R. Slatyer, INTEGRAL constraints on primordial black holes and particle dark matter. Phys. Rev. D **101**(12), 123514 (2020)
64. L. Bouchet, E. Jourdain, J. Roques, A. Strong, R. Diehl, F. Lebrun, R. Terrier, INTEGRAL SPI all-sky view in soft gamma rays: study of point source and galactic diffuse emissions. Astrophys. J. **679**, 1315 (2008)

Hot Big-Bang Cosmology

A

A.1 Friedmann Equations

The universe at large scales is homogenous and isotropic with no preferred location (The Cosmological Principle). The metric which describes the universe at large scale is the Friedmann-Robertson–Walker (FRW) metric given by the line element[1]

$$ds^2 = dt^2 - a(t)^2 \left(\frac{dr^2}{1 - kr} + r^2(d\theta^2 + \sin^2\theta d\phi^2) \right) \tag{A.1}$$

where $k = 0$ for flat 3-space, $k = 1$ for 3-space with positive curvature and $k = -1$ for negative curvature 3-space. This describes a space-time where the spatial sections are have constant curvature (positive, negative or zero) which are expanding with time proportional to a scale factor $a(t)$. The curvature and the expansion rate depends upon the energy-momentum tensor which by the Cosmological Principle is also assumed to be homogenous and isotropic at large scale.

The energy density $\rho(t)$ and pressure $p(t)$ are the components of the perfect fluid isotropic energy-momentum tensor

$$T^\mu{}_\nu = \text{diagonal}(\rho, -P, -P, -P) \tag{A.2}$$

The dynamics of the scale factor $a(t)$ can be derived from the Einstein equation

$$G^\mu{}_\nu = 8\pi G\, T^\mu{}_\nu \tag{A.3}$$

---

[1] In the FRW metric with any non-zero value of k one can make the transformation $a/\sqrt{k} = \tilde{a}$ and $\sqrt{k}r = \tilde{r}$ to obtain an FRW metric in terms of $(\tilde{a}, \tilde{r})$ and $k = \pm 1$.

© Springer Nature Switzerland AG 2020
S. Mohanty, *Astroparticle Physics and Cosmology*,
Lecture Notes in Physics 975, https://doi.org/10.1007/978-3-030-56201-4

The FRW metric in Cartesian coordinates is described by

$$ds^2 = dt^2 - a^2\gamma_{ij}dx^i dx^j$$

$$= dt^2 - a^2\left(dx^{i2} + k\frac{x^{i2}dx^{i2}}{1 - kx^{i2}}\right). \tag{A.4}$$

The nonzero components of the Christoffel connection are

$$\Gamma_{ij}^0 = \dot{a}a\gamma_{ij}, \quad \Gamma_{0j}^i = \frac{\dot{a}}{a}\delta_j^i, \quad \Gamma_{jk}^i = \frac{k}{a^2}x^i\gamma_{jk}. \tag{A.5}$$

For the FRW metric the non-zero components of the Ricci tensor $R_{\mu\nu} = g_{\rho\sigma}R^\rho{}_{\sigma\mu\nu}$ are

$$R_{00} = -3\frac{\ddot{a}}{a}, \quad R_{0i} = 0$$

$$R_{ij} = \frac{1}{a^2}\left(\ddot{a}a + 2\dot{a}^2 + 2k\right)\gamma_{ij} \tag{A.6}$$

and the Ricci scale $R = g^{\mu\nu}R_{\mu\nu}$ is

$$R = \frac{6}{a^2}\left(\ddot{a}a + \dot{a}^2 + k\right) \tag{A.7}$$

To compute the Einstein's equations for the FRW metric, we start with 00 component of $G^\mu{}_\nu$

$$G^0{}_0 = \frac{3}{a^2}(\dot{a}^2 + k) \tag{A.8}$$

which gives us the 00 component of the Einstein equation as

$$\left(\frac{\dot{a}}{a}\right)^2 + \frac{k}{a^2} = \frac{8\pi G}{3}\rho \tag{A.9}$$

The ij component of the Einstein tensor is

$$G^i{}_j = \left(2a\ddot{a} + \dot{a}^2 + k\right)\delta_j^i \tag{A.10}$$

which gives us the ij component of Einstein's equation as

$$2\frac{\ddot{a}}{a} + \left(\frac{\dot{a}}{a}\right)^2 + \frac{k}{a^2} = -8\pi G\,p \tag{A.11}$$

Subtracting (A.9) from (A.11) we obtain the equation for the acceleration of the scale factor,

$$\left(\frac{\ddot{a}}{a}\right) = -\frac{4\pi G}{3}(\rho + 3p) \tag{A.12}$$

The pair of Eqs. (A.9) and (A.12) are called the Friedman equations which describe how the scale factor evolves in a homogenous isotropic universe.

To consider different matter contributions to the total energy density of the universe. The density fraction is defined as the ratio of the energy density of a particular component to the critical density,

$$\Omega_i \equiv \frac{\rho_i}{\rho_c} \tag{A.13}$$

where the critical density is

$$\rho_c = \frac{3H^2}{8\pi G} = 1.9 \times 10^{-29} h^2 \frac{\text{gm}}{\text{cm}^3}. \tag{A.14}$$

The critical density is a measure of the Hubble parameter whose present value from PLANCK-2018 [1] is $H_0 = 100\, h\, \text{km s}^{-1}\, \text{Mpc}^{-1} = 67.36 \pm 0.54\, \text{km s}^{-1}\, \text{Mpc}^{-1}$, (68%CL). If we divide Friedmann equation (A.9) by H^2, it can be written as

$$\sum_i \Omega_i + \Omega_\kappa = \Omega_m + \Omega_r + \Omega_\Lambda + \Omega_\kappa = 1 \tag{A.15}$$

where

$$\Omega_\Lambda = \frac{\Lambda}{3H^2} = \frac{8\pi G}{3H^2}\rho_\Lambda \quad \text{and} \quad \Omega_\kappa = -\frac{\kappa}{H^2 a^2} \tag{A.16}$$

are the dark energy and curvature density parameters respectively. The matter density parameter consists of baryonic matter and non-relativistic cold dark matter (CDM), i.e. $\Omega_m = \Omega_b + \Omega_c$. The present observations suggest that our universe is composed of nearly 4.9% atoms (or baryons), 26.8% (cold) dark matter and 68.3% of dark energy which adds up to approximately 1 in the total density parameter. According to Planck combined with BAO data [1] $\Omega_\kappa = 0.0007 \pm 0.0037$ (95% CL) i.e., the observations suggest that the intrinsic geometry of our universe is (very close to) flat i.e. $\kappa \simeq 0$ or the universe is at the critical density. Why the universe is so close to flat geometry or at its critical density is known as *flatness problem* which is solved by Inflation.

The time evolution of the stress tensor components $(\rho(t), p(t))$ is given by the conservation of the energy-momentum tensor

$$\nabla_\mu T^\mu{}_\nu = \partial_\mu T^\mu{}_\nu + \Gamma^\mu_{\mu\lambda} T^\lambda{}_\nu - \Gamma^\lambda_{\mu\nu} T^\mu{}_\lambda = 0 \tag{A.17}$$

Taking $\nu = 0$ and the diagonal T_ν^μ and the non-zero components of the Christoffel connection for the FRW metric, we obtain

$$\partial_0 T_0^0 + \Gamma_{\mu 0}^\mu T_0^0 - T_{\mu 0}^\lambda T_{\mu 0}^\mu = 0$$

$$\Rightarrow \quad \dot\rho + 3\frac{\dot a}{a}(\rho + p) = 0 \tag{A.18}$$

This equation also implies that the expansion of the universe is adiabatic i.e. the entropy is conserved during expansion as we will see in the next section. If the equation of state is $\rho = \omega p$ then from (A.18) we have

$$\dot\rho = -3\frac{\dot a}{a}(1 + \omega)\rho$$

$$\Rightarrow \rho \propto a^{-3(1+\omega)} \tag{A.19}$$

In the radiation era $\omega = 1/3$ and $\rho \propto a^{-4}$. In the matter era $\omega = 0$ and $\rho \propto a^{-3}$. In the cosmological constant dominated era $\omega = -1$ and $\rho = $ constant during the cosmological expansion.

Form the first Friedmann equation (A.9) and (A.19) we can solve for the time variation of the scale factor for any equation of state characterised by ω,

$$\frac{\dot a}{a} \propto \sqrt{\rho} \propto a^{-(3/2)(1+\omega)}$$

$$\Rightarrow a \sim \begin{cases} t^{(2/3)(1+\omega)^{-1}} & \text{if } \omega \neq -1 \\ e^{Ht} & \text{if } \omega = -1. \end{cases} \tag{A.20}$$

In the radiations era $a \sim t^{1/2}$, matter era $a \sim t^{2/3}$ and in the cosmological constant dominated era $a \sim e^{Ht}$ where $H = (8\pi G\rho/3)^{1/2} = $ constant.

A.2   Thermodynamics of the FRW Universe

The probability distribution function of fermions and bosons at equilibrium with a heat bath at temperature T and with chemical potential μ_i is

$$f(p, T) = \frac{1}{\exp\left(\frac{E_i - \mu_i}{T}\right) \pm 1}, \quad \text{fermions (+1), bosons (-1)} \tag{A.21}$$

with $E_i(p) = \sqrt{\mathbf{p}^2 + m_i^2}$ and μ_i the chemical potential. The distribution function then determines the number density

$$n = g_i \int \frac{d^3p}{(2\pi)^3} f(p, T) \tag{A.22}$$

the energy density

$$\rho = g_i \int \frac{d^3p}{(2\pi)^3} f(p, T) E_i(p) \tag{A.23}$$

and the pressure

$$p = g_i \int \frac{d^3p}{(2\pi)^3} f(p, T) \frac{\mathbf{p}^2}{3E_i(p)}. \tag{A.24}$$

Here g_i is the degree of freedom or the number of spin states of a given particle species i. When the chemical potential $\mu_i \ll T$ the expressions for n, ρ, p in terms of the temperature can be computed by evaluating the integrals in the limit $E = m + p^2/(2m)$ for non-relativistic particles ($m \geq T$) and taking $E = p$ for relativistic particles ($m \ll T$).

A.2.1   Relativistic Particles

At high temperatures $f(p) = (e^{p/T} \pm 1)^{-1}$ and the expression for number density of bosons and fermions number density as a function of T is[2]

$$\begin{aligned}
n &= \frac{g}{(2\pi)^3} \int_0^\infty \frac{4\pi p^2}{e^{p/T} \pm 1} dp \\
&= \frac{1}{2\pi^2} g T^3 \int_0^\infty \frac{x^2}{e^x \pm 1} dx, \quad x = p/T
\end{aligned} \tag{A.25}$$

$$= \begin{cases} g \frac{\zeta(3)}{\pi^2} T^3 & \text{bosons,} \\[2mm] \frac{3}{4} g \frac{\zeta(3)}{\pi^2} T^3 & \text{fermions.} \end{cases} \tag{A.26}$$

---

[2] $\int_0^\infty \frac{x^2}{e^x-1} dx = 2\zeta(3)$, $\int_0^\infty \frac{x^2}{e^x+1} dx = \frac{3}{2}\zeta(3)$.

And the expressions for the energy density of relativistic bosons and fermions is

$$\rho = \frac{g}{(2\pi)^3} \int_0^\infty \frac{4\pi p^3}{e^{p/T} \pm 1} dp$$

$$= \frac{1}{2\pi^2} g T^4 \int_0^\infty \frac{x^3}{e^x \pm 1} dx, \quad x = p/T$$

(A.27)

$$= \begin{cases} g \frac{\pi^2}{30} T^4 & \text{bosons,} \\ \\ \frac{7}{8} g \frac{\pi^2}{30} T^4 & \text{fermions.} \end{cases}$$

(A.28)

When there are a number of species of particles then the total energy density can be expressed in terms of an effective degree of freedom for all species as

$$\rho = \frac{\pi^2}{30} T_\gamma^4 \sum_{i=bosons} g_i \left(\frac{T_i}{T_\gamma}\right)^4 + \frac{7}{8} \sum_{j=fermions} g_j \left(\frac{T_j}{T_\gamma}\right)^4$$

$$\equiv g_\rho \frac{\pi^2}{30} T_\gamma^4$$

(A.29)

where the effective degrees of freedom contributing to the energy density is

$$g_\rho = \sum_{i=bosons} g_i \left(\frac{T_i}{T_\gamma}\right)^4 + \frac{7}{8} \sum_{j=fermions} g_j \left(\frac{T_j}{T_\gamma}\right)^4$$

(A.30)

When all species are in thermal equilibrium with the same temperature then the effective degrees of freedom is simply

$$g_\rho = \sum_{i=bosons} g_i + \frac{7}{8} \sum_{j=fermions} g_j$$

(A.31)

At $T > 175\,\text{GeV}$ (the top quark mass), all the standard model particles are in equilibrium. The total degrees of freedom which contribute to the energy density can be counted as follows. The degrees of freedoms of quarks is (6 flavor × 3 color × 2 helicity × 2 antiparticle); the charged leptons is (3 flavors × 2 helicity × 2 antiparticle); and for the neutrinos (3 flavors × 1 helicity × 2 antiparticle) which makes the fermionic degrees of freedom $g_f = 90$. For the bosons, photons (2); $W^\pm$, Z (3 × 3 spin polarizations); gluons (8 colors × 2 spin) and Higgs (1) the total bosonic degrees of freedom is $g_B = 28$. The value of g_* for the standard model is therefore $28 + (\frac{7}{8}) \times 90 = 106.75$.

The Hubble expansion rate depends on the effective degrees of species in the radiation bath and is given by

$$H = \left(\frac{8\pi G}{3}\frac{\pi^2}{30}g_\rho T^4\right)^{1/2}$$

$$= 1.66\sqrt{g_\rho}\frac{T^2}{M_P} \tag{A.32}$$

The reaction rates during the BBN era constrain the Hubble expansion rate which in turn limits the number of extra relativistic particles which can be relativistic at the time of BBN which is ~ 1 MeV.

A.2.2 Non-relativistic Particles

When the temperature drops below some particle mass then annihilates with its antiparticle in the bath and the number density is exponentially suppressed. The distribution function with $E = m + p^2/(2m)$ and with $m \gg T$ and $\mu \ll T$ reduces to the form

$$f(p) = \frac{1}{e^{(m+p^2/2m-\mu)/T}} \tag{A.33}$$

for both bosons and fermions (as long as they are not degenerate $\mu \ll T$). The expression for the number density of relativistic particles is

$$n = \frac{g}{(2\pi)^3}e^{-(m-\mu)/T}\int_0^\infty 4\pi p^2 e^{-p^2/(2mT)}dp$$

$$= \frac{g}{2\pi^2}e^{-(m-\mu)/T}(2mT)^{3/2}\int_0^\infty x^2 e^{-x^2}dx$$

$$= g\left(\frac{mT}{2\pi}\right)^{3/2}e^{-(m-\mu)/T} \tag{A.34}$$

where we used the result $\int_0^\infty x^2 e^{-x^2}dx = \frac{\sqrt{\pi}}{4}$. The energy density for non-relativistic particles is $\rho = mn + (3/2)nT$ and the pressure is $p = nT$.

If the chemical potential μ is non-zero and $T \gg m$ then the particle-antiparticle number difference in thermal equilibrium is

$$n - \bar{n} = \frac{g}{2\pi^2}\int dp p^2 \left(\frac{1}{e^{(p-\mu)/T}+1} - \frac{1}{e^{(p+\mu)/T}+1}\right)$$

$$= \frac{1}{6\pi^2}gT^3\left[\pi^2\left(\frac{\mu}{T}\right) + \left(\frac{\mu}{T}\right)^2\right]$$

If the chemical potential μ is non-zero and $T < m$ then the particle-antiparticle number difference for non-relativistic particles in thermal equilibrium is

$$n - \bar{n} = 2g \left(\frac{mT}{2\pi}\right)^{3/2} e^{-m/T} \sinh\left(\frac{\mu}{T}\right)$$

The number density, energy density and pressure for bosons and fermions at thermal equilibrium with temperature T are summarized in Table 1.

	Relativistic bosons	Relativistic fermions	Non-relativistic
n	$g_i \frac{\xi(3)}{\pi^2} T^3$	$\frac{3}{4} g_i \frac{\xi(3)}{\pi^2} T^3$	$g_i \left(\frac{mT}{2\pi}\right)^{3/2} e^{-m/T}$
ρ	$g_i \frac{\pi^2}{30} T^4$	$\frac{7}{8} g_i \frac{\pi^2}{30} T^4$	$mn + \frac{3}{2} nT$
p	$\frac{\rho}{3}$	$\frac{\rho}{3}$	$nT \ll \rho$
$\langle E \rangle = \frac{\rho}{n}$	$2.701T$	$3.151T$	$m + \frac{3}{2}T$

The expression for the stress tensor of a thermodynamic distribution of fermions or bosons is

$$T_\nu^\mu = g_i \int \frac{d^3 p}{(2\pi)^3} f(p, T) \frac{p^\mu p_\nu}{p^0}. \tag{A.35}$$

A.2.3   Entropy of Relativistic Particles

The general thermodynamic relation for the change in entropy in terms of pressure and energy density is

$$T dS = d(\rho V) + p dV \tag{A.36}$$

which defines the entropy S in a given volume V. In the context of expanding universe take $V = a^3 V_0$ in terms of the reference comoving volume V_0. We have $d(\rho V) = a^3 V_0 d\rho + 3\rho a^2 V_0 da$ and $p dV = 3p a^2 V_0 da$. The change in entropy is therefore

$$T dS = a^3 V_0 d\rho + 3(\rho + p) a^2 V_0 da \tag{A.37}$$

From the energy-momentum conservation relation (A.18) we have

$$d\rho = -3 \frac{da}{a} (\rho + p) \tag{A.38}$$

Substituting (A.38) in (A.37) we see that $dS = 0$ and total entropy is conserved in the expanding universe.

Entropy density is related to the pressure by the Maxwell relation

$$s = \frac{dp}{dT} \tag{A.39}$$

When the fermions and bosons have identical temperatures the entropy density is then given by

$$s = \frac{2\pi^2}{45} g_* T^3 \tag{A.40}$$

where the effective degrees of relativistic degrees of freedom contributing to entropy

$$g_* = \sum_{i=bosons} g_i + \frac{7}{8} \sum_{j=fermions} g_j \tag{A.41}$$

Conservation of total entropy implies

$$d(g_* T^3 a^3) = 0 \tag{A.42}$$

This implies that in the expanding universe

$$T \propto \frac{1}{g_*^{1/3} u}. \tag{A.43}$$

The value of $g_*(T)$ at $T > 175\,\text{GeV}$ when all the standard model particles are relativistic 106.75. As the universe cools below the particle masses they annihilate into the lighter relativistic particles of species already present in the heat bath and the effective degrees of freedom g_* goes down at T the annihilating particle mass. The $g_*(T)$ at different temperatures is given in Table 2 and a plot of $g_*(T)$ as a function of temperature is given in Chap. 2, Fig. 2.3.

When the species are out of equilibrium with the remaining particles in the thermal bath then the effective degrees of freedom which contribute to the entropy is

$$g_* = \sum_{i=bosons} g_i \left(\frac{T_i}{T_\gamma}\right)^3 + \frac{7}{8} \sum_{j=fermions} g_j \left(\frac{T_j}{T_\gamma}\right)^3 \tag{A.44}$$

Temperature	Annihilating particles	$g_*(T)$
$T > 170\,\text{GeV}$	All SM particles present	106.75
$120\,\text{GeV} < T < 170\,\text{GeV}$	t	96.25
$80\,\text{GeV} < T < 120\,\text{GeV}$	Higgs boson	95.25
$4\,\text{GeV} < T < 80\,\text{GeV}$	$W^\pm, Z$	86.25
$1\,\text{GeV} < T < 4\,\text{GeV}$	b	75.75
$200\,\text{MeV} < T < 1\,\text{GeV}$	c, τ	61.75
$100\,\text{MeV} < T < 200\,\text{MeV}$	$u, d, g \to \pi^\pm, \pi^0$	17.25
$0.5\,\text{MeV} < T < 100\,\text{MeV}$	$\pi^\pm, \pi^0$	10.75
$T < 0.5\,\text{MeV}$	$e^\pm \to \gamma$	3.36

A.2.4 T_ν / T_γ Temperature Ratio

When the degrees of freedom change then the temperature will change not just due to expansion but also due to the change in g_* so as to conserve the total entropy $(g_* a^3 T^3)$. For example at temperatures $T > m_e$ the entropy gets contribution from $\gamma, e^\pm$ and neutrinos.

$$S(T > m_e) = \frac{2\pi^2}{45} a^3 T^3 \left(2 + \frac{7}{8}(4+6) \right) \tag{A.45}$$

when the temperature is $T < m_e$ the $e^+ + e^- \to 2\gamma$ annihilation increases the photon $a^3 T_\gamma^3$ for photons but for neutrinos which have decoupled at $T \sim 1\,\text{MeV}$ the value of $a^3 T$ remains constant. During $e^+ e^-$ annihilation the photon temperature drops slower than $1/a$. After the electron-positron annihilation the entropy will be

$$S(T < m_e) = \frac{2\pi^2}{45} a^3 T^3 \left(2 \left(\frac{T_\gamma}{T} \right)^3 + \frac{7}{8} 6 \right) \tag{A.46}$$

Equating the entropies before and after electron-positron annihilation, we obtain the relation between the photon temperature T_γ and the neutrino temperature $T_\nu = T$,

$$\left(\frac{T_\gamma}{T} \right) = \left(\frac{11}{4} \right)^{1/3} \tag{A.47}$$

A.2.5 ρ_r and s_0 in the Present Era

The temperature of the photon background in the present universe is $T_\gamma = T_0 = 2.725\,\text{K} = 0.2348 \times 10^{-6}\,\text{eV}$. The photon number density is $n_\gamma = 410.5\,\text{photons/cm}^3$ and the energy density of photons is $\rho_\gamma = 0.2607\,\text{eV/cm}^3$. The total energy density in radiation (neutrinos and photons) in the present universe is

$$\rho_r = \frac{\pi^2}{30}T_0^4\left(2 + \frac{7}{8} \times 2 \times 3 \times \left(\frac{4}{11}\right)^{4/3}\right) = 0.424 \ \text{eV/cm}^3 \qquad (A.48)$$

The entropy density in the present universe is

$$s_0 = \frac{2\pi^2}{45}T_0^3\left(2 + \frac{7}{8} \times 2 \times 3 \times \left(\frac{4}{11}\right)\right) = 2891.8 \ \text{cm}^{-3}. \qquad (A.49)$$

This numbers is important for determination of the relic densities of particles in the present epoch.

Reference

1. N. Aghanim, et al. [Planck Collaboration], Planck 2018 results. VI. Cosmological parameters (2019). arXiv:1807.06209 [astro-ph.CO]

Neutrino Oscillations

B

B.1   Vacuum Oscillations

Neutrinos are produced and detected with weak interaction processes in gauge or 'flavour' eigenstates $\nu_\alpha = \{\nu_e, \nu_\mu, \nu_\tau\}$ but propagate as mass eigenstates $\nu_i = \{\nu_1, \nu_2, \nu_3\}$. The transformation between the mass and flavour neutrino fields by a unitary matrix called the PMNS (Pontecorvo-Maki-Nakagawa-Sakata) matrix $U_{\alpha i}$,

$$\nu_\alpha = \sum_i^3 U_{\alpha i} \nu_i \tag{B.1}$$

Neutrinos produced in nuclear fusion (like in the sun) or beta-decays of nuclei are the electron type neutrinos ν_e or antineutrinos $\bar{\nu}_e$. However when they propagate they do so in the mass basis with energy and momentum related as $E_i = \sqrt{p_i^2 + m_i^2}$. This change in basis of the production and propagation eigenstates results in the process of flavour oscillations of neutrinos.

Consider a neutrino of flavour ν_α created in a weak interaction process. Its state at time $t = 0$ is

$$|\nu_\alpha, t = 0\rangle = \nu_\alpha^\dagger |0\rangle . \tag{B.2}$$

In time it evolves to the state

$$|\nu, t\rangle = e^{-iHt} \nu_\alpha^\dagger |0\rangle = e^{-iHt} \sum_j U_{\alpha j}^* \nu_j^\dagger |0\rangle = \sum_j e^{-E_j t} U_{\alpha j}^* |\nu_j\rangle \tag{B.3}$$

© Springer Nature Switzerland AG 2020
S. Mohanty, *Astroparticle Physics and Cosmology*,
Lecture Notes in Physics 975, https://doi.org/10.1007/978-3-030-56201-4

Note that the states are related as $|\nu_\alpha\rangle = \sum_j U_{\alpha j}^* |\nu_j\rangle$ while the fields are related as $\nu_\alpha = \sum_j U_{\alpha j} \nu_j$. The amplitude for it to be detected as flavour state ν_β is

$$A(\alpha \rightarrow \beta) = \langle \nu_\beta | \nu, t \rangle = \sum_{kj} \langle \nu_k | U_{\beta k} e^{-E_j t} U_{j\alpha}^* | \nu_j \rangle = \sum_j U_{\beta j} U_{\alpha j}^* e^{-E_j t} \quad (\text{B.4})$$

using the orthogonality of the mass eigenstates $\langle \nu_k | \nu_j \rangle = \delta_{kj}$. The probability of flavour oscillation over a distance L is therefore given by

$$P(\alpha \rightarrow \beta)(L) = |A(\alpha \rightarrow \beta)|^2 = \left| \sum_{ij} U_{\alpha i} U_{\beta j}^* \exp\left(\frac{-i(m_i^2 - m_j^2)L}{2E}\right) \right|^2 \quad (\text{B.5})$$

where we have used the relativistic expansion $E_i \simeq p + m_i^2/2p \simeq p + m_i^2/2E$ and taken $L \simeq ct$ (with $c = 1$). We have also assumed that different mass eigenstates have the same momentum but different energies.[1]

The PMNS matrix is unitary $U^\dagger U = UU^\dagger = \mathcal{I}$. This can also be written as the unitarity relations

$$\sum_i U_{\alpha i} U_{\beta i}^* = \delta_{\alpha \beta}, \quad \sum_\alpha U_{\alpha i}^* U_{j\alpha} = \delta_{ij}. \quad (\text{B.6})$$

The oscillation formula (B.5) can be written is a form to make its CP properties clearer

$$P(\alpha \rightarrow \beta)(L) = \delta_{\alpha\beta} - 4 \sum_{j<k} Re\left[U_{\alpha j} U_{\beta j}^* U_{\alpha k}^* U_{\beta k} \right] \sin^2\left(\frac{\Delta m_{jk}^2 L}{4E}\right)$$

$$+ 2 \sum_{j<k} Im\left[U_{\alpha j} U_{\beta j}^* U_{\alpha k}^* U_{\beta k} \right] \sin\left(\frac{\Delta m_{jk}^2 L}{2E}\right) \quad (\text{B.7})$$

where $\Delta m_{jk}^2 \equiv m_j^2 - m_k^2$. For the oscillation probability of antineutrinos $P(\bar{\alpha} \rightarrow \bar{\beta})$ we replace U by U in (B.7). When we replace U by U^* the term $Im\left[U_{\alpha j} U_{\beta j}^* U_{\alpha k}^* U_{\beta k} \right]$ changes sign. This term is non-zero if the if CP violating phase δ_{CP}. For Dirac neutrinos the 3×3 PMNS matrix has 3 mixing angles $(\theta_{12}, \theta_{13}, \theta_{23})$ and one CP violation phase δ_{CP}. In neutrinos are Majorana then the PMNS matrix has two additional phases which cannot be removed by field

---

[1]The assumption that the neutrinos can be localised in position and momentum is clearly incorrect. The correct treatment takes neutrino wave-functions to have a spread in both position and momentum which depends upon the production process of the neutrinos. For all situations of practical interest the wave-packet treatment gives the same result as the plane wave assumption made here [1].

redefinitions. These Majorana phases do not appear in the oscillation formula and one cannot tell the difference whether the neutrino mass is Dirac or Majorana from neutrino oscillation experiments. Neutrino Majorana mass violate lepton number by two units so a non-zero signal of neutrino-less double beta decay would be a sign of neutrino mass and this observation will put constraints on the Majorana phases.

The standard (PDG) parameterisation of the PMNS matrix is

$$
U = \begin{pmatrix} c_{12}c_{13} & s_{12}c_{13} & s_{13}e^{-i\delta} \\ -s_{12}c_{23} - c_{12}s_{23}s_{13}e^{i\delta} & c_{12}c_{23} - s_{12}s_{23}s_{13}e^{i\delta} & s_{23}c_{13} \\ s_{12}s_{23} - c_{12}c_{23}s_{13}e^{i\delta} & -c_{12}s_{23} - s_{12}c_{23}s_{13}e^{i\delta} & c_{23}c_{13} \end{pmatrix} \tag{B.8}
$$

where $c_{ij} = \cos\theta_{ij}$, $s_{ij} = \sin\theta_{ij}$. In case of Majarano neutrinos U gets multiplied by the Majorana phase matrix $P = \text{diagonal}(1, e^{i\alpha_2}, e^{i\alpha_3})$ here $\{\alpha_2, \alpha_3\}$ are the two Majorana phases, however these do not enter the equation for neutrino oscillations.

One can take a two neutrino mixing limit of the oscillation formula depending upon the neutrino energy and the baseline and the oscillation probability is

$$
P(\alpha \to \beta) = \sin^2 2\theta \sin^2 \left(\frac{\Delta m^2 L}{4E} \right) = \sin^2 2\theta \sin^2 \left(\frac{\pi L}{L_{osc}} \right) \tag{B.9}
$$

where the oscillation length is minimum length at which a complete oscillation takes place for a given Δm^2 and E,

$$
L_{osc}(\text{km}) = \frac{2.48 \, E(\text{GeV})}{\Delta m^2(\text{eV}^2)} \tag{B.10}
$$

When the length scale $L \gg L_{osc}$ the $\sin(\Delta m^2 L/4E)$ factor averages to 1/2 and $P(\alpha \to \beta) = (1/2)\sin^2 2\theta$. In the 2 flavour vacuum oscillation formula there is a degeneracy in the sign $\Delta m^2 \to -\Delta m^2$ and a degeneracy $\theta \to (\pi/2 - \theta)$ (the octant of the mixing angle).

Cosmic rays protons produce pions and (fewer) kaons on when they collide with atoms at the top of the atmosphere. Pion decays produce $\pi^+ \to \mu^+ \nu_\mu$. Their decay into the channel $\pi^+ \to e^+ \nu_e$ is kinematically possible but is suppressed by angular momentum conservation. The muons further decay $\mu^+ \to e^+ \nu_e \bar{\nu}_\mu$. So in the decay of π^+ and π^- followed by muon decays the flux ratio

$$
R = \frac{\nu_\mu + \bar{\nu}_\mu}{\nu_e + \bar{\nu}_e}. \tag{B.11}
$$

is expected to be $R \simeq 2$. This ratio of muon to electron neutrinos is ubiquitous and is expected in any hydronic source of neutrinos, for example in proton collisions in Blazer jets which produce the PeV neutrinos observed at IceCube. In the Super-Kamiokande experiment [2] a water Čerenkov detector was used to observe both upward and downward going muons and electrons which were produced in the detector by atmospheric ν_μ and ν_e. They observed that for the upward going events the ratio $R/R_{MC} = 0.63 \pm 0.05$ while for downward pointing events $R/R_{MC} \simeq 1$. They also observed that in the electrons were not in excess compared to Monte-Carlo prediction while the muons were less by a factor of 0.5. The atmospheric neutrino events had $E_\nu \sim$ GeV and from the zenith angle dependence of the neutrino fluxes they concluded that neutrinos going through distances larger than 400 km were undergoing $\nu_\mu \rightarrow \nu_\tau$ oscillations. The best fit values of the mixing angle and mass are $\sin^2 2\theta_{23} = 1$ and $\Delta m_{23}^2 = 2.2 \times 10^{-3}\,\text{eV}^2$.

Deficit of neutrinos from the sun as measurements made by Homestake [3] (based on the reactions $\nu_e + {}^{37}Cl \rightarrow {}^{37}Ar + e^-$), GALLEX [4] and SAGE [5] (using $\nu_e + {}^{71}Ga \rightarrow {}^{71}Ge + e^-$ respectively) compared to the predictions of the solar models were the first evidence that neutrino flavour conversions were taking place. Super-Kamiokande observed neutrinos from the ${}^8B$ in the solar fusion cycle. Super-Kamiokande [6] measured the ${}^8B$ and the *hep* solar neutrinos by electron scattering ($\nu_x + e^- \rightarrow \nu_x + e^-$) in water Cerenkov detector. SNO [7] measured the ${}^8B$ neutrinos through deuterium splitting through both the charged current ($\nu_e + d \rightarrow 2p + e^-$) and the neutral current process ($\nu_x + d \rightarrow \nu_x + p + n$) which is equally sensitive to all flavours of neutrinos coming from the sun. These experiments confirmed that of the total flux of ν_e from the core of the sun only fraction survives and the remaining ν_e converted to ν_μ and/or ν_τ. The total ${}^8B$ neutrino flux was $\Phi_\nu = (5.045 \pm 0.16) \times 10^6\,\text{cm}^{-2}\,\text{s}^{-1}$ which was in agreement if the predictions from the solar models [8]. In the sun there is a conversion of ν_e to ν_μ by the matter effect known as the MSW mechanism [9, 10].

B.2   MSW Mechanism

The MSW effect occurs when neutrinos propagate through matter of varying density. The neutrino energy eigenstates are a combination of neutrino mass and the neutrino effective potential in matter. In propagating through matter with varying density neutrinos evolution of a neutrino flavour eigenstate can be described by the evolution equation

$$i\frac{d}{dt}|\nu_\alpha(t)\rangle = \sum_\beta H_{\alpha\beta}|\nu_\beta(t)\rangle \tag{B.12}$$

we describe the Hamiltonian in the flavour basis as the interaction with matter is diagonal in the flavour basis. We can write $\hat{H} = \hat{H}_0 + \hat{V}$ as a sum of the free particle Hamiltonian and a matter potential. The potential is due to forward scattering of the neutrinos by e^-, p, n in matter due to CC and NC interactions. In the flavour basis the potential is diagonal $\langle\alpha|\hat{V}|\beta\rangle = V_\alpha\delta_{\alpha\beta} = (V_{CC,\alpha} + V_{NC,\alpha})\delta_{\alpha\beta}$. The charged current potential is

$$V_{CC,\nu_e} = \sqrt{2}G_F n_e(x)\,, \qquad V_{CC,\nu_\mu/\nu_\tau} = 0\,. \tag{B.13}$$

The neutral current potential gets opposite contributions from e^- and p and in a charged neutral medium it only get contribution from the neutrons,

$$V_{NC,\alpha} = -\frac{G_F}{\sqrt{2}} n_n(x) \,, \qquad \alpha = \nu_e, \nu_\mu, \nu_\tau \,. \tag{B.14}$$

Since V_{NC} is the same for all flavours it can be subtracted from the Hamiltonian without affecting the flavour oscillations. Hence in normal flavour oscillations of neutrinos the neutral current interactions do not contribute. If there was an additional sterile neutrino which has zero potential with matter, the neutral current will have to be kept when considering oscillations of the four flavours $\nu_\alpha = \{\nu_e, \nu_\mu, \nu_\tau, \nu_s\}$ as $(V_{NC})_{\alpha\beta} = (-G_F n_n/\sqrt{2})$diagonal$(1, 1, 1, 0)$ is no longer proportional to the identity matrix.

The free Hamiltonian $(H_0)_{ij} = (p + m_i^2/(2E))\delta_{ij}$ is diagonal in the mass basis. In the flavour basis it is

$$(H_0)_{\alpha\beta} = \sum_{i,j} \langle \nu_\alpha | \nu_i \rangle \langle \nu_i | \hat{H} | \nu_j \rangle \langle \nu_j | \nu_\beta \rangle$$

$$= \sum_i U_{\alpha i} U_{\beta i}^* \left(\frac{m_i^2}{2E} \right) \tag{B.15}$$

where we have subtracted the constant term proportional to the identity matrix as it will not affect oscillations.

So the total Hamiltonian in the flavour basis is

$$H_{\alpha\beta} = \sum_i U_{\alpha i} U_{\beta i}^* \left(\frac{m_i^2}{2E} \right) + (V_{CC,\beta})\delta_{\beta \nu_e} \,. \tag{B.16}$$

For the case of anti-neutrino propagation through matter we replace $U \to U^*$ and $V \to -V$.

The matter potential is linear in G_F as the refractive index is proportional to the forward scattering amplitude [11]. The optical theorem gives a relation between the refractive index $n = E/p$ and the forward scattering amplitude $f(0)$,

$$n - 1 = \frac{2\pi}{E} n_e f(0) \tag{B.17}$$

For the charged current scattering the forward scattering amplitude is

$$f(0) = \frac{G_F E}{\sqrt{2}\pi} \tag{B.18}$$

from which we see that

$$V_{CC} = E - p = (n - 1)p = \sqrt{2}G_F n_e(x) \tag{B.19}$$

where in the final step we have taken $p \simeq E$. For a field theoretic calculation of neutrino potentials at finite temperature and density see [12–15].

We are able to treat the matter effect as Hamiltonian evolution in a medium with refractive index as the coherence length—the distance the neutrino travels before scattering is much larger than the oscillation length which is the case in most situations (solar neutrino conversion and atmospheric neutrino regeneration in the earth) where the matter effect is applied.

We can solve the two neutrino mixing case analytically. Consider the case of $\nu_e - \nu_\mu$ mixing in the sun. The matrix connecting mass and gauge eigenstates is

$$\begin{pmatrix} \nu_e \\ \nu_\mu \end{pmatrix} = \begin{pmatrix} \cos\theta & \sin\theta \\ -\sin\theta & \cos\theta \end{pmatrix} \begin{pmatrix} \nu_1 \\ \nu_2 \end{pmatrix}$$

Now we can derive the Hamiltonian in the flavour basis,

$$H_{\alpha\beta} = U \begin{pmatrix} 0 & 0 \\ 0 & \frac{\Delta m^2}{2E} \end{pmatrix} U^\dagger + \begin{pmatrix} \sqrt{2}G_F n_e & 0 \\ 0 & 0 \end{pmatrix} \tag{B.20}$$

where $\Delta m^2 = m_2^2 - m_1^2$ and have subtracted $(m_1^2/(2E))I$ from the free Hamiltonian. The Hamiltonian in the flavour basis can then be written in the form

$$H_{\alpha\beta} = \left(\frac{\Delta m^2}{4E}\right) \begin{pmatrix} -\cos 2\theta + A & \sin\theta \\ \sin 2\theta & \cos 2\theta - A \end{pmatrix}$$

where

$$A = \frac{2\sqrt{2}G_F n_e E}{\Delta m^2} \tag{B.21}$$

This Hamiltonian can be diagonalised by a matrix U_m

$$U_m^\dagger H_m U_m = \begin{pmatrix} E_1 & 0 \\ 0 & E_2 \end{pmatrix}, \qquad U_m = \begin{pmatrix} \cos\theta_m & \sin\theta_m \\ -\sin\theta_m & \cos\theta_m \end{pmatrix}$$

where the r dependent eigenvalues and mixing angles in matter are

$$E_1(r) - E_2(r) = \Delta m^2 \sqrt{(\cos 2\theta - A(r))^2 + \sin^2 2\theta} \,,$$

$$\sin 2\theta_m(r) = \frac{\sin 2\theta}{\sqrt{(\cos 2\theta - A(r))^2 + \sin^2 2\theta}} \,. \tag{B.22}$$

If the matter density is constant then if A is large the mixing angle $\sin 2\theta_m \to 0$ and a large potential suppresses the oscillations. In the limit $n_e \to 0$ we get back the vacuum oscillation formulas for flavour oscillations. If neutrinos propagate through a region of changing $n_e(r)$ then if at some point $A(r) = \cos 2\theta$ then at that point the mixing angle in matter $\sin 2\theta_m \simeq 1$. In that case we can have a resonant conversion of neutrino flavour even when the vacuum mixing angle $\theta \ll \pi/4$.

The flavour eigenstates are related to the energy eigenstates $|v_1^m\rangle$ and $|v_2^m\rangle$ through the position dependent mixing angles θ_m as

$$|v_e\rangle = \sin\theta_m |v_1^m\rangle + \cos\theta_m |v_2^m\rangle$$

$$|v_\mu\rangle = -\sin\theta_m |v_2^m\rangle + \cos\theta_m |v_1^m\rangle \tag{B.23}$$

In the high density core of the sun where neutrinos are produced $A \gg 1$ and $\sin\theta_m^c \simeq 0$. At the point where $A = \cos 2\theta$ there is a resonance in the sense that $E_1 - E_2$ attains its smallest value $E_1 - E_2 = \Delta^2 \sin^2 2\theta$ and $\theta_m = \pi/4$. Outside the sun $\theta_m = \theta$ and $E_1 - E_2 = \Delta m^2$. Neutrinos produced in the core are $|v_e\rangle$ and have a probability $P_1 = \cos^2\theta_m$ to be in state $|v_1^m\rangle$ and a probability $P_2 = \sin^2\theta_m$ to be in state $|v_2^m\rangle$. If the matter density varies slowly then a neutrino in state $|v_1^m\rangle_c$ in the core of the sun evolves into $|v_1\rangle$ outside. And similarly a neutrino in the state $|v_2^m\rangle_c$ in the sun evolves into $|v_2\rangle$ outside. In this case of adiabatic evolution of the energy eigenstates we can calculate the flavour conversion probability as follows. The probability of v_e produced in the solar core to be observed as v_e outside is

$$P^{ad}(v_e \to v_e) = \cos^2\theta \; P_1 + \sin^2\theta \; P_2$$

$$= \cos^2\theta \; \cos^2\theta_m^c + \sin^2\theta \; \sin^2\theta_m^c$$

$$= \frac{1}{2} + \frac{1}{2}\cos 2\theta_m^c \cos 2\theta \tag{B.24}$$

Here θ_m^c is value of the mixing angle at the point of production of v_e.

In case the matter density varies rapidly there is a finite probability of a quantum transition between $|v_1^m\rangle \leftrightarrow |v_2^m\rangle$ in the region of resonance when the energy levels come closest. By taking into account this Landau-Zener crossing probability P_{LZ}, we can give a more accurate formula for the flavour conversion in the sun.

Taking into account the non-adiabatic transitions between the energy levels near resonance the v_e survival probability can be calculated as follows. The probability of $|v_e\rangle$ being in the state $|v_1^m\rangle$ after the point of resonance is the sum of the probabilities of two probabilities. The probability of starting as $|v_1^m\rangle$ and not crossing at the after resonance and the probability of starting as $|v_2^m\rangle$ and crossing to $|v_1^m\rangle$. The total probability of $|v_e\rangle$ being in state $|v_1^m\rangle$ on emerging from the sun is

$$P_1 = \cos^2\theta_m^c \; (1 - P_{LZ}) + \sin^2\theta_m^c \; P_{LZ} \tag{B.25}$$

Similarly the probability of $|v_e\rangle$ to emerge from the sun in the state $|v_2^m\rangle$ is

$$P_2 = \sin^2 \theta_m^c \ (1 - P_{LZ}) + \cos^2 \theta_m^c \ P_{LZ} \tag{B.26}$$

The probability of detecting $|v_e\rangle$ outside the sun is therefore

$$P(v_e \rightarrow v_e) = \cos^2 \theta \ P_1 + \sin^2 \theta \ P_2$$

$$= \frac{1}{2} + \left(\frac{1}{2} - P_{LZ}\right) \cos 2\theta_m^c \ \cos 2\theta . \tag{B.27}$$

This is the well used Parke's formula [16] for neutrino conversion in a matter potential. The Landau-Zener formula for the crossing probability[2] is $P_c \simeq e^{-(\pi/2)\gamma}$ with the non-adiabaticity parameter

$$\gamma = \frac{E_1 - E_2}{2 \frac{d\theta_m}{dr}}\bigg|_{r=res} = \frac{\Delta m^2 \sin 2\theta}{2E \cos 2\theta \frac{d \ln n_e}{dr}\big|_{res}} \tag{B.29}$$

In the sun the central density is $\rho_c \simeq 150 \, \text{gm cm}^{-3}$ and $n_e(0) = 1.6 \times 10^{26} \, \text{cm}^{-3}$. The electron number profile in the sun is $n_e(r) = n_e(r_c) \exp(-10.54r/R_\odot)$ and $V_{CC}(r_c) \simeq 10^{-11} \, \text{eV}$. The best fit solution with all the observations is the LMA solution $\Delta m^2 = 7.5 \times 10^5 \, \text{eV}^2$ and $\tan \theta = 0.4$. The adiabaticity parameter $\gamma \gg 1$ and $P_{LZ} \sim 0$ and the adiabatic solution (B.24) holds. This solution depends only on θ_m^c or $n_e(r_c)$. We can define an energy E_c for which the resonance point will occur at the core of the sun

$$E_c = \frac{\Delta m^2 \cos 2\theta}{2\sqrt{2} G_F n_e(r_c)} \tag{B.30}$$

for the LMA parameters $E_c = 2.2$ MeV.

Since $n(r) < n(r_c)$ neutrinos with energy $E > E_c$ will satisfy the resonance condition at some $r > r_c$. The matter mixing angle for these neutrinos goes from $\theta_m = \theta$ outside the sun to $\theta_m = \pi/4$ at the resonance point to $\theta_m^c = \pi/2$ at the core. So for the neutrinos with $E > E_c$ the v_e survival probability (B.24) is

$$P^{ad}(v_e \rightarrow v_e) = \sin^2 \theta, \quad E > E_c \tag{B.31}$$

---

[2] A better approximation relevant for very non-adiabatic conditions is [17]

$$P_{LZ} = \frac{e^{-b} - e^{-b/\sin^2 \theta}}{1 - e^{-b/\sin^2 \theta}} \tag{B.28}$$

where $b = (\pi/2)(F\gamma)$ and F depends upon the n_e profile with $F = 1 - \tan^2 \theta$ for the exponential profile relevant to the sun.

On the other hand neutrinos with $E < E_c$ never satisfy the resonance condition at any position in the sun as $n(r)$ is a monotonically decreasing function of r. For these neutrinos $A < \cos 2\theta$ and $\theta_m \sim \theta$ everywhere. The survival probability of these neutrinos is from (B.24),

$$P^{ad}(\nu_e \to \nu_e) = 1 - \frac{1}{2}\sin^2 2\theta, \quad E < E_c. \qquad (B.32)$$

This bimodal pattern is seen in the solar neutrino observations. The Chlorine, Gallium experiments which measure the low energy ($E_\nu < 0.5\,\text{MeV}$) observed the survival probability $P(\nu_e \to \nu_e) \sim 0.6$. Borexino (which observes the ^7Be ($E_\nu = 0.862\,\text{MeV}$) [18] and *pep* neutrinos ($E_\nu = 1\text{--}1.5\,\text{MeV}$) [19] measured the ν_e survival probability more precisely to be 0.51 ± 0.007. The higher energy ($E_\nu > 5\,\text{MeV}$) observations of Super-Kamiokande and SNO give $P(\nu_e \to \nu_e) \simeq 0.32$ which gives the mixing angle from the solar neutrino experiments to be $\sin^2 \theta_{12} = 0.32$. The change in the spectrum of the survival probability occurs at $E_c \simeq 2.2$ from which one can infer the mass difference $\Delta m_{12}^2 = 7.5 \times 10^{-5}\,\text{eV}^2$.

The sign of the potential is opposite for antineutrinos therefore if the neutrinos have a resonance at some at some point in matter the antineutrinos will not have a resonance there. For $\nu_e e^-$ CC interaction $V_{CC} > 0$ the solar neutrino observations therefore establish the sign $\Delta_{sol} \equiv \Delta m_{12}^2 = m_2^2 - m_1^2 > 0$. The sign of the mass difference established from atmospheric neutrino oscillations is however undetermined $\Delta m_{atm} = \pm(m_3^2 - m_2^2)$ as both signs give the same vacuum oscillations probability. So for the three neutrino masses (m_1, m_2, m_3) we can have two possible orderings of masses. The first called normal hierarchy (NI) is $m_3| > m_2 > m_1$ and the second is called the inverted hierarchy (IH) $m_2 > m_1 > m_3$. So far experiments have not established which is the correct neutrino mass hierarchy.

This mass and mixing angle was confirmed by Kamland [20] which observed the $\bar{\nu}_e$ from nuclear reactors at an average flux weighted distance from the reactors $L \sim 180\,\text{km}$ with neutrino energies $E_{\bar{\nu}_e} \sim 1$ through the reaction $\bar{\nu}_e + p \to n + e^+$.

Finally the remaining mixing matrix element U_{e3} was determined by Daya Bay experiment [21] using reactor neutrinos to be $\sin^2(2\theta_{13}) = 0.084 \pm 0.005$. The importance of the non-zero U_{e3} is that the CP violation parameter the Jarlskog determinant of PMNS matrix $J = s_{13}\, c_{13}^2\, s_{12}\, c_{12}\, s_{23}\, c_{23} \sin delta_{CP}$ can be non-zero if $\sin \delta_{CP}$ is non-zero and the CP phase can be measured by comparing the oscillations in the neutrino versus antineutrino channels.

B.3   Ultra-High Energy Neutrinos

Neutrino observations at IceCube provide a good test of neutrino properties like neutrino decay over cosmological distances or oscillations into mirror states with mass difference as small as $\Delta m^2 = 10^{-15}\,\text{eV}^2$ [22]. The most stringent bounds on the invisible decays of neutrinos (decays to Majoron $\nu_i \to \bar{\nu}_j + J$ or $\nu \to 3\nu$) of the three mass eigenstates are as follows. Bound on ν_1 decay comes from

SN1987a observation [23], and is $\tau_1/m_1 > 10^5$ s/eV. Bound on ν_2 lifetime comes from solar neutrinos, $\tau_2/m_2 > 10^{-4}$ s/eV [24–27]. The bound on ν_3 decay comes from atmospheric neutrino observations, $\tau_3/m_3 > 10^{-10}$ s/eV [28].

In IceCube the non-observation of Glashow resonance $\bar{\nu}_e + e^- \rightarrow W^-$ at $E_\nu = 6.3$ PeV whose cross section is 200 times larger than the observed νN processes is still a puzzle. Recently, there is a possibility that the Glashow resonance has been seen in a contained shower event between 4–8 PeV range [29]. If the Glashow resonace is confirmed that would mean that neutrinos have not decayed from $\nu_1 \rightarrow \nu_3$ or $\nu_2\nu_3$ as if the final state of decay is ν_3 its ν_e component would be very small as $|U_{e3}|^2 < 0.05$. Therefore observation of Glashow resonace would establish the lifetimes $\tau_1/m_1 > 2.91 \times 10^{-3}$ s/eV and $\tau_2/m_1 > 1.26 \times 10^{-3}$ s/eV [30].

Neutrino Flavour Ratios Measurement of the flavour ratios of the high energy neutrinos can give a lot of information about the source as well as neutrino properties [31–34].

Neutrinos are produced in AGN by pp collisions which produce pions. The decay $\pi^+ \rightarrow \mu^+ + \nu_\mu$ is followed by muon decay $\mu^+ \rightarrow \bar{\nu}_\mu + e^+ + \nu_e$ which yields 2 ν_μ's for each ν_e. The flux of each neutrino flavour at the source is expected to be in the ratio $f_\alpha^S = (f_e : f_\mu : f_\tau) = (1 : 2 : 0)$ (where f_α is the flux of $\nu_\alpha + \bar{\nu}_\alpha$). For the high energy neutrinos observed at IceCube coming from cosmological distances $\sim$100 Mpc, the oscillations are averaged over and the oscillation probability of $\nu_\alpha \rightarrow \nu_\beta$ is then, $P_{\alpha\beta} = \sum_i |U_{\alpha i}|^2 |U_{\beta i}|^2$. The flavour ratio of neutrinos observed on the Earth is $f_\alpha^O = \sum_\alpha P_{\alpha\beta} f_\beta^S$. For the values of neutrino mixing angles $\left(\sin^2(2\theta_{12}), \sin^2(2\theta_{23}), \sin^2(2\theta_{13}) \right) = (0.85, 0.93, 0.1)$ the flux ratio on Earth for hadronic sources is $f_\alpha^O \simeq (1 : 1 : 1)$. IceCube measures ν_μ from the muon tracks produced and ν_e and ν_τ from the cascade of charged particles. Form the ratio of the track events to cascade events and assuming $f_e = f_\tau$, IceCube gives a flux ratio which is consistent with $f_\alpha^O = (1 : 1 : 1)$ [35].

If at the source the muon gets absorbed before decaying then for such muon damped sources the flavour ratio is $f_\alpha^S = (0, 1, 0)$. The prediction for the flavour ratio on Earth for the muon damped source is $f_\alpha^O = (1/2, 1, 1)$. This changes the track to cascade ratio compared to the $(1 : 1 : 1)$ flux.

Now consider the possibility of neutrinos decaying to the lowest mass eigenstate by the time of arrival at Earth [32]. In the normal hierarchy the final state consists of the lowest mass eigenstate ν_1. Then the flavour ratio of observed neutrinos will be $f_\alpha^O = \left(|U_{e1}|^2 : |U_{\mu1}|^2 : |U_{\tau1}|^2 \right) \simeq (6 : 1; 1)$. This is a big deviation from the observed ratio of (1:1:1). This puts a bound on $\tau_2/m_2, \tau_3/m_3 > 10^{-3}$ s/eV. For inverted hierarchy where the final state is ν_3 the flavour ratio is $f_\alpha^O = \left(|U_{e3}|^2 : |U_{\mu3}|^2 : |U_{\tau3}|^2 \right) \simeq (0 : 1; 1)$. This decay mode can explain the lack of Glashow resonance as the ν_e fraction of the final state is $|U_{e3}|^2 \simeq 0$.

References

1. W. Grimus, Revisiting the quantum field theory of neutrino oscillations in vacuum (2020). [arXiv:1910.13776 [hep-ph]].
2. Y. Fukuda, et al. [Super-Kamiokande], Evidence for oscillation of atmospheric neutrinos. Phys. Rev. Lett. **81**, 1562–1567 (1998)
3. B.T. Cleveland, T. Daily, R. Davis, Jr., J.R. Distel, K. Lande, C.K. Lee, P.S. Wildenhain, J. Ullman, Measurement of the solar electron neutrino flux with the Homestake chlorine detector. Astrophys. J. **496**, 505 (1998)
4. W. Hampel, et al., GALLEX solar neutrino observations: results for GALLEX IV. Phys. Lett. **B447**, 127 (1999)
5. J.N. Abdurashitov, et al., Solar neutrino flux measurements by the Soviet-American Gallium Experiment (SAGE) for half the 22 year solar cycle. J. Exp. Theor. Phys. **95**, 181–193 (2002)
6. S. Fukuda, et al., Solar ^8B and hep neutrino measurements from 1258 days of super-kamiokande data. Phys. Rev. Lett. **86**, 5651 (2001)
7. Q. Ahmad, et al. [SNO], Measurement of the rate of $v_e + d \rightarrow p + p + e^-$ interactions produced by ^8B solar neutrinos at the sudbury neutrino observatory. Phys. Rev. Lett. **87**, 071301 (2001)
8. J.N. Bahcall, G. Shaviv, Solar models and neutrino fluxes. Astrophys. J. **153** 113 (1968)
9. L. Wolfenstein, Neutrino oscillations in matter. Phys. Rev. D **17**, 2369–2374 (1978)
10. S. Mikheyev, A. Smirnov, Resonance amplification of oscillations in matter and spectroscopy of solar neutrinos. Sov. J. Nucl. Phys. **42**, 913–917 (1985)
11. P. Langacker, J. Liu, Standard model contributions to the neutrino index of refraction in the early universe. Phys. Rev. D **46**, 4140–4160 (1992)
12. D. Notzold, G. Raffelt, Neutrino dispersion at finite temperature and density. Nucl. Phys. B **307**, 924–936 (1988)
13. P.B. Pal, T.N. Pham, Field-theoretic derivation of Wolfenstein's matter-oscillation formula. Phys. Rev. D **40**, 259 (1989)
14. J.F. Nieves, Neutrinos in a medium. Phys. Rev. D **40**, 866 (1989)
15. K. Enqvist, K. Kainulainen, J. Maalampi, Refraction and oscillations of neutrinos in the early universe. Nuclear Phys. B, **349** 754 (1991)
16. S.J. Parke, Nonadiabatic level crossing in resonant neutrino oscillations. Phys. Rev. Lett. **57**, 1275–1278 (1986)
17. T.K. Kuo, J. Pantaleone, Nonadiabatic neutrino oscillations in matter. Phys. Rev. D **39**, 1930 (1989)
18. G. Bellini, et al., Precision measurement of the 7Be solar neutrino interaction rate in Borexino. Phys. Rev. Lett. **107**, 141302 (2011)
19. G. Bellini, et al. [Borexino], First evidence of pep solar neutrinos by direct detection in Borexino. Phys. Rev. Lett. **108**, 051302 (2012)
20. S. Abe, et al. [KamLAND], Precision measurement of neutrino oscillation parameters with KamLAND. Phys. Rev. Lett. **100**, 221803 (2008)
21. B.Z. Hu [Daya Bay], Recent results from daya bay reactor neutrino experiment (2015). [arXiv:1505.03641 [hep-ex]]
22. S. Pakvasa, A. Joshipura, S. Mohanty, Explanation for the low flux of high energy astrophysical muon-neutrinos. Phys. Rev. Lett. **110**, 171802 (2013)
23. K. Hirata, et al. [Kamiokande-II], Observation of a neutrino burst from the Supernova SN 1987a. Phys. Rev. Lett. **58**, 1490–1493 (1987)
24. A.S. Joshipura, E. Masso, S. Mohanty, Constraints on decay plus oscillation solutions of the solar neutrino problem. Phys. Rev. D **66**, 113008 (2002)
25. J.F. Beacom, N.F. Bell, Do solar neutrinos decay? Phys. Rev. D **65**, 113009 (2002)
26. A. Bandyopadhyay, S. Choubey, S. Goswami, Neutrino decay confronts the SNO data. Phys. Lett. B **555**, 33–42 (2003)
27. J.M. Berryman, A. de Gouvea, D. Hernandez, Solar neutrinos and the decaying neutrino hypothesis. Phys. Rev. D **92**(7), 073003 (2015)

28. M. Gonzalez-Garcia, M. Maltoni, Status of oscillation plus decay of atmospheric and long-baseline neutrinos. Phys. Lett. B **663**, 405–409 (2008)
29. L. Lu, Presented at UHECR 2018, October 10, 2018, Paris. https://indico.in2p3.fr/event/17063/
30. M. Bustamante, New limits on neutrino decay from the Glashow resonance of high-energy cosmic neutrinos (2020). [arXiv:2004.06844 [astro-ph.HE]]
31. J.G. Learned, S. Pakvasa, Detecting tau-neutrino oscillations at PeV energies. Astropart. Phys. **3**, 267–274 (1995)
32. J.F. Beacom, N.F. Bell, D. Hooper, S. Pakvasa, T.J. Weiler, Decay of high-energy astrophysical neutrinos. Phys. Rev. Lett. **90**, 181301 (2003)
33. S. Pakvasa, Neutrino properties from high energy astrophysical neutrinos. Mod. Phys. Lett. A **19**,1163 (2004)
34. S. Pakvasa, W. Rodejohann, T.J. Weiler, Flavor ratios of astrophysical neutrinos: implications for precision measurements. J. High Energy Phys. **02**, 005 (2008)
35. M. Aartsen, et al. [IceCube], Flavor ratio of astrophysical neutrinos above 35 TeV in IceCube. Phys. Rev. Lett. **114**(17), 171102 (2015)

Thermal Field Theory

C

C.1   Partition Function of Bosons and Fermions

Thermodynamic averages of observables are obtained starting from the grand canonical partition function

$$Z = tr\left(e^{-\beta(H-\mu N)}\right) \tag{C.1}$$

where $\beta = 1/(kT)$ and H is the Hamiltonian, μ is the chemical potential and N is the number operator..

For bosons the free particle Hamiltonian is $H = \omega a_k^\dagger a_k$, $N = a_k^\dagger a_k$ and the grand canonical partition function (for a given momentum state $\mathbf{k}$ can be evaluated by taking the trace over the occupation number states $|n_k\rangle$, defined as $a_k^\dagger a |n_k\rangle = n_k|n_k\rangle$,

$$Z_k = \sum_{n_k=0}^{\infty} \langle n_k|e^{-\beta(H-\mu N)}|n_k\rangle$$

$$= \sum_{n_k=0}^{\infty} e^{-\beta(E_k-\mu)n_k}$$

$$= \frac{1}{1 - e^{-\beta(E_k-\mu)}} . \tag{C.2}$$

where $E_k = \sqrt{\mathbf{k}^2 + m^2}$.

For fermions the occupation number $n_k = 0, 1$ and the partition function is

$$Z_k = \sum_{n_k=0}^{1} \langle n_k|e^{-\beta(H-\mu N)}|n_k\rangle$$

$$= 1 + e^{-\beta(E_k-\mu)} . \tag{C.3}$$

© Springer Nature Switzerland AG 2020
S. Mohanty, *Astroparticle Physics and Cosmology*,
Lecture Notes in Physics 975, https://doi.org/10.1007/978-3-030-56201-4

For non-interacting bosons and fermions the partition function can be obtained by summing over all the momentum modes

$$
Z = tr\left(e^{-\beta \sum_k (E_k - \mu N_k)}\right)
$$

$$
= \prod_k Z_k \tag{C.4}
$$

Which implies that

$$
\ln Z = \sum_k \ln Z_k
$$

$$
= V \int \frac{d^3 k}{(2\pi)^3} \ln Z_k
$$

$$
= V \int \frac{d^3 k}{(2\pi)^3} \ln\left(1 \pm e^{-\beta(E_k - \mu)}\right)^{\pm} \tag{C.5}
$$

where the $+$ sign is for fermions and the $-$ sign is for bosons.

Using the partition function we obtain the pressure, number density and energy density as follows,

$$
P = \frac{T}{V} \ln Z = T \int \frac{d^3 k}{(2\pi)^3} \ln\left(1 \pm e^{-\beta(E_k - \mu)}\right)^{\pm} ,
$$

$$
\frac{N}{V} = \frac{1}{V} \frac{\partial T \ln Z}{\partial \mu} = \int \frac{d^3 k}{(2\pi)^3} \frac{1}{e^{\beta(E_k - \mu)} \pm 1} ,
$$

$$
\frac{E}{V} = \frac{-1}{V} \frac{\partial \ln Z}{\partial \beta} = \int \frac{d^3 k}{(2\pi)^3} \frac{E_k}{e^{\beta(E_k - \mu)} \pm 1} , \tag{C.6}
$$

C.2   Thermal Two Point Functions

The thermal averaged two point correlation functions of bosons and fermions is

$$
\langle \phi(\mathbf{x}, t)\phi(\mathbf{y}, 0)\rangle_\beta = \frac{1}{Z} tr\left[e^{-\beta H} \phi(\mathbf{x}, t)\phi(\mathbf{y}, 0)\right]. \tag{C.7}
$$

From this we can derive a relation,

$$
\langle \phi(\mathbf{x}, t)\phi(\mathbf{y}, 0)\rangle_\beta = \frac{1}{Z} tr\left[\phi(\mathbf{x}, t)e^{-\beta H} e^{i(-i\beta H)}\phi(\mathbf{y}, 0)e^{-i(-i\beta H)}\right]
$$

$$
= \frac{1}{Z} tr\left[\phi(\mathbf{x}, t)e^{-\beta H}\phi(\mathbf{y}, -i\beta)\right]
$$

$$
= \langle \phi(\mathbf{y}, -i\beta)\phi(\mathbf{x}, t)\rangle_\beta . \tag{C.8}
$$

Now we can define a imaginary time variable $t = -i\tau$ and in the imaginary time, and the equation above becomes,

$$\langle\phi(\mathbf{x}, -i\tau)\phi(\mathbf{y}, 0)\rangle_\beta = \langle\phi(\mathbf{y}, -i\beta(-i))\phi(\mathbf{x}, -i\tau)\rangle_\beta \tag{C.9}$$

which can be written as

$$\langle\phi(\mathbf{x}, \tau)\phi(\mathbf{y}, 0)\rangle_\beta = \langle\phi(\mathbf{y}, \beta)\phi(\mathbf{x}, \tau)\rangle_\beta. \tag{C.10}$$

This is known as the Kubo-Martin-Schwinger (KMS) relation.

Using the KMS relation we see that

$$\langle\phi(\mathbf{x}, \tau)\phi(\mathbf{y}, 0)\rangle_\beta = \pm\langle\phi(\mathbf{y}, 0)\phi(\mathbf{x}, \tau)\rangle_\beta$$
$$= \langle\phi(\mathbf{y}, \beta)\phi(\mathbf{x}, \tau)\rangle_\beta, \tag{C.11}$$

where the $+(-)$ sign is due to the exchange of bosons (fermions) and the second line is the KMS relation. From this it follows that $\phi(\mathbf{y}, 0) = \pm\phi(\mathbf{y}, \beta)$, which implies that, bosons (fermions) are periodic (anti-periodic) under the shift $t \to t - i\beta$,

$$\phi(\mathbf{y}, t) = \pm\phi(\mathbf{y}, t - i\beta). \tag{C.12}$$

Due to the periodicity constraint on bosonic and fermionic fields the Fourier transform the fields can be written as

$$\phi(\mathbf{x}, \tau) = \sqrt{\frac{\beta}{V}} \sum_{n=-\infty}^{\infty} \int \frac{d^3p}{(2\pi)^3} \phi(\omega_n, \mathbf{p}) e^{i(\omega_n\tau + \mathbf{p}\cdot\mathbf{x})} \tag{C.13}$$

where n are integers and ω_n are the Matsubara frequencies

$$\omega_n = \frac{2\pi n}{\beta} \qquad \text{bosons},$$

$$\omega_n = \frac{2\pi(n+1)}{\beta} \qquad \text{fermions}, \tag{C.14}$$

which ensure that the bosonic (fermionic) fields are periodic (anti-periodic) under the shift $\tau \to \tau + \beta$.

C.2.1 Partition Function of Scalar Fields

To evaluate the partition function for a scalar field, we note that the partition function can be expressed as the path integral of the action in Euclidean time $\tau = -it$,

$$Z = tr\left(e^{-\beta H}\right) = \int D[\phi]e^{S_E} \tag{C.15}$$

where the path integral includes the fields which are periodic in τ, $\phi(\mathbf{x}, \tau) = \phi(\mathbf{x}, \tau + \beta)$ and where the Euclidean action is

$$
\begin{aligned}
S_E &= -\frac{1}{2} \int_0^\beta d\tau \int d^3x \left[\left(\frac{\partial \phi}{\partial \tau} \right)^2 + (\nabla \phi)^2 + m^2 \phi^2 \right] \\
&= -\frac{1}{2} \int_0^\beta d\tau \int d^3x \, \phi \left(-\frac{\partial^2}{\partial \tau^2} - \nabla^2 + m^2 \right) \phi \\
&= -\frac{1}{2} \beta^2 \sum_n \int \frac{d^3p}{(2\pi)^3} (\omega_n^2 + \omega^2) \phi(\omega_n, \mathbf{p}) \phi^*(\omega_n, \mathbf{p}) \quad\quad \text{(C.16)}
\end{aligned}
$$

where we substituted (C.13) for $\phi(\mathbf{x}, \tau)$ and used the orthogonality relations[1] and the property $\phi(-\omega_n, \mathbf{p}) = \phi^*(\omega_n, \mathbf{p})$ for real $\phi(\mathbf{x}, \tau)$ The partition function (C.15) can be evaluated by using (C.16),

$$
\begin{aligned}
Z &= \int D|\phi_n| e^{-\frac{1}{2}\beta^2 \sum_n \sum_p (\omega_n^2 + E_p^2)|\phi_n|^2} \\
&= \prod_n \prod_p \int d|\phi_n| e^{-\frac{1}{2}\beta^2 \sum_n \sum_p (\omega_n^2 + E_p^2)|\phi_n|^2} \\
&= \mathcal{N} \prod_n \prod_p \left[\beta^2 (\omega_n^2 + E_p^2) \right]^{-1/2} \quad\quad \text{(C.17)}
\end{aligned}
$$

For a scalar field we therefore obtain, after dropping the temperature independent normalization constant,

$$
\ln Z = -\frac{1}{2} V \sum_n \int \frac{d^3p}{(2\pi)^3} \ln \left[\beta^2 (\omega_n^2 + E_p^2) \right] \quad\quad \text{(C.18)}
$$

The summation over the bosonic Matsubara frequencies $\omega_n = 2\pi n/\beta$ can be performed by using the following mathematical identities. We write

$$
\ln \left[\beta^2 \left(\omega_n^2 + E_p^2 \right) \right] = \int_1^{\beta^2 E_p^2} dy \frac{1}{(2\pi n)^2 + y^2} . \quad\quad \text{(C.19)}
$$

Use the identity

$$
\sum_{n=-\infty}^{\infty} \frac{1}{(2\pi n)^2 + y^2} = \frac{1}{2y} + \frac{1}{y} \frac{1}{e^y - 1} . \quad\quad \text{(C.20)}
$$

---

[1]

$$
\int_0^\beta d\tau e^{-i(\omega_n + \omega_m)\tau} = \beta \, \delta(\omega_m + \omega_n), \quad \int d^3x e^{-i(\mathbf{p}+\mathbf{k})\cdot x} = V \delta^3(\mathbf{p} + \mathbf{k}).
$$

Substituting the rhs of (C.20) as the integrand in (C.19) and performing the integral we obtain

$$\int_1^{\beta^2 E_p^2} dy \left(\frac{1}{2y} + \frac{1}{y} \frac{1}{e^y - 1} \right) = \beta E_p + 2 \ln \left(1 - e^{-\beta E_p} \right) \tag{C.21}$$

where we have dropped the constants independent β and E_p. Using these results in (C.18) we obtain the expression for the partition function of real scalar fields.

$$\ln Z = V \int \frac{d^3 p}{(2\pi)^3} \left[-\frac{1}{2} \beta E_p - \ln \left(1 - e^{-\beta E_p} \right) \right] \tag{C.22}$$

For complex scalars the will be an extra factor two which will account for the separate contribution of particles and antiparticles.

The thermodynamic potential or free energy is the negative of pressure and is given by

$$F = -P = -\frac{T}{V} \ln Z$$

$$= \int \frac{d^3 p}{(2\pi)^3} \left[\frac{1}{2} E_p + \frac{1}{\beta} \ln \left(1 - e^{-\beta E_p} \right) \right] \tag{C.23}$$

where $E_p = (m^2 + \mathbf{p}^2)^{1/2}$.

In a theory where the scalar mass is a function of the background filed, for example if $V_0 = (1/2)m_0^2 \phi_c^2 + (\lambda/4)\phi_c^4$ where $m^2(\phi_c) = m_0^2 + (\lambda/2)\phi_c^2$, the thermodynamic potential a function of the background scalar. Thermal equilibrium is obtain by maximizing pressure or minimizing the thermodynamic potential with ϕ_c.

C.2.2 Partition Function of Spin 1/2 Fields

The partition function of Dirac fields is the path integral of the Euclidean action

$$Z_F = \int D[i\psi^\dagger] D[\psi] \exp \left[\int_0^\beta d\tau \int d^3 x \, \psi^\dagger \gamma_0 \left(-i\gamma_0 \frac{\partial}{\partial \tau} + i\boldsymbol{\gamma} \cdot \nabla - m_f \right) \psi \right] \tag{C.24}$$

where the path integral is over Grossman variables ϕ and $i\psi^\dagger = \Pi$ which are treated as independent variables and the anti-periodic condition $\psi(\mathbf{x}, \tau) = \psi(\mathbf{x}, \tau + \beta)$ is imposed.

We go to Fourier space

$$\psi_\alpha(\mathbf{x}, \tau) = \frac{1}{\sqrt{V}} \sum_{n=-\infty}^{\infty} \int \frac{d^3 p}{(2\pi)^3} e^{i(\mathbf{p} \cdot \mathbf{x} + \omega_n \tau)} \tilde{\psi}_{\alpha,n}(\mathbf{p}) \tag{C.25}$$

where α is the spinorial index and $\omega_n = (2n + 1)\pi/\beta$.

The partition function expressed in Fourier modes is

$$Z_F = \prod_n \prod_p \int d[i\,\tilde{\psi}^\dagger_{\alpha,n}(\mathbf{p})]\,d[\tilde{\psi}_{\sigma,n}(\mathbf{p})]\exp\left[\sum_n \sum_p -i\,\tilde{\psi}^\dagger_{\alpha,n}(\mathbf{p})\,D_{\alpha\sigma}\,\tilde{\psi}_{\sigma,n}(\mathbf{p})\right]$$

(C.26)

where

$$D_{\alpha\sigma} = i\beta\left[\left(i\omega_n\gamma^0\boldsymbol{\gamma}\cdot\mathbf{p}\gamma_0\,m_f\right)\right]_{\alpha\sigma}.$$

(C.27)

The Grassmann integrations gives us,

$$Z_F = \prod_n \prod_p \det D.$$

(C.28)

We can write (C.27) explicitly as a 4×4 matrix

$$D = i\beta\begin{pmatrix} i\omega_n - \mathbf{p} & 0 & m & 0 \\ 0 & i\omega_n + \mathbf{p} & 0 & m \\ m & 0 & i\omega_n + \mathbf{p} & 0 \\ 0 & m & 0 & i\omega_n - \mathbf{p} \end{pmatrix}$$

using which we calculate the determinant

$$\det D = \beta^4\left(\omega_n^2 + E_p^2\right)^2$$

(C.29)

Using (C.29) in (C.28) we obtain

$$\ln Z_F = \sum_n \sum_p 2\ln\left[\beta^2\left(\omega_n^2 + E_p^2\right)\right]$$

$$= \sum_n V\int \frac{d^3 p}{(2\pi)^3}\,2\ln\left[\beta^2\left(\omega_n^2 + E_p^2\right)\right]$$

(C.30)

The summation over the fermionic Matsubara frequencies $\omega_n = 2\pi(n+1)/\beta$ can be performed by using the following mathematical identities. We write

$$\ln\left[\beta^2\left(\omega_n^2 + E_p^2\right)\right] = \int_1^{\beta^2 E_p^2} dy\,\frac{1}{(2\pi(n+1))^2 + y^2}.$$

(C.31)

Use the identity

$$\sum_{n=-\infty}^{\infty} \frac{1}{(2\pi (n+1))^2 + y^2} = \frac{1}{2y} - \left(\frac{1}{y} \frac{1}{e^y + 1}\right). \tag{C.32}$$

Substituting the rhs of (C.32) as the integrand in (C.31) and performing the integral we obtain

$$\int_{1}^{\beta^2 E_p^2} dy \left(\frac{1}{2y} - \frac{1}{y} \frac{1}{e^y + 1}\right) = \beta E_p + 2\ln\left(1 + e^{-\beta E_p}\right) \tag{C.33}$$

where we have dropped the constants independent β and E_p. Using these results we can write the partition function of Dirac fields (C.30) as

$$\ln Z_F = 4\beta V \int \frac{d^3 p}{(2\pi)^3} \left[\frac{E_p}{2} + \frac{1}{\beta}\ln\left(1 + e^{-\beta E_p}\right)\right] \tag{C.34}$$

The factor of 4 accounts for the two helicity states of both particles and anti-particles of states represented by Dirac fermions. For chiral fermions (like neutrinos) the partition function will have an overall factor of 2 instead of 4. The first term of the integrand represents the zero-point energy and comes has a relative negative sign for fermions compared to scalars (C.22).

The free energy is for Dirac fermions is

$$F = -P = -\frac{T}{V}\ln Z_F$$

$$= 4\int \frac{d^3 p}{(2\pi)^3}\left[-\frac{1}{2}E_p - \frac{1}{\beta}\ln\left(1 + e^{-\beta E_p}\right)\right] \tag{C.35}$$

where $E_p = (m_f^2 + \mathbf{p}^2)^{1/2}$.

C.2.3   Thermal Correlation Function

The thermal average of the number operator for bosonic fields is

$$\langle a(\mathbf{p})^\dagger a(\mathbf{k})\rangle_\beta = \frac{1}{Z_k}\sum_{n_k=0}^{\infty} n_k e^{-\beta(E_k - \mu)n_k}\delta^3(\mathbf{p} - \mathbf{k})$$

$$= \frac{1}{Z_k}\frac{e^{-\beta(E_k - \mu)}}{\left(1 - e^{-\beta(E_k - \mu)}\right)^2}\delta^3(\mathbf{p} - \mathbf{k})$$

$$= n_B(E_k)\delta^3(\mathbf{p} - \mathbf{k}), \tag{C.36}$$

where

$$n_B(E_k) = \frac{1}{e^{\beta(E_k - \mu)} - 1} \tag{C.37}$$

is the distribution function of bosons at finite temperature. Using the commutation relations, $a(\mathbf{p})a(\mathbf{k})^\dagger = (\delta^3(\mathbf{p} - \mathbf{k}) + a(\mathbf{k})^\dagger a(\mathbf{p}))$ we have

$$\langle a(\mathbf{p})a(\mathbf{k})^\dagger \rangle_\beta = (1 + n_B(E_k))\,\delta^3(\mathbf{p} - \mathbf{k})\,. \tag{C.38}$$

For fermionic fields the thermal average of the number operator is

$$\langle b(\mathbf{p})^\dagger b(\mathbf{k}) \rangle_\beta = \frac{1}{Z_k} \sum_{n_k=0}^{1} n e^{-\beta(E_k - \mu)n_k} \delta^3(\mathbf{p} - \mathbf{k})$$

$$= \frac{1}{Z_k} e^{-\beta(E_k - \mu)} \delta^3(\mathbf{p} - \mathbf{k})$$

$$= n_F(E_k)\delta^3(\mathbf{p} - \mathbf{k})\,, \tag{C.39}$$

where

$$n_F(E_k) = \frac{1}{e^{\beta(E_k - \mu)} + 1} \tag{C.40}$$

is the distribution function of fermions at finite temperature. From the anti-commutation relations $b(\mathbf{p})b(\mathbf{k})^\dagger = (\delta^3(\mathbf{p} - \mathbf{k}) - b(\mathbf{k})^\dagger b(\mathbf{p}))$ we have

$$\langle b(\mathbf{p})b(\mathbf{k})^\dagger \rangle_\beta = (1 - n_F(E_k))\,\delta^3(\mathbf{p} - \mathbf{k})\,. \tag{C.41}$$

Index

© Springer Nature Switzerland AG 2020
S. Mohanty, *Astroparticle Physics and Cosmology*,
Lecture Notes in Physics 975, https://doi.org/10.1007/978-3-030-56201-4

Printed in the United States
By Bookmasters